AF693663

THERMOMECHANICS OF VISCOPLASTICITY

For other titles published in this series, go to
www.springer.com/series/5613

Advances in Mechanics and Mathematics

VOLUME 20

Series Editors
David Y. Gao (Virginia Polytechnic Institute and State University)
Ray W. Ogden (University of Glasgow)

Advisory Board
Ivar Ekeland (University of British Columbia, Vancouver)
Tim Healey (Cornell University, USA)
Kumbakonam Rajagopal (Texas A&M University, USA)
Tudor Ratiu (École Polytechnique Fédérale, Lausanne)
David J. Steigmann (University of California, Berkeley)

Aims and Scope

Mechanics and mathematics have been complementary partners since Newton's time, and the history of science shows much evidence of the beneficial influence of these disciplines on each other. The discipline of mechanics, for this series, includes relevant physical and biological phenomena such as: electromagnetic, thermal, quantum effects, biomechanics, nanomechanics, multiscale modeling, dynamical systems, optimization and control, and computational methods.

Driven by increasingly elaborate modern technological applications, the symbiotic relationship between mathematics and mechanics is continually growing. The increasingly large number of specialist journals has generated a complementarity gap between the partners, and this gap continues to widen. *Advances in Mechanics and Mathematics* is a series dedicated to the publication of the latest developments in the interaction between mechanics and mathematics and intends to bridge the gap by providing interdisciplinary publications in the form of monographs, graduate texts, edited volumes, and a special annual book consisting of invited survey articles.

THERMOMECHANICS OF VISCOPLASTICITY

Fundamentals and Applications

By

MILAN V. MIĆUNOVIĆ
University of Kragujevac, Serbia

Springer

Milan V. Mićunović
Faculty of Mechanical Engineering
University of Kragujevac
Sestre Janjića 6a
34000 Kragujevac
Serbia
mmicun@kg.ac.rs

Series Editors:

David Y. Gao
Department of Mathematics
Virginia Polytechnic Institute
Blacksburg, VA 24061
gao@vt.edu

Ray W. Ogden
Department of Mathematics
University of Glasgow
Glasgow, Scotland, UK
rwo@maths.gla.ac.uk

ISSN: 1571-868
ISBN: 978-0-38 SBN: 978-0-387-89490-4
DOI: 10.1007/9

Library of Congress Control Number: PCN applied for

Mathematics Subject Classification (2000): 74C20, 74D10, 74M25

© Springer Science+Business Media, LLC 2009
All rights reserved. This work may not be translated or copied in whole or in part without the written permission of the publisher (Springer Science+Business Media, LLC, 233 Spring Street, New York, NY 10013, USA), except for brief excerpts in connection with reviews or scholarly analysis. Use in connection with any form of information storage and retrieval, electronic adaptation, computer software, or by similar or dissimilar methodology now known or hereafter developed is forbidden.

The use in this publication of trade names, trademarks, service marks, and similar terms, even if they are not identified as such, is not to be taken as an expression of opinion as to whether or not they are subject to proprietary rights.

Printed on acid-free paper

springer.com

This book is dedicated to my two teachers

Rastko Stojanović and Henryk Zorski

Series Preface

As any human activity needs goals, mathematical research needs problems.
—David Hilbert

Mechanics is the paradise of mathematical sciences.
—Leonardo da Vinci

Mechanics and mathematics have been complementary partners since Newton's time, and the history of science shows much evidence of the beneficial influence of these disciplines on each other. Driven by increasingly elaborate modern technological applications, the symbiotic relationship between mathematics and mechanics is continually growing. However, the increasingly large number of specialist journals has generated a duality gap between the partners, and this gap is growing wider.

Advances in Mechanics and Mathematics (AMMA) is intended to bridge the gap by providing multidisciplinary publications that fall into the two following complementary categories:

1. An annual book dedicated to the latest developments in mechanics and mathematics;
2. Monographs, advanced textbooks, handbooks, edited volumes, and selected conference proceedings.

The AMMA annual book publishes invited and contributed comprehensive research and survey articles within the broad area of modern mechanics and applied mathematics. The discipline of mechanics, for this series, includes relevant physical and biological phenomena such as: electromagnetic, thermal, and quantum effects, biomechanics, nanomechanics, multiscale modeling, dynamical systems, optimization and control, and computation methods. Especially encouraged are articles on mathematical and computational models and methods based on mechanics and their interactions with other fields. All contributions will be reviewed so as to guarantee the highest possible scientific standards. Each chapter will reflect the most recent achievements in the area. The coverage should be conceptual, concentrating on the methodological thinking that will allow the nonspecialist reader to understand it. Discussion of possible future research directions in the area is welcome. Thus, the annual volumes will provide a continuous documentation of the most recent developments in these active and important interdisciplinary fields. Chapters published in this series could form bases from which possible AMMA monographs or advanced textbooks could be developed.

Volumes published in the second category contain review/research contributions covering various aspects of the topic. Together these will provide an overview of the state-of-the-art in the respective field, extending from an introduction to the subject right up to the frontiers of contemporary research. Certain multidisciplinary topics, such as duality, complementarity, and symmetry in mechanics, mathematics, and physics are of particular interest.

The *Advances in Mechanics and Mathematics* series is directed to all scientists and mathematicians, including advanced students (at the doctoral and postdoctoral levels) at universities and in industry who are interested in mechanics and applied mathematics.

David Y. Gao
Ray W. Ogden

Preface

Very often, theoreticians and experimentalists have disjoint ways of thinking. Typically, a theoretician begins an analysis with "let us assume" and anything following this phrase is valid. The procedure leading from a theorem to its proof is untouchable and in some way sacrosanct. On the other hand, an experimentalist often says that a theory is nothing but a servant of experimental facts. Anything more than this is like a chess game replacing a real battle with canons and people. In the opinion of the author, the just path lies between the two extremes. Namely, each sophisticated experiment needs a lot of thought and a sound theory in planning. One must not be blind when preparing it, expecting miracles. On the other hand, a theory which predicts a behaviour contrary to physical experience is certainly wrong and must be improved. This discussion holds especially for complex shape specimens with a testing campaign aimed to characterize the influence of previous history. When you add thermal, electromagnetic, and other nonmechanical fields, everything starts to resemble a nightmare or a novel written by Agatha Christie. On the other hand, these nonmechanical fields could tell you a lot about characteristic singular points in inelastic material behaviour. Typical examples are plasticity commencement and initiation of a magistral crack.

If you use synergetic behaviour in a proper way, then such couplings become highly valuable. One measurable quantity, like temperature, can replace some other quantities either not measurable directly or measurable with difficulties due to specific shapes of specimens. If your testing is aimed at a better understanding of an inelastic behaviour, then your path to a more proper description could be illustrated by a three-dimensional helix theory-experiment-numerics-theory... etc.

In reading some classical papers advocating normality of plastic strain rate to the yield surface, I have wondered how weak are logical foundations of such a proposition. Its great advantage is simplicity in application, especially when material data available are far from plentiful. In short, this monograph is intended to be an alternative to the commonly applied and used (visco)plasticity papers based on the so-called associative flow rules, i.e., yield surface normality and in the most special case J_2-theory.

This book contains geometrical and thermodynamical issues indispensable for development of a rational theory of thermoviscoplasticity. A geometrical picture of coupled thermomagnetomechanical histories of damaged solids is built both by means of Kröner's incompatibility approach as well as by Eshelbian implanting eigenstrains. Duality of Euclidean anholonomic and non-Euclidean natural state space is emphasized. Damaged inelastic materials of differential type, discrete and infinitesimal memory are obtained from the principle of thermo-inelastic memory. The issue of plastic spin is considered. A postulate of minimal plastic work and corresponding nonassociativity 4-tensor are then used to show whether associativity of flow rule holds. Postulates of Drucker, Iliushin and Hill are discussed. Thermodynamics of inelasticity is extensively discussed in classical, rational, extended and endochronic version taking into account statistical thermodynamics. A nonsteady aging is used in endochronic thermodynamics to cover creep-plasticity coupled inelastic histories. Multiaxial dynamic experiments with cylindrical, "bicchierino" and cruciform specimen from austenitic stainless steels are analyzed. Quasi rate independence and Rabotnov's plastic delay is combined with tensor representation. Inelastic ferromagnetics are treated by means of extended as well as endochronic thermodynamics. For low-cycle fatigue, the experimentally observed displacement of magnetic induction history with respect to stress history is analyzed. Self-consistent method applied to inelastic polycrystals is based on constrained micro-rotations and free meso-rotations. Special attention is devoted to slight disorder of polycrystal grains. The theory is confronted with classical J_2-theory. Different inelastic multiaxial stress histories are analyzed and corresponding active slip systems determined. For numerical results micro quasi rate independence and relaxed Taylor's model are used. The theory of inelastic micromorphic polycrystals with couple stresses needs a very small number of necessary material constants. Nonproportionality of strain history as well as intergranular continuity are related to antisymmetry of stress tensor.

The last three chapters are devoted to applications of the theory to plastic waves (in the tension type split Hopkinson bar), ratchetting phenomenon as well as the issue of metal forming of orthotropic sheets. These examples serve to enlighten the advantages and applicability of the approach.

It is intended that this book could serve not only researchers in the field of continuum mechanics and material science but also engineers dealing with metal forming in car body design and construction. The concepts and results employed here are readily applicable to the rapidly developing field of biomechanics (hard and soft tissues and bones). Some recent papers in biomechanics use the concept of natural state intermediate local configuration. In order to widen the audience of potential readers, the level of mathematical apparatus used throughout the book does not exceed that applied in a classical book of continuum mechanics (like in Eringen, 1962). Also, the differential geometry needed for understanding the issue of eigen-stresses is condensed to be as simple as possible. Thus, the text tries to avoid unnecessary mathematics.

My sincere thanks go to many people who have helped in this work. First, to my late teachers, Rastko Stojanović and Henryk Zorski, for their support and helpful criticism whenever necessary. Discussions on "Belgrade's Rheology Group" meetings as well as in meetings at the Institute of Fundamental Technological Research PAN in Warsaw were always friendly and stimulating. To David Gao and Ray Ogden, I owe very much for their support during publishing procedure. The Springer editorial team led by Elizabeth Loew has been highly professional, efficient, and encouraging to me. Comments of the reviewers are also much appreciated. Last, but most important, my wife Ludmila has given me sincere care and blessed understanding during this work.

Kragujevac, Serbia
August 2008

Milan Mićunović

Contents

List of Figures

List of Tables

Part I

Theoretical and Experimental Aspects

Introduction

There are many books and a huge amount of papers devoted to viscoplasticity. Still, there are many things to be clarified and underlined in this vast field.

Theoretical consideration of viscoplasticity has become an important item for finite element codes, which pretend to perform calculations of complex structures with a high precision. In a majority of these codes the evolution equation for plastic strain rate is oversimplified and of associate type, i.e., it is perpendicular to yield surface in stress space. Anisotropy induced by plastic strains is very rarely properly taken into account. It should be noted that usually the yield function is detected from tension tests and then applied to calculation during arbitrary stress–strain histories appearing in real structures. Such a procedure could produce significant errors destroying geometrical accuracy, which FEM codes offer. Thus a need for new theories still exists.

Let us start this book by giving first a list of difficulties that must be faced and cannot be escaped.[1]

- At the very beginning it seems logical to say something about experimental evidence, which should serve as a final judgement whether the theoretical approach is correct or not. This is exactly the point of view taken by James Bell in his huge review of experimental foundations of solid mechanics [Bel65]. Here influence of strain rate, temperature, damage and nonmechanical effects like magnetic fields and irradiation are crucial. However, we must be cautious since there is no single experiment without constitutive assumptions being involved. Every experiment provides a partial picture of the process with many holes to be filled by patches provided by some constitutive theory. For instance, even if we restrict consideration to pure mechanical thermo-inelasticity without other nonmechanical effects,

[1] The prefixes "macro," "meso" and "micro" are reserved for properties of the whole body, of the volume element composed of many grains and of an individual grain, respectively. This distinction is important only in the last two chapters dealing with polycrystals.

what we are able to measure are just displacement and temperature. Stress must be calculated by Hooke's or some other nonquestionable law. Moreover, shape and size of a specimen are very important. As an example - a cylindrical specimen is very convenient due to its axial symmetry but since measurements are limited to its outer surface we must assume something about spatial homogeneity. If a cruciform specimen is chosen, then it provides very well a homogeneous two-dimensional straining picture but the stress analysis is very complex.

For fast processes (detected by a small specimen like in testing by Hopkinson bar technique) measurements are performed outside the specimen and plastic waves are almost always neglected. Such an assumed homogeneity inside the specimen is a very rough assumption[2] and it influences our knowledge about such "facts" like upper and lower yield point.

- Microstructural topological changes either for single crystals or for polycrystals are extremely hard to be adequately modeled due to the very large number of structural defects and cracks. Their motion is so complex that in crystalline materials dislocations often behave and move like crazy snakes taking peculiar forms extending over more volume elements (cf. [Ant92]). Nevertheless, their micro-motion is the main cause of plastic straining.

  On the other hand, plastic slips cause breaking of symmetry if defined in the traditional way. For single crystals the frame of reference must be connected to crystal frame. If such a frame is chosen, then corresponding invariants in a defective single crystal[3] are not affected by elastic straining. With regard to symmetry the problem is how to properly define symmetry in a polycrystal volume element (often termed *representative volume element*) composed of many grains with random orientations and diverse inter and intragranular slips.[4]

  Here the non-Euclidean approach promoted by the classic works [Eck48, Kon58, Kro60, Bil60, Sto62] must be learned prior to any serious attempt to give a correct explanation of inelastic straining processes. This is even emphasized when we try to cover magneto-inelastic coupling or other nonmechanical causes of inelastic behaviour.
- If polycrystals are treated, then intergranular residual stresses are such a serious problem that the usual attempt is to assume some kind of self-consistent method. In other words, such a method often considers behaviour of one grain as isolated and immersed into otherwise continuous effective matrix. How do we find adequate properties of such a matrix? Is it necessary to take into account couple stresses or not?
- Next difficulty is a synergetic coupling of viscoplastic behaviour with nonmechanical processes like magnetization, irradiation, etc. Damage induced

[2] The situation is explained in detail in the paper [MB-02]. It is shown there that a homogeneity is reached only at advanced strains.

[3] This has been analyzed in [DP-91].

[4] A promising and original approach to identification of symmetry type has been given by Rychlewski and his followers [Ryc83, KO-04].

either by neutron irradiation or by thermal creep is also present. The coupling must be treated properly since an approach by the so-called principle of equipresence is too general and cannot tell us anything about the order of magnitude of diverse couplings. Anyway such couplings are often a treasure, which helps us to detect where characteristic points like yield onset are situated. Such information is mostly valuable since there are many definitions of yield commencement. Minimum of measured temperature gives precious information about such a point at multiaxial tension tests.

- As Naghdi mentioned in his review [Nag90] a number of issues in plasticity of finite strains are controversial and major disagreements do exist since the field has been rapidly developing. Notwithstanding, there exists a consensus about the decisive role of thermodynamics in proper treatment of constitutive as well as evolution equations. However, which version of thermodynamics to apply—classical, rational, extended or endochronic—when we do not know even how to define nonequilibrium temperature and entropy?

A collection of all these difficulties seems very discouraging for prospective authors. The question of "why a generally accepted framework for plasticity is still lacking" posed in [Ber98] is too optimistic according to our taste. Anyway, something has to be done. The most important issues that should not be missed in a book like this, as listed in the subsequent chapters, are:

- Straining geometry with explanation of continuous distribution of dislocations deserves special attention since there is often a misunderstanding of this delicate issue. Mistakes often lead to misinterpretation of such important things like plastic spin.
- Does associativity of flow rule hold in general and if the answer is "yes," is it supported by thermodynamics?
- Is it better to use a memory approach by means of functionals or internal variables? If memory is preferred, should we then take the history of total or plastic strain? In other words, does a viscoplastic material have a total or plastic memory?
- A brief review of thermodynamics of inelastic processes is inevitable if we want to see how to correctly impose restrictions on constitutive and evolution equations.
- Experiments covering a broad range of strain rates and multiaxial stress and strain histories are indispensable for calibration of a theory and check of its validity.
- A brief treatment of soft ferromagnetics given here is aimed, first of all, at an explanation of nondestructive testing like low-cycle fatigue. In the prospective we hope that it could serve fast inelastic processes.[5]

[5] Among the extensive list of references in this field the comprehensive book [Mau88] is the nearest to a conventional approach to continuum mechanics and gives almost everything needed for a more profound insight into the subject.

- How to treat polycrystals? Is it necessary to take into account couple stresses or not? If the answer is "yes," how to do that in the simplest way? It is known that the self-consistent method for polycrystals is an approximation. It would be worthwhile to include in such a concept more sophisticated geometric concepts of continuum theory of dislocations.

The very limited number of references given here is incomplete but chosen in such a way to support the exposition.

1

Physical and Geometrical Background

This chapter is devoted to fundamental issues common to coupled fields inside a crystalline body deforming plastically. A lot of figures, some of them being very simple, serve to make the text more comprehensive. It is basic for almost all the subjects in the subsequent chapters of this monograph. When necessary, in each of the following chapters a brief kinematical addendum is given.

1.1 Basic notions and simple concepts

In order to understand clearly the fundamental difference between elastic and plastic strains consider the following two deformations given either by $(\kappa) \rightarrow (\chi')$, or by $(\kappa) \rightarrow (\chi'')$ depicted in Fig. 1.1.

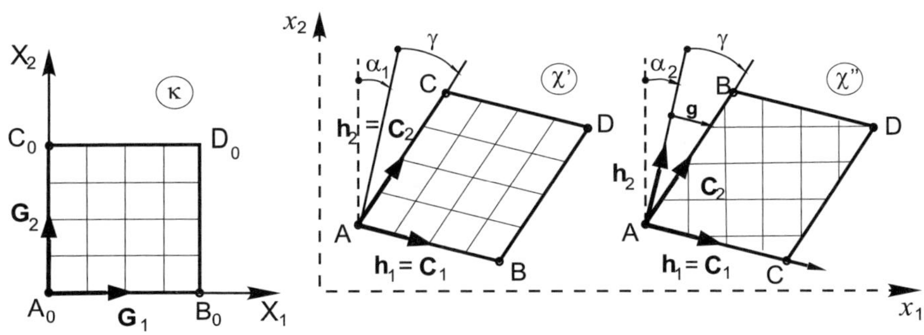

Fig. 1.1. The difference between elastic and plastic strain

Comparing (κ) and (χ') we see that they have different shapes and different intrinsic (i.e., crystalline) arrangements. Let (κ) be a natural state configuration which means a configuration at a uniform temperature field and without any mechanical loads. Then it is quite clear that (χ') can be maintained only

M.V. Mićunović, *Thermomechanics of Viscoplasticity: Fundamentals and Applications*, Advances in Mechanics and Mathematics 20, DOI: 10.1007/978-0-387-89490-4_1,
© Springer Science + Business Media, LLC 2009

by means of some shearing forces realizing the shear angle γ. After removal of these forces we would get (κ) configuration completely recovered. Hence, the mapping $(\kappa) \to (\chi')$ is an elastic deformation.

On the other hand, shapes of (κ) and (χ'') are different but their lattices are identical.

Thus, (χ'') has been permanently deformed and no removal of external forces is needed to get a relaxed state since (χ'') itself is already a natural state configuration. Therefore, the mapping $(\kappa) \to (\chi'')$ is a plastic deformation.

This simple figure is very enlightening due to its significant information about the very nature of elastic and plastic deformations. From a "black box" approach (i.e., an approach which disregards the intrinsic structure) only the total deformation defined by the shape of the considered body could be estimated. However, for $\alpha_1 = \alpha_2$ shapes and positions of (χ') and (χ'') are identical but their properties are fundamentally different.

Consequently, in order to give a precise meaning to the terms: total, elastic and plastic deformation, not only the shape change but also the intrinsic structure change must be taken into account.

In the above example all the configurations have been made of crystals without defects. On the contrary, in the following figure (κ) is a piece of an ideal crystal, while (χ) is a real crystal configuration.

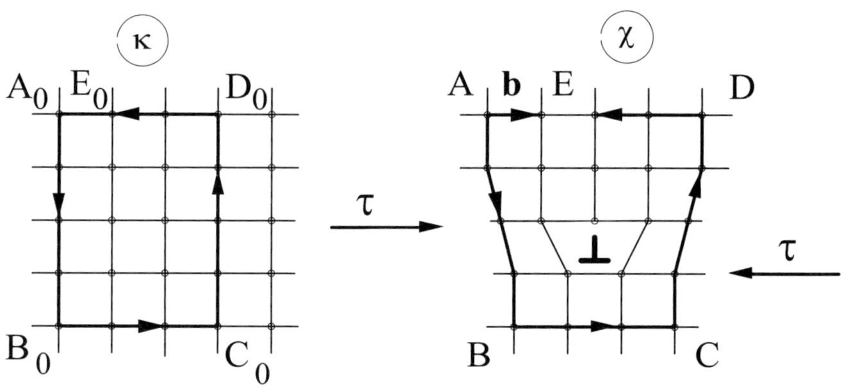

Fig. 1.2. Elementary edge dislocation

The deflection of the real crystal structure of (χ) with respect to an ideal crystal (κ) is defined by the mapping of the closed contour $A_0B_0\ C_0D_0E_0$ in (κ) onto the corresponding open contour $ABCDE$ in (χ). For a chosen sequence of atoms in (κ) and the positive walk direction around the loop, the closure failure vector in (χ) is always the same and amounts to $\mathbf{b}$, which is referred to as the Burgers vector. Examining (χ) more closely we see that imperfection of (χ) is caused by insertion of a half-plane of additional atoms. Its boundary is called the dislocation line. Obviously, the magnitude of Burgers

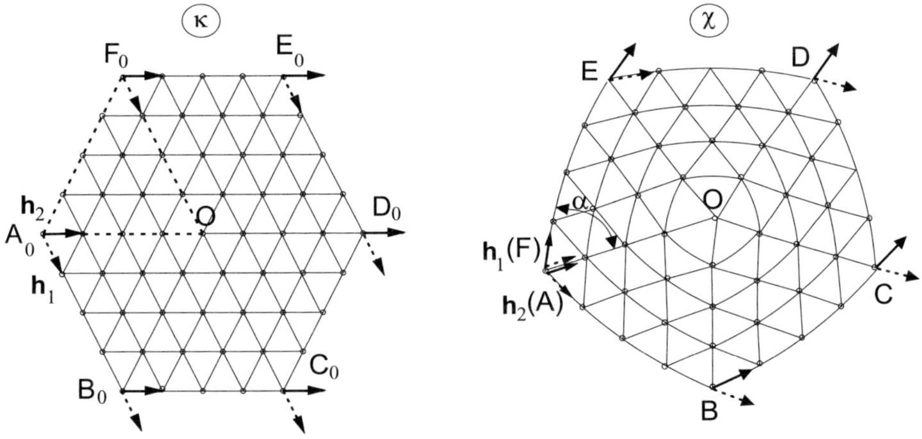

Fig. 1.3. Elementary wedge disclination

vector depends on the number of dislocations encircled, as well as on their sign. Usually, dislocation density is defined by dividing the Burgers vector with the encircled area, i.e.,

$$\alpha = \frac{b}{\triangle A}, \tag{1.1}$$

which will be stated more precisely (in tensorial notation) in the following section.

The most distinctive feature of the above real crystal is that the glide in the direction of **b** may be caused by very small shearing forces due to the fact that atoms nearest to the dislocation line are in an instable equilibrium. A jump to the nearest stable position would be an elementary plastic deformation as a slip along preferred crystallographic planes and directions. The shown dislocation where **b** is perpendicular to its line is termed an edge while a screw dislocation has its line and **b** mutually parallel.

The other type of closure failure is represented in Fig. 1.3 (cf. [KA-75]).

In ideal crystal configuration (κ), let us choose a closed contour A_0 $B_0C_0D_0E_0F_0A_0$, defined by the corresponding atomic spacings, and subsequently let us move two crystallographic vectors (say $\mathbf{h}_1$ and $\mathbf{h}_2$) along the contour by parallel displacements. Then $\mathbf{h}_1$ and $\mathbf{h}_2$ in A_0 will coincide after the whole encirclement. Imagine now that the wedge A_0OF_0 is removed and edges A_0O and OF_0 are subsequently joined together to form configuration (χ). Repeating the encirclement along $ABCDEF$ contour we note that $\mathbf{h}_1$ and $\mathbf{h}_2$ at A are not coincident any more with those at F and that a closure failure angle δ shows the imperfection of the lattice of (χ). The above crystal defect, for the obvious reason, is called a *wedge disclination* and its density could by represented by

$$D = \frac{\delta}{\triangle A}, \tag{1.2}$$

where $\triangle A$ is the encircled area. Again, the more general tensorial formula will be given in the next section. The other type of crystal disclination possible is referred to as a twist disclination.

Comparing a disclination with a dislocation it is seen that a disclination line is not moved so easily inside a crystal and that for its creation a high energy would be necessary. Thus, only partial disclinations could exist in a real solid crystal ([KA-75]).

Other types of defects are vacancies (a missing atom from the lattice) or interstitial atoms (extra atoms immersed into the lattice) and they prevent or enhance plastic deformation accomplished by a dislocation motion.

Concluding this section it is possible to say that a real crystal is made of crystal grains which encompass dislocations whereas different orientations of grains are due to the existence of intergranular dislocations. Their highly complicated motion inside and between the grains is caused by external forces or activated by a corresponding temperature field.

1.2 Intermediate configurations, continuum geometry and kinematics

Consider, now, a real crystal body, $\mathcal{B}$, in its configuration (χ_t) (which is time-dependent) acted upon by external forces and possessing an inhomogeneous time-dependent temperature field $\theta(X^K,t) \equiv T(X^K,t) - T_0$, where X^K $(K \in \{1,2,3\})$ are material coordinates of $\mathcal{B}$ and t is the time. Here T and T_0 are absolute current and room temperatures, respectively, whereas $\theta(X^K,t)$ is the corresponding temperature increment. This configuration has been obtained from the initial real crystal configuration (χ_0) of $\mathcal{B}$ by means of the mentioned external agencies (cf. Fig. 1.4). During the loading process the shape of $\mathcal{B}$ as well as its intrinsic structure are changed (in general) so that when external agencies are taken away (χ_t) does not relax into (χ_0) but into a so-called "unloaded" configuration (χ_t^{res}).

The difference between (χ_0) and (χ_t^{res}) is usually referred to as plastic deformation while the difference between (χ_t) and (χ_t^{res}) is identified with elastic deformation. In the initial elastic region (χ_0) and (χ_t^{res}) coincide while in subsequent elastic regions (χ_t^{res}) does not change for fixed (χ_t). At first sight this picture is quite logical and, moreover, fits perfectly well with the experimental evidence being termed here the engineering picture for this reason.

However, looking more carefully at (χ_0) and (χ_t^{res}) we note that these configurations enclose a great many line and point defects causing residual stresses related to elastic residual strains. The only way to relax the body completely is to cut it isothermally into small elements free of neighbors getting in this way from (χ_t) a mosaic of (ν_t) local natural state elements at constant temperature T_0. If they are small enough, their crystals are ideal and the tearing process is "perfect". In the terminology of [Kon58] it is called a

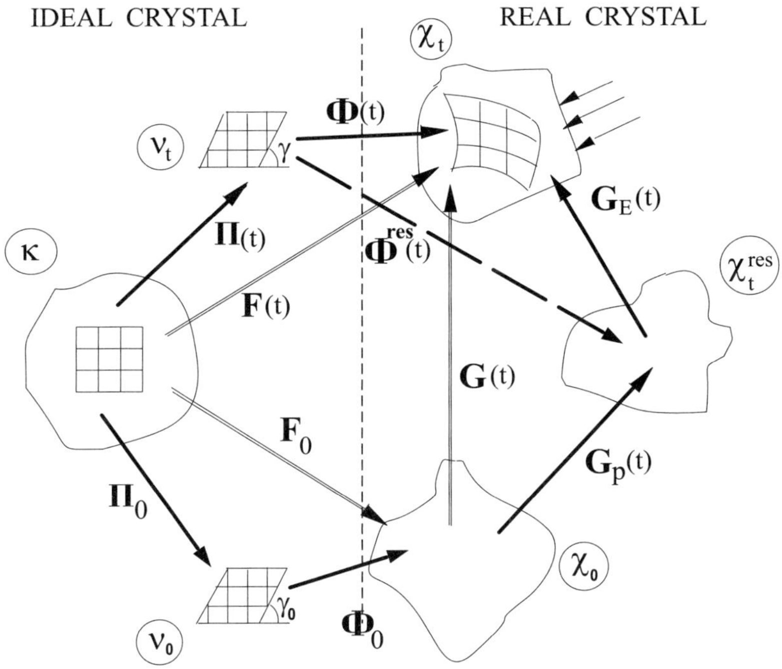

Fig. 1.4. Basic configurations and distortions

perfect tearing. Otherwise, they would still contain a small (negligible) number of defects. The difference between (ν_t) and the corresponding particle in (χ_t) shows now the whole elastic strain, the difference between (ν_t) and (χ_t^{res}) amounts to the residual elastic strain while the difference in (χ_t) and (χ_t^{res}) gives the engineering elastic strain.

The whole plastic strain is gained by comparing a global (ideal crystal) reference configuration (κ) at (θ_0) with the local elements (ν_t). The picture obtained in this way could be termed the intrinsic structure picture.

Obviously, a picture with inverse direction (cf. [Bil60]) is equally valid. Namely, we contemplate that (κ) is isothermally cut into small elements, these subsequently being plastically deformed in the way presented in Fig. 1.1 (mappings of the type $(\kappa) \to (\chi''))$ and joined together into a configuration (χ_t) by means of external forces and an inhomogeneous temperature field.

It should be remarked that elements (ν_t) could be joined together into a continuous stress-free entity (without forces) only in some non-Euclidean space (e.g., [Eck48, Sto62, Ant70, Mic74b]) giving a vast space for interesting and useful mathematical interpretations omitted here for the sake of brevity. Only some of its most important aspects will be reviewed in the sequel keeping simplicity of exposition.

Translating the above considerations into the language of continuum mechanics the following formulae are acquired:

$$2\mathbf{E} = \mathbf{F}^T\mathbf{F} - \mathbf{1}, \tag{1.3}$$

$$2\mathbf{e} = \mathbf{1} - \mathbf{F}^{-T}\mathbf{F}^{-1}, \tag{1.4}$$

$$2\mathbf{E}_E = \boldsymbol{\Phi}^T\boldsymbol{\Phi} - \mathbf{1}, \tag{1.5}$$

$$2\mathbf{e}_E = \mathbf{1} - \boldsymbol{\Phi}^{-T}\boldsymbol{\Phi}^{-1}, \tag{1.6}$$

$$2\mathbf{E}_P = \boldsymbol{\Pi}^T\boldsymbol{\Pi} - \mathbf{1}, \tag{1.7}$$

$$2\mathbf{e}_P = \mathbf{1} - \boldsymbol{\Pi}^{-T}\boldsymbol{\Pi}^{-1}, \tag{1.8}$$

where $\{\mathbf{E}, \mathbf{E}_E, \mathbf{E}_P\}$ and $\{\mathbf{e}, \mathbf{e}_E, \mathbf{e}_P\}$ are, respectively, Lagrangian and Eulerian, total, elastic and plastic strain tensors in intrinsic picture.

These tensors are obtained by comparing lengths of the infinitesimal vectors $d\mathbf{r}_\kappa$, $d\mathbf{r}_\nu$ and $d\mathbf{r}$ connecting the same adjacent points M and N in (κ), (ν_t) and (χ_t) configurations, respectively. For instance,

$$ds^2 - ds_\kappa^2 \equiv d\mathbf{r} \cdot d\mathbf{r} - d\mathbf{r}_\kappa \cdot d\mathbf{r}_\kappa = 2d\mathbf{r}\ \mathbf{e}\ d\mathbf{r} = 2d\mathbf{r}_\kappa\ \mathbf{E}\ d\mathbf{r}_\kappa, \tag{1.9}$$

etc. The corresponding engineering strain tensors are easily met by means of

$$2\mathbf{E}^{\text{eng}} = \mathbf{G}^T\mathbf{G} - \mathbf{1}, \tag{1.10}$$

$$2\mathbf{e}^{\text{eng}} = \mathbf{1} - \mathbf{G}^{-T}\mathbf{G}^{-1}, \tag{1.11}$$

$$2\mathbf{E}_E^{\text{eng}} = \mathbf{G}_E^T\mathbf{G}_E - \mathbf{1}, \tag{1.12}$$

$$2\mathbf{e}_E^{\text{eng}} = \mathbf{1} - \mathbf{G}_E^{-T}\mathbf{G}_E^{-1}, \tag{1.13}$$

$$2\mathbf{E}_P^{\text{eng}} = \mathbf{G}_P^T\mathbf{G}_P - \mathbf{1}, \tag{1.14}$$

$$2\mathbf{e}_P^{\text{eng}} = \mathbf{1} - \mathbf{G}_P^{-T}\mathbf{G}_P^{-1}. \tag{1.15}$$

Obviously, the multiplicative decompositions

$$\mathbf{F} = \boldsymbol{\Phi}\boldsymbol{\Pi}, \tag{1.16}$$

as well as

$$\mathbf{G} = \mathbf{G}_E\mathbf{G}_P, \tag{1.17}$$

are to be taken into account.

A few important notes are now in order.

Remark 1.1. (Additive decomposition)

First of all, the very popular additive decompositions of strains (e.g., [RK-88, Per71]) like

$$\mathbf{E} = \mathbf{E}_E + \mathbf{E}_P, \quad \mathbf{e} = \mathbf{e}_E + \mathbf{e}_P \tag{1.18}$$

are only approximately valid for very small plastic strains which harshly restricts the field of their application. Instead, the correct formulae read

$$\mathbf{E} = \mathbf{E}_P + \mathbf{\Pi}^T \mathbf{E}_E \mathbf{\Pi}, \quad \mathbf{e} = \mathbf{e}_E + \mathbf{\Phi}^{-T} \mathbf{e}_P \mathbf{\Phi}^{-1}, \tag{1.19}$$

and these can be easily derived from (1.16) and (1.3)–(1.8). The same nonlinearity (in decomposition) holds for engineering strains.

As an elementary illustration of the difference between the above linear and nonlinear decompositions let us consider the case of uniaxial deformation of a straight rod having lengths $L_0, L^{\rm res}, L$ in engineering configurations $(\chi_0), (\chi^{res}), (\chi)$, respectively. From this sequence it is elementary to calculate that

$$E_{11}^{\rm eng} = E_{P_{11}}^{\rm eng} + E_{E_{11}}^{\rm eng}(1 + 2E_{P_{11}}^{\rm eng}) \neq E_{P_{11}}^{\rm eng} + E_{E_{11}}^{\rm eng}. \tag{1.20}$$

A similar formula holds true in the intrinsic picture.

Remark 1.2. (Kröner's decomposition)

Very often, decomposition (1.16) is believed to have introduced by E. H. Lee for the first time (cf. [Lee69]). This is not true and E. Kröner was apparently the first who wrote this decomposition in [Kro60]. A lot of others, like Bilby, Stojanovic, Kondo, etc., have made extensive use of this decomposition[1] before Lee.

Remark 1.3. (Elastic and plastic rotations)

A great deal of confusion is introduced by misunderstanding of plastic and elastic rotations. This could be explained in simple terms in Fig. 1.5. Introducing the polar decomposition (cf. [TN-65]) for the plastic deformation "gradient" tensor

$$\mathbf{\Pi} = \mathbf{R}_P \mathbf{U}_P = \mathbf{V}_P \mathbf{R}_P, \quad \mathbf{U}_P^2 = \mathbf{\Pi}^T \mathbf{\Pi}, \quad \mathbf{R}_P^T = \mathbf{R}_P^{-1}, \tag{1.21}$$

its stretch part $\mathbf{U}_P$ shows the deformation of a sphere in (κ) into an ellipsoid in (ν_t) while its rotation part $\mathbf{R}_P$ describes an arbitrary rotation α and principal axes rotation β caused by shear γ. The fundamental fact that (κ) and (ν_t) are intrinsically indistinguishable allows us to fix α and most naturally to put $\alpha = 0$. Then crystal directions in (κ) and (ν_t) become parallel (*isoclinic* in the terminology of [Man81]). In this way plastic rotation is defined only by shear and not by an arbitrary rotation analogous to objective global rotations of spatial frames (cf. [Lee69, Nem79]).

A formula similar to (1.21) holds also for elastic deformation:

$$\mathbf{\Phi} = \mathbf{R}_E \mathbf{U}_E = \mathbf{V}_E \mathbf{R}_E, \quad \mathbf{U}_E^2 = \mathbf{\Phi}^T \mathbf{\Phi}, \quad \mathbf{R}_E^T = \mathbf{R}_E^{-1}. \tag{1.22}$$

Its rotation part is set by Lee to be unity [LL-81]. Since there exists only a global rotation of spatial coordinate frame in (χ_t) this would mean that we

[1] The first notions about a non-Euclidean interpretation of plastic strains were made in [Eck48].

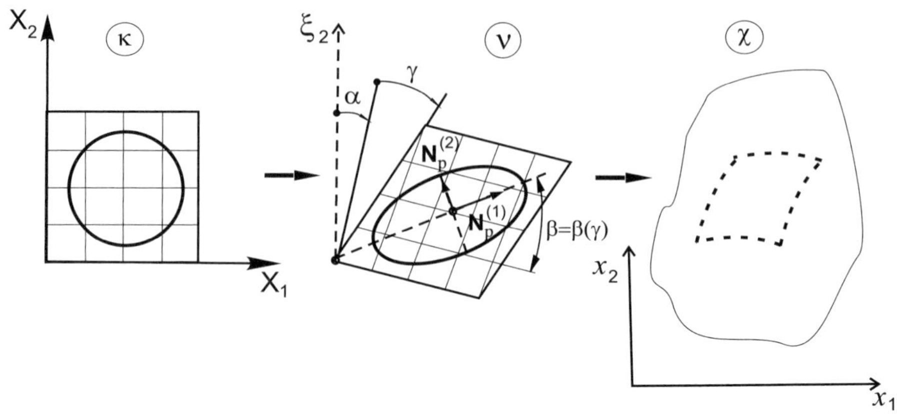

Fig. 1.5. Plastic and elastic rotations

would have to adapt orientations of (ν_t) elements in order to get $\mathbf{R}_E = \mathbf{1}$. This choice is very artificial and allows for a confusion between an arbitrary angle α (introduced by an observer) and crystal shear angle γ which is by its very definition a physically significant feature.

In terms of isoclinic (κ) and (ν_t) configurations the plastic distortion $\mathbf{\Pi}$ can always be represented by a product of elementary slip deformation tensors

$$\left\{\begin{matrix} 1 & \pi_{(1)} & 0 \\ 0 & 1 & 0 \\ 0 & 0 & 1 \end{matrix}\right\}, \quad \left\{\begin{matrix} 1 & 0 & 0 \\ \pi_{(2)} & 1 & 0 \\ 0 & 0 & 1 \end{matrix}\right\}, \quad \ldots, \quad \left\{\begin{matrix} 1 & 0 & \pi_{(6)} \\ 0 & 1 & 0 \\ 0 & 0 & 1 \end{matrix}\right\}.$$

They are introduced in [Mic86b] and called plastic slip generators. Their importance lies in the fact that nine components of $\mathbf{\Pi}$ may be expressed by six elementary plastic slips

$$\mathbf{\Pi} = \prod_{\alpha \in I} \mathbf{\Pi}(\pi_{(\alpha)}), \tag{1.23}$$

where $I \subset \mathcal{I}$ and $\mathcal{I}$ is the set of all permutations of the indices $\{1, \ldots, 6\}$. For each I the (ν_t) configuration is locally isomorphic with the global ideal crystal (κ) (cf. Fig. 1.4).[2] This is completely in accordance with the above discussion.

Remark 1.4. (Principal stretches)

Besides generally accepted strain tensors, $\mathbf{E}$ and $\mathbf{e}$ (cf. (1.3)–(1.15)) it is possible to introduce (cf. [Hil68, Ogd03]) generalized strain measures by

$$\mathbf{E}^{(m)} = \sum_a \frac{\lambda_a^{2m} - 1}{2m} \mathbf{N}^a \otimes \mathbf{N}^a, \tag{1.24}$$

[2] Given the resulting distortion $\mathbf{\Pi}$, the decomposition (1.23) is not unique. Moreover as illustrated in [Ant92] different choices of I could give very strange and unexpected isomorphic local faces inside a volume element.

where λ_a ($a \in \{1,2,3\}$) are principal extension ratios (along principal directions) and $\mathbf{N}^a$ are principal direction unit vectors. For $m = 1$ we get (1.3) or (1.10), i.e., Lagrangian strain tensor, $m = -1$ corresponds to Almansi strain tensor, whereas for $m = 0$ and $m = \frac{1}{2}$ Hill's logarithmic and Biot's strain tensors are acquired, respectively.

All these tensors vanish when $\lambda_1 = \lambda_2 = \lambda_3 = 1$, which means that for $ds = ds_\kappa$ all of them are equal to zero. Similar formulae are available for elastic and plastic tensors as well.

Here, one property is worth mentioning. By means of the relations

$$\mathbf{E}_E^{\text{gen}} = \sum_a \frac{\lambda_{Ea}^{2m} - 1}{2m} \mathbf{N}_E^a \otimes \mathbf{N}_E^a, \quad \mathbf{E}_P^{\text{gen}} = \sum_a \frac{\lambda_{Pa}^{2m} - 1}{2m} \mathbf{N}_P^a \otimes \mathbf{N}_P^a, \tag{1.25}$$

it is possible to show that

$$\lambda_a \neq \lambda_{Ea} \lambda_{Pa} \tag{1.26}$$

since proper directions of total, elastic and plastic deformations do not correspond to the same material fibres (curves along the same material points).

Choosing m to be equal to zero would lead to *Hill's natural logarithmic strains*[3] as follows:

$$\mathbf{E}^{\text{Hill}} = \sum_a \ln(\lambda_a) \mathbf{N}^a \otimes \mathbf{N}^a, \tag{1.27}$$

$$\mathbf{E}_E^{\text{Hill}} = \sum_a \ln(\lambda_{Ea}) \mathbf{N}_E^a \otimes \mathbf{N}_E^a, \quad \mathbf{E}_P^{\text{Hill}} = \sum_a \ln(\lambda_{Pa}) \mathbf{N}_P^a \otimes \mathbf{N}_P^a. \tag{1.28}$$

At first sight this definition would lead to additivity. However, lack of coincidence of principal vectors leads to (1.26) such that for a general strain history the inequality

$$\mathbf{E}^{\text{Hill}} \neq \mathbf{E}_E^{\text{Hill}} + \mathbf{E}_P^{\text{Hill}} \tag{1.29}$$

holds. It may be satisfied only in some very special cases like coaxial plastic as well as elastic tension, etc.

Thus, application of (1.27)–(1.28) is not convenient for general considerations in plasticity.[4]

Remark 1.5. From all important relationships at this point only those appearing in the first and the second law of thermodynamics are quoted. First of all the plastic strain time rate $D_t\mathbf{E}_P$[5] amounts to

$$D_t \mathbf{E}_P = \mathbf{N} D_t p, \tag{1.30}$$

[3] In subsequent sections we will use other notations: either $\boldsymbol{\varepsilon}_P$ or $\boldsymbol{\varepsilon}^{pl}$ commonly explored in engineering practice.

[4] A lot of other important remarks about geometry of deformation could be made (e.g., [Mic74b]) but have to be omitted for the sake of brevity.

[5] The notation $D_t(\bullet) \equiv \frac{\partial(\bullet)}{\partial t} + \mathbf{v}\,\text{grad}(\bullet)$ stands for material time derivative.

with

$$D_t p^2 \stackrel{def}{=} D_t \mathbf{E}_P : D_t \mathbf{E}_P \equiv \mathrm{tr}\{D_t \mathbf{E}_P^2\} \geq 0 \tag{1.31}$$

and

$$p(t) = \int_{t_0}^{t} D_t p(\tau)\, d\tau. \tag{1.32}$$

Here $\mathbf{N}$ is the symmetric second-order unit length tensor of plastic strain rate direction, whereas the scalar p represents the nondecreasing plastic deformation path length (cumulated plastic strain in terminology of [CR-83]).

Alternatively, with the representation

$$D_t \mathbf{\Pi} = \mathbf{M} D_t p = \sum_{\alpha=1}^{6} D_t \pi_{(\alpha)} \partial_{\pi_{(\alpha)}} \mathbf{\Pi}, \tag{1.33}$$

where $\mathbf{M}$ is an asymmetric nonunit tensor, the relationship

$$2\mathbf{N} = \mathbf{M}^T \mathbf{M} + \mathbf{M}\mathbf{M}^T \tag{1.34}$$

is obtained readily.

On the other hand, the total stretching tensor [TN-65] is by definition the symmetric part of the velocity gradient tensor ($2\mathbf{D} = \mathbf{L} + \mathbf{L}^T$):

$$\mathbf{L} := D_t \mathbf{F}\, \mathbf{F}^{-1}, \tag{1.35}$$

whose elastic as well as plastic counterparts amount to

$$\mathbf{L}_E := D_t \mathbf{\Phi}\, \mathbf{\Phi}^{-1}, \tag{1.36}$$

$$\mathbf{L}_P := D_t \mathbf{\Pi}\, \mathbf{\Pi}^{-1} = D_t p\, \mathbf{M} \mathbf{\Pi}^{-1}, \tag{1.37}$$

respectively. Again the corresponding decomposition is nonlinear,

$$\mathbf{L} = \mathbf{L}_E + \mathbf{\Phi} \mathbf{L}_P \mathbf{\Phi}^{-1} \neq \mathbf{L}_E + \mathbf{L}_P. \tag{1.38}$$

unless $\mathbf{\Phi} - \mathbf{1}$, $D_t \mathbf{\Phi}$ and $D_t \mathbf{\Pi}$ are infinitesimal (i.e., approximately very small of the same order). Very often this fact has been misunderstood by many people dealing with theoretical, experimental and numerical plasticity.

1.3 Characteristics of the intrinsic structure. Continuous dislocations

Let us concentrate now more carefully on dislocation densities explained roughly in the first section of this chapter.

For this purpose consider a natural state element (ν_t) and the corresponding material particle in (χ_t)—Fig. 1.6.

If we choose a closed contour c in (χ_t), then an open contour c_ν in (ν_t) corresponds to it in general, where $d\mathbf{r}, d\mathbf{r}_\nu$ as well as $d\mathbf{h}_\alpha(\chi)$, $d\mathbf{h}_\alpha(\nu), \alpha \in \{1,2,3\}$ are line element vectors along the contours as well as lattice vectors in the mentioned configurations (χ_t) and (ν_t), respectively.

According to the explanation accompanying Fig. 1.2 *Burgers vector* (the closure failure vector) may be presented by

$$\triangle\mathbf{b} := \oint_c d\mathbf{r}_\nu = \oint_c d\mathbf{r}\,\mathbf{\Phi}^{-T} = \int_{\triangle s} \mathbf{n}\,(\nabla \times \mathbf{\Phi}^{-T})\,ds, \tag{1.39}$$

where $\nabla \equiv \nabla_\chi$ is the differential operator defined in (χ_t) configuration, $\triangle s$ is the area of the surface enclosed by c and $\mathbf{n}\,ds$ is the outer surface material element ($\mathbf{n}$ being the unit vector while ds is the area of the element). More precisely, when $ds \to 0$ the integral sign is not necessary any more so that (1.39) becomes

$$d\mathbf{b} = \mathbf{n}\,(\nabla \times \mathbf{\Phi}^{-T})\,ds \equiv \mathbf{n}\,\mathbf{A}_E\,ds, \tag{1.40}$$

where

$$\mathbf{A}_E = \nabla \times \mathbf{\Phi}^{-T} \equiv \ \mathrm{curl}\mathbf{\Phi}^{-T} \tag{1.41}$$

is the *true dislocation density* [Teo70, Mic74b]. This is an asymmetric second-order tensor representing orientations and average Burgers vector for all dislocations piercing a (volume defined) material particle. For example, in the case of parallel screw dislocations along the x_1-axis all the components of $\mathbf{A}_E$ except $(\mathbf{A}_E)_{11}$ equal zero, whereas in the case of parallel edge dislocations along the x_1-axis its only nonzero components are $(\mathbf{A}_E)_{21}$ and $(\mathbf{A}_E)_{31}$.

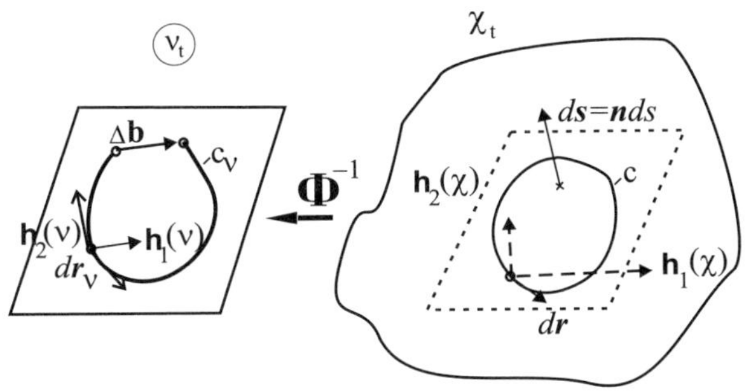

Fig. 1.6. Burgers vector mapping

It is worth noting here that $\mathbf{A}_E$ (like stress tensor) is independent while $d\mathbf{b}$ (like stress vector) is dependent on orientation of some infinitesimal closed curve c encircling the corresponding surface element $\mathbf{n}\,ds$. This is very important because at each point there exists a bundle of infinite number of diverse orientations.

In the relationship (1.39) the Burgers vector is determined in (ν_t) configuration. If it is mapped by means of an elastic distortion to (χ_t) configuration, then we get *local dislocation density* (cf. [Teo70, Mic74b]) as follows:

$$\mathbf{A}_E^{\text{local}} = \mathbf{A}_E \,\mathbf{\Phi}^T = (\nabla \times \mathbf{\Phi}^{-T})\,\mathbf{\Phi}^T = (\text{curl}\mathbf{\Phi}^{-T})\,\mathbf{\Phi}^T. \tag{1.42}$$

On the other hand, as shown in Fig. 1.3 a disclination is introduced and described by lattice vectors in configurations (ν_t) and (χ_t). They are different due to elastic deformation so that (cf. Fig. 1.6)

$$\mathbf{h}_\alpha(\chi_t) = \mathbf{\Phi}\,\mathbf{h}_\alpha(\nu_t), \qquad \alpha \in \{1,2,3\}. \tag{1.43}$$

If the real crystal in (χ_t) possesses disclinations, then the integral

$$\oint_c \delta\mathbf{h}_\alpha(\chi_t) \;=\; \triangle\mathbf{h}_\alpha(\chi_t) \;=\; \Big(\oint_c \delta\mathbf{\Phi}\Big)\,\mathbf{h}_\alpha(\nu_t) \tag{1.44}$$

is not equal to zero (compare to right-hand side of Fig. 1.3), which means that $\delta\mathbf{\Phi}$ is not a total differential but may be expressed by third-order tensor $\mathcal{A}_E$ in the following way:

$$\delta\mathbf{\Phi} \;=\; d\mathbf{r}\,\mathcal{A}_E \neq\; d\mathbf{r}\,\text{grad}\mathbf{\Phi}. \tag{1.45}$$

Thus, transforming the line integral appearing in (1.44) by means of the corresponding surface integral

$$\begin{aligned}\triangle\mathbf{h}_\alpha(\chi_t) \;&=\; \Big(\oint_c d\mathbf{r}\,\mathcal{A}_E\Big)\,\mathbf{h}_\alpha(\nu_t) = \Big(\int_{\triangle s} \mathbf{n}\,(\nabla \times \mathcal{A}_E)ds\Big)\,\mathbf{h}_\alpha(\nu_t)\\ &=\; \Big(\int_{\triangle s} \mathbf{n}\,(\nabla \times \mathcal{A}_E)\,ds\Big)\,\mathbf{\Phi}^{-1}\,\mathbf{h}_\alpha(\chi_t)\end{aligned} \tag{1.46}$$

and letting the area $\triangle s$ tend to zero we would acquire

$$\triangle\mathbf{h}_\alpha(\chi_t) \;= \mathbf{n}\,(\nabla \times \mathcal{A}_E)\,\mathbf{\Phi}^{-1}\,\mathbf{h}_\alpha(\chi_t)\,ds. \tag{1.47}$$

The third-order tensor

$$\mathcal{D}_E := (\nabla \times \mathcal{A}_E)\;\;\mathbf{\Phi}^{-1} = (\text{curl}\mathcal{A}_E)\;\;\mathbf{\Phi}^{-1} \tag{1.48}$$

vanishes whenever $\delta\mathbf{\Phi}$ is a total differential for in that case

$$\text{curl}\,\mathcal{A}_E \;=\; \text{curl grad}\,\mathbf{\Phi} \;\equiv\; \nabla \times \nabla \otimes \mathbf{\Phi} \;=\; 0.$$

Again, it is worth noting that $\mathcal{D}_E$ does not depend on the orientation chosen and that it describes closure failures for all three lattice vectors in (χ_t) after encirclement of c. For these reasons it is termed the *true disclination density* (cf. [Mic74a]) whose simplest representative is (1.2). As an example in the case of a wedge disclination line parallel to the x_1-axis the only nonzero components of $\mathcal{D}_E$ would be $(\mathcal{D}_E)_{221}$ and $(\mathcal{D}_E)_{231}$.

The above formulae for dislocation and disclination densities are stated in terms of elastic deformation "gradient" tensor. By making use of the fundamental decomposition (1.16) and the fact that the total deformation is compatible, i.e.,

$$\nabla \times \mathbf{F}^{-T} \equiv \mathrm{curl}\ \mathbf{F}^{-T} = 0, \quad \nabla_0 \times \mathbf{F}^{T} \equiv \mathrm{CURL}\ \mathbf{F}^{T} = 0, \tag{1.49}$$

(which follow from $F^k_K = \partial x^k / \partial X^K$, where x^k are the spatial and X^K the material coordinates), we could reach the *reference dislocation density*

$$\mathbf{A}_P = \mathrm{CURL}\boldsymbol{\Pi}^T \equiv \nabla_0 \times \boldsymbol{\Pi}^T = (\mathrm{det}\mathbf{F})\ \mathbf{A}_E\ \mathbf{F}^{-T}, \tag{1.50}$$

where $\nabla_0 \equiv \nabla_\kappa$ is the differential operator defined in (κ) configuration.

Due to the fact that $\mathrm{CURL}\boldsymbol{\Pi}^T$ and curl $\mathbf{F}^{-T}$ are different from zero, $\boldsymbol{\Pi}$ and $\boldsymbol{\Phi}$ cannot be deformation gradient tensors. Instead, they are often termed *distortion* tensors or incompatible deformation tensors. Physically this fact means that (ν_t) elements cannot be fitted into a Euclidean stress-free global configuration but they could be joined into a continuous (χ_t) configuration (filled with defects) only by means of external forces, temperature and residual stress—the last being caused by defects ([Mic74a]).

Expression similar to (1.50) could be derived for the corresponding reference disclination density in terms of the plastic distortion tensor (details are given in [Mic74a]). For the sake of brevity it is not given here.

Furthermore, in the sequel it is supposed that disclinations can exist only as partial disclinations (cf. [KA-75]) wherein it is possible to represent them by $\mathbf{A}_E$ letting $\mathcal{D}_E$ vanish. This has as an advantage that at each material point X^K and each time t the distortion tensors $\boldsymbol{\Phi}$ and $\boldsymbol{\Pi}$ are uniquely defined.

At this point, the following notes should be taken into account:

Remark 1.6. (Burgers or glide vector?)

In some papers (e.g., [Pec83]) Burgers vector is misinterpreted by the glide vector $\mathbf{g}$ (as shown by mapping $(\kappa) \to (\chi'')$ in Fig. 1.1) without a mapping of a closed contour to the corresponding open contour. This has as a consequence that instead of CURL $\boldsymbol{\Pi}^T$ the plastic distortion $\boldsymbol{\Pi}$, itself, is taken as a measure of dislocation density. However, there exist total, elastic and plastic glides (for instance in Fig. 1.1 mapping $(\kappa) \to (\chi')$ is an elastic glide) so that this is not a good measure. While the tensor $\boldsymbol{\Pi}$ is not able to indicate explicitly number, orientations and glide directions of dislocation lines, the true dislocation density $\mathbf{A}_E$ can do all of this. Unacceptability of such a choice for a measure of intrinsic structure irregularities will be discussed in more detail, below.

Remark 1.7. (Grain disorientations)

To account for lattice grain disorientations presented in the left side of Fig. 1.7 or, more generally, in its right side, it would be possible to deal with some volume defined dislocation density measure. One possible definition is proposed as follows.

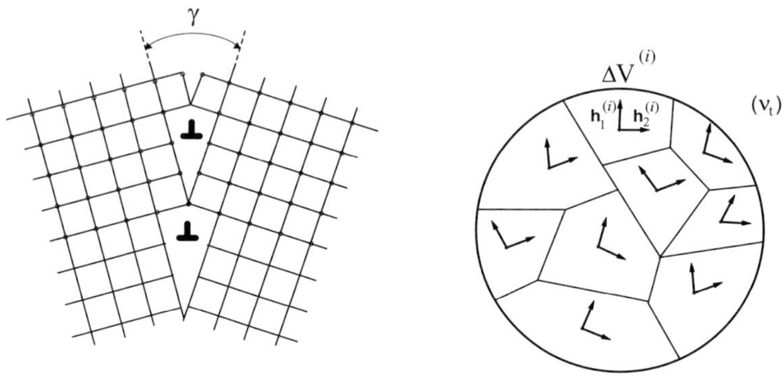

Fig. 1.7. Grain disorientations

In a polycrystalline volume element of natural state (ν_t) configuration let us have some number of single crystal grains with volumes $\triangle V^{(i)}$ and *grain "concentration" factors* $c^{(i)} \equiv \triangle V^{(i)}/\triangle V$ as well as the major grain $\triangle V^{(1)} \equiv \max \triangle V^{(i)}$. Suppose that orthonormal lattice vectors of individual grains are denoted by $\mathbf{h}_\alpha^{(i)}$ with $\mathbf{h}_\alpha^{(1)} \equiv \mathbf{h}_\alpha^{\max}$ corresponding to the major grain. In this case rotations, i.e., disorientations of individual grains with respect to the major grain orientation are given by rotation tensors $\mathbf{R}^{(i)}$, i.e.,

$$\mathbf{h}_\alpha^{(i)} = \mathbf{R}^{(i)}\mathbf{h}_\alpha^{\max}, \quad \mathbf{R}^{(i)}\mathbf{R}^{(i)T} = \mathbf{1}, \quad \alpha \in \{1, 2, 3\}.$$

The average disorientation of the whole element with regard to its major grain orientation is then represented by the tensor

$$\mathbf{R}_m = \sum_i \mathbf{R}^{(i)} c^{(i)}, \tag{1.51}$$

defined on the basis of average lattice vectors $\mathbf{h}_\alpha$

$$\triangle V \ \mathbf{h}_\alpha = \sum_i \triangle V(i) \ \mathbf{h}_\alpha^{(i)}, \tag{1.52}$$

as well as the definition

$$\mathbf{h}_\alpha \equiv \mathbf{R}_m \ \mathbf{h}_\alpha^{\max}. \tag{1.53}$$

Let us now turn to the concept of plastic distortion tensor $\mathbf{\Pi}$. Most naturally, choosing $\mathbf{h}_\alpha$ to be isoclinic with $\mathbf{G}_K$ (cf. Fig. 1.1), i.e.,

$$\mathbf{h}_\alpha = \delta_\alpha^K \, \mathbf{G}_K, \tag{1.54}$$

for double tensor components ([Eri60]) of $\boldsymbol{\Pi}$ we acquire

$$\boldsymbol{\Pi}(h) = \Pi_K^\alpha \, \mathbf{h}_\alpha \otimes \mathbf{G}^K. \tag{1.55}$$

In other words, the whole plastic distortion tensor

$$\boldsymbol{\Pi}(h) = \mathbf{R}_m \, \boldsymbol{\Pi}(h^{\max}) \tag{1.56}$$

takes into account not only single crystal slips described by $\boldsymbol{\Pi}(h^{\max})$ but also disorientations inside a polycrystal described by $\mathbf{R}_m$. Therefore for considerable polycrystal disorientations Kröner's fundamental decomposition (1.16) could be replaced by the more detailed formula

$$\mathbf{F} = \boldsymbol{\Phi} \, \mathbf{R}_m \, \boldsymbol{\Pi} \tag{1.57}$$

with

$$\boldsymbol{\Pi} \equiv \boldsymbol{\Pi}(h^{\max}). \tag{1.58}$$

A noteworthy feature of the above definition for $\mathbf{R}_m$ as a measure of grain disorientations inside a polycrystal is that it does not depend on a surface element orientation inside the polycrystal. The condition for this analysis is that the considered polycrystal possesses a major grain. If on the contrary there is no such dominant grain, then the corresponding analysis would follow the line presented in the last two chapters of the second part of this monograph, where a more detailed analysis is given.

Another approach has been proposed in [Pec83] such that disorientation should be described by the second-rank tensor $\mathbf{D}_{\text{Pech}}$ (cf. (1.48)):

$$d\mathbf{h}_\alpha(\chi) = (\mathbf{D}_{\text{Pech}} - \mathbf{1}) \, \mathbf{h}_\alpha(\chi) = \mathbf{n} \, \mathcal{D}_E \, \mathbf{h}_\alpha(\chi) \, ds, \tag{1.59}$$

describing closure failures of lattice vectors after an encirclement of a closed contour in (χ_t) configuration given by $\mathbf{n}\, ds$. However, this measure (a) depends explicitly on the orientation of the contour (by means of $\mathbf{D}^T_{\text{Pech}} = \mathbf{1} + \mathbf{n} \, \mathcal{D}_E \, ds$), which is a considerable shortcoming, and (b) vanishes whenever disclination density $\mathcal{D}_E$ becomes zero. This is always the case in real solid crystals due to excessive stresses—according to the present state of knowledge in the field [KA-75].

Remark 1.8. (Replacement invariance)

Let us now turn again to the picture depicted in Fig. 1.4. Suppose that instead of the global reference configuration (κ) another global reference configuration (κ^*) is chosen such that both configurations are isoclinic and differ for a homogeneous relative plastic distortion $\boldsymbol{\Pi}_D$ (cf. Fig. 1.8). In such a case, the relation between the two sets of deformation measures is given as

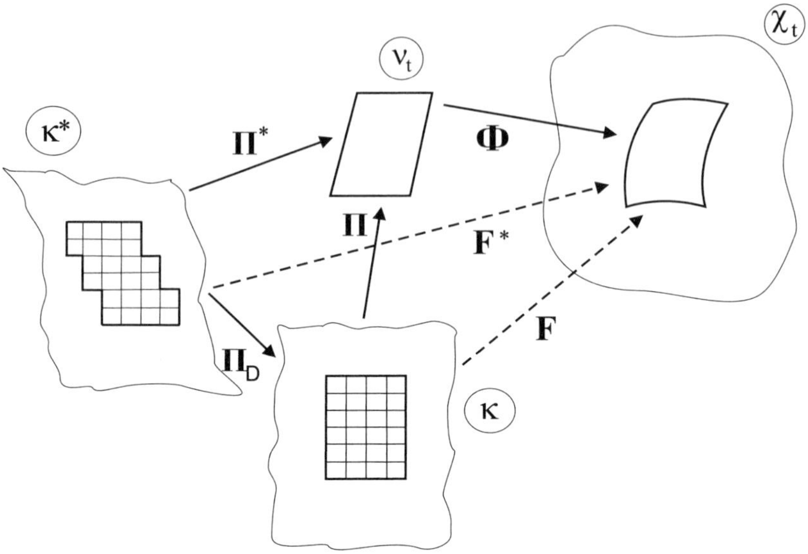

Fig. 1.8. Illustration of replacement invariance requirement

$$
\begin{aligned}
\mathbf{\Pi}^* &= \mathbf{\Pi}\mathbf{\Pi}_D,\ \mathbf{F}^* = \mathbf{F}\mathbf{\Pi}_D,\ \mathbf{\Phi}^* = \mathbf{\Phi}, \\
\mathbf{A}_E^* &= \mathbf{A}_E,\ \mathcal{D}_E^* = \mathcal{D}_E,\ \mathbf{A}_P^* = \mathbf{A}_P\mathbf{\Pi}_D^{-T}, \\
\mathbf{L}^* &= \mathbf{L},\ \mathbf{L}_P^* = \mathbf{L}_P,\ \mathbf{L}_E^* = \mathbf{L}_E, \\
D_t\mathbf{E}_P^* &= \mathbf{\Pi}_D^T D_t\mathbf{E}_P\mathbf{\Pi}_D,\ D_t\mathbf{E}_E^* = D_t\mathbf{E}_E^*,\ D_t\mathbf{E}^* = \mathbf{\Pi}_D^T D_t\mathbf{E}\mathbf{\Pi}_D,
\end{aligned}
\tag{1.60}
$$

where asterisk-labeled measures are aimed to describe the chain mapping $(\kappa^*) \to (\nu_t) \to (\chi_t)$. It should be noted that local reference dislocation density (defined in the same way as (1.50)) is replacement invariant. Obviously,

$$
\mathrm{GRAD}\mathbf{\Pi}_D = 0, \tag{1.61}
$$

$$
\mathbf{\Pi}_D = \prod_{k=1}^{6} \left(\mathbf{1} + \pi_{D(k)}\ \mathbf{g}_{(k)} \otimes \mathbf{n}_{(k)}\right), \quad \det\mathbf{\Pi}_D = 1 \tag{1.62}
$$

have been taken into account to derive (1.56).

Physically, configurations (κ^*) and (κ), being global ideal crystals with the same intrinsic structure, are equivalent[6] so that both descriptions—with or without asterisk—corresponding to the same instant deformed configuration (χ_t) are correct and must be equivalent. Since there exists infinity of constant tensors fulfilling (1.61) and (1.62), plastic distortion tensor $\mathbf{\Pi}$ is determined only up to the set of constant plastic shear tensors (cf. also (1.23)):

$$
\mathcal{P} = \{\mathbf{\Pi}\mathbf{\Pi}_D \mid \mathbf{\Pi}_D \text{ satisfies (1.61) and (1.62)}\}, \tag{1.63}
$$

[6] In other more precise words, they are locally isomorphic.

which emphasizes the fact that dislocations could be introduced into an ideal crystal during solidification only if *inhomogeneous* plastic shears with $\mathbf{A}_P \neq 0$ are applied to it.

The equivalence of global reference configurations, namely, (κ^*) and (κ), has been called *replacement invariance* in [Das86] when discussing application of (1.60) to constitutive equations.

This ambiguity may be eliminated as follows. Namely, it is worth noting that replacement of equivalent global reference configurations has no effect at all if engineering picture of elastic–plastic deformation is chosen. In fact, if we choose mapping $\mathbf{G} : (\chi(t_0)) \Rightarrow (\chi(t))$ as the total deformation gradient, then the corresponding plastic distortion may be defined as the mapping $\mathbf{F}_P :$ $(\nu(t_0)) \Rightarrow (\nu(t))$(cf. Fig. 1.4). In other words:

$$\mathbf{F}_P := \mathbf{\Pi}(\mathbf{t})\mathbf{\Pi}(\mathbf{t_0})^{-1}, \tag{1.64}$$

being the same as $\mathbf{F}^*_P = \mathbf{\Pi}^*(\mathbf{t})\mathbf{\Pi}^*(\mathbf{t_0})^{-1}$. This agrees with the approach by [Teo70]. The key point is that in such a case the global reference configuration is not a hypothetical ideal crystal but the initial configuration $(\chi(t_0))$ composed of real crystal. Therefore,

$$\mathbf{G}(t) = \mathbf{\Phi}(t)\mathbf{F}_P(t)\mathbf{\Phi}(t_0)^{-1} \tag{1.65}$$

is insensitive to plastic shears in global ideal crystal (κ). On the other hand, this definition is not always practical as we will see in forthcoming sections devoted to applications. Namely, if we use Hill's logarithmic elastic and plastic strains (cf. (1.29)) then three constituents in Teodosiu's definition make the analysis even more cumbersome.

1.3.1 Quasi-plastic strain and anholonomic coordinates

Until now we have assumed that the instant deformed configuration (χ_t) contains dislocations and, eventually, disclinations. In this way cutting it into infinitesimal pieces and bringing them to a constant reference temperature T_0 (by means of "perfect tearing" along the line of thought of [Kon58]) would give us a mosaic of ideal crystalline pieces with same structure. Suppose, now, that inside one of them we insert some extra atoms (black dots in Fig. 1.9). Then Kröner's decomposition rule may be extended to read

$$\mathbf{F} = \mathbf{\Phi}\,\mathbf{\Pi}^{\omega}\,\mathbf{\Pi}^{p} \tag{1.66}$$

where a new *quasi-plastic* distortion tensor (cf. [Ant70]) $\mathbf{\Pi}^{\omega} : (\nu_t) \rightarrow (\nu_t^{\omega})$ describes the local strain induced by the extra matter (or holes as negative extra matter). The new local reference configuration is called in [Mic93] the *damaged natural state*.

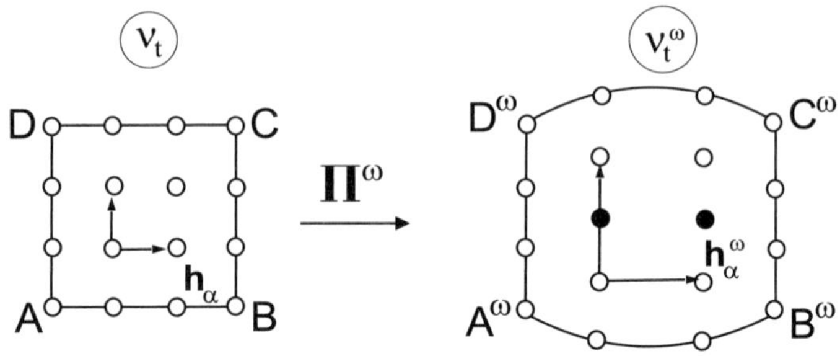

Fig. 1.9. Extra matter illustration and the origin of nonmetric connection in local reference configuration

Therefore, here inelastic distortion is composed of quasi-plastic and pure plastic distortions by the rule

$$\mathbf{\Pi} = \mathbf{\Pi}^{\omega}\,\mathbf{\Pi}^{p}. \tag{1.67}$$

According to Fig. 1.9 structural vectors in the damaged natural state (ν_t^{ω}) are given by

$$\mathbf{h}_{\alpha}^{\omega} = \mathbf{\Pi}^{\omega}\mathbf{h}_{\alpha} \tag{1.68}$$

and the corresponding *quasi-plastic* strain is defined in the same way as above (cf. (1.3)–(1.8)):

$$(ds_{\nu}^{\omega})^2 - (ds_{\nu})^2 = 2\,d\mathbf{r}_{\nu}\,\mathbf{E}_P^{\omega}\,d\mathbf{r}_{\nu}, \qquad 2\,\mathbf{E}_P^{\omega} = \mathbf{\Pi}^{\omega T}\mathbf{\Pi}^{\omega} - \mathbf{1}_{\nu}. \tag{1.69}$$

By means of structural vectors in (ν_t) it is possible to define anholonomic structural coordinates:

$$(d\xi)^{\alpha} = (\mathbf{\Pi})_K^{\alpha}\,dX^K. \tag{1.70}$$

Accordingly, if quasi-plastic strain is inhomogeneous—depending on structural coordinates—then (ν_t)-defined gradient of quasi-plastic strain does not vanish:

$$\mathcal{Q} = \mathrm{grad}_{\nu}\mathbf{E}_P^{\omega} \equiv \nabla_{\nu} \otimes \mathbf{E}_P^{\omega}. \tag{1.71}$$

This tensor will be used in the following subsection for general definition of geometric objects in (ν_t^{ω})-space.

1.3.2 Torsion, curvature and nonmetricity in natural state

For anholonomic coordinates in a non-Euclidean 3-space coefficients of connection have the following form (cf. [Sch54]):

$$\Gamma_{\alpha\beta}^{\cdot\,\cdot\,\gamma} = \mathcal{H}_{\alpha\beta}^{\cdot\,\cdot\,\gamma} + \mathcal{T}_{\alpha\beta}^{\cdot\,\cdot\,\gamma} \tag{1.72}$$

where the first term on the right-hand side is the Christoffel symbol of the second kind given by

$$\mathcal{H}_{\alpha\beta}^{\cdot\cdot\,\gamma} = h^{\gamma\delta}\mathcal{H}_{\alpha\beta\delta} \equiv \frac{1}{2}h^{\gamma\delta}(-\partial_\delta h_{\alpha\beta} + \partial_\alpha h_{\beta\delta} + \partial_\beta h_{\alpha\delta}), \tag{1.73}$$

whereas the second term being so-called *co-torsion* has three constituents, namely,

$$\begin{aligned} \mathcal{T}_{\alpha\beta}^{\cdot\cdot\,\gamma} &= \mathcal{S}_{\alpha\beta}^{\cdot\cdot\,\gamma} + h^{\gamma\delta}(\mathcal{S}_{\delta\alpha\beta} + \mathcal{S}_{\delta\beta\alpha}) \\ &+ \frac{1}{2}h^{\gamma\delta}(\mathcal{Q}_{\alpha\beta\delta} + \mathcal{Q}_{\beta\alpha\delta} - \mathcal{Q}_{\delta\alpha\beta}) \\ &- \Omega_{\alpha\beta}^{\cdot\cdot\,\gamma} - h^{\gamma\delta}(\Omega_{\delta\alpha\beta} + \Omega_{\delta\beta\alpha}). \end{aligned} \tag{1.74}$$

The skew symmetric *anholonomic object* is responsible for lack of global transformation either from material to structural coordinates or from structural to spatial coordinates. It is given by

$$\Omega_{\alpha\beta}^{\cdot\cdot\,\gamma} = (\boldsymbol{\Pi}^{-1})^{A}_{.\alpha}(\boldsymbol{\Pi}^{-1})^{B}_{.\beta}\partial_{[A}(\boldsymbol{\Pi})^{\gamma}_{B]} \tag{1.75}$$

$$= \Phi^{m}_{.\alpha}\Phi^{n}_{.\beta}\partial_{[m}(\boldsymbol{\Phi}^{-1})^{\gamma}_{n]}, \tag{1.76}$$

where the notations $2(skw\mathbf{A})_{mn} \equiv A_{mn} - A_{nm} \equiv 2A_{[mn]}$ and $2(sym\mathbf{A})_{mn} \equiv A_{mn} + A_{nm} \equiv 2A_{(mn)}$ are used throughout this book to indicate symmetrization of the chosen indices.

Torsion and curvature tensors in anholonomic coordinates $(d\xi)^\alpha$ of the natural state space (ν_t^ω) read [Sch54]

$$\mathcal{S}_{\alpha\beta}^{\cdot\cdot\,\gamma} = \Gamma_{[\alpha\beta]}^{\cdot\cdot\,\gamma} + \Omega_{\alpha\beta}^{\cdot\cdot\,\gamma}, \tag{1.77}$$

$$(\mathbb{R})_{\alpha\beta\gamma}^{\cdot\cdot\cdot\,\delta} = 2(\partial_\alpha\Gamma_{\beta\gamma}^{\cdot\cdot\,\delta} + \Gamma_{\alpha\lambda}^{\cdot\cdot\,\delta}\Gamma_{\beta\gamma}^{\cdot\cdot\,\lambda}\Omega_{\alpha\beta}^{\cdot\cdot\,\lambda}\Gamma_{\lambda\gamma}^{\cdot\cdot\,\delta})_{[\alpha\beta]}. \tag{1.78}$$

Both tensors are antisymmetric in the first two indices.

It should be noted that an alternative approach to deformation geometry of a body with defects was given in [Ant70], where a metric structural tensor and corresponding structural coefficients of connection are introduced into the instant deformed configuration (χ_t). In this paper co-torsion is termed *hyper-deformation*. The results of his analysis are the same as here. In the mentioned paper nonmetricity tensor is explicitly given for inhomogeneous temperature field in an isotropic body as well as spontaneous magnetization of a cubic crystal.

The above expressions for coefficients of connection are too general for a proper description of continuous distribution of structural defects in (χ_t) configuration and corresponding geometric structure of damaged natural state space, i.e., (ν_t^ω) configuration. Let us restrict our attention to disclination-free distribution of defects. For the sake of illustration we give the following two examples.

Anholonomic coordinates in E_3

Suppose that there exists only quasi-plastic strain due to inhomogeneous temperature field in (χ_t). Then, it is enough for an adequate description to consider damaged natural space elements as a mosaic in Euclidean 3-space with anholonomic coordinates. Therefore, $\mathcal{S} = 0$, $\mathbb{R} = 0$ and coefficients of connection reduce to

$$\mathit{\Gamma}_{\alpha\beta}^{\cdot\,\cdot\,\gamma} = \mathcal{H}_{\alpha\beta}^{\cdot\,\cdot\,\gamma} + \mathcal{T}_{\alpha\beta}^{\cdot\,\cdot\,\gamma}(\mathcal{Q}, \mathit{\Omega}). \tag{1.79}$$

If the considered body is thermally anisotropic with extremal values of coefficients of thermal expansion $\alpha_1, \alpha_2, \alpha_3$, then quasi-plastic Lagrangian strain in proper directions frame reads

$$\{E_P^\omega(\theta)\} = \begin{pmatrix} \alpha_1(\theta-\theta_0) & 0 & 0 \\ 0 & \alpha_2(\theta-\theta_0) & 0 \\ 0 & 0 & \alpha_3(\theta-\theta_0) \end{pmatrix} \tag{1.80}$$

such that the corresponding thermal nonmetricity tensor is

$$\mathcal{Q}(\theta)_{\lambda\alpha\beta} = \alpha_\lambda\, \delta_{\alpha\beta}\, (\nabla_\nu)_\lambda \theta. \tag{1.81}$$

Here $\delta_{\alpha\beta}$ is the Kronecker delta symbol, whereas the thermal quasi-plastic distortion for small values of thermal expansion coefficients is equal to

$$(\mathbf{\Pi}^\theta)^{\lambda}_{\cdot\mu} \equiv (\mathbf{\Pi}^\omega(\theta))^{\lambda}_{\cdot\mu} = \delta^\lambda_\mu\big(1 + \alpha_\mu(\theta-\theta_0)\big). \tag{1.82}$$

With the above values of $\mathbf{E}_P^\omega(\theta)$ and $\mathcal{Q}(\theta)$ it is easy to find explicit values of coefficients of connection.

Holonomic coordinates in L_3

If the body possesses dislocations and some nonmetric defect, then again $\mathbb{R} = 0$ and there is no need for anholonomic coordinates, i.e., $\mathit{\Omega} = 0$. In this case coefficients of connection simplify to

$$\mathit{\Gamma}_{\alpha\beta}^{\cdot\,\cdot\,\gamma} = \mathcal{H}_{\alpha\beta}^{\cdot\,\cdot\,\gamma} + \mathcal{T}_{\alpha\beta}^{\cdot\,\cdot\,\gamma}(\mathcal{Q}, \mathcal{S}). \tag{1.83}$$

The Riemann–Christoffel curvature tensor disappears but the torsion tensor is different from zero, i.e.,

$$\mathcal{S}_{\alpha\beta}^{\cdot\,\cdot\,\gamma} = \mathit{\Gamma}_{[\alpha\beta]}^{\cdot\,\cdot\,\gamma} \neq 0, \tag{1.84}$$

which means that the natural state (ν_t^ω)-space has the distant parallelism structure. In this case structural vectors in (ν_t^ω)-space are uniquely defined (cf. also Fig. 1.3) and components of $\mathcal{S}$ are the same as components of the anholonomic object in (1.75).

A comparison

Comparing the above two examples we may conclude that mosaic of anholonomic Euclidean elements means simply a collection of tangent Euclidean spaces at the non-Euclidean curved natural space with holonomic coordinates.

By the same approach it is possible to find quasi-plastic magnetization strain, magnetization nonmetricity tensor as well as the corresponding coefficients of connection (based on formula (32) in [Ant70]).[7]

1.4 Somigliana dislocations and eigen-strains of Eshelby

Let us imagine that a hollow cylinder (cf. Fig. 1.10) is cut along an axial plane, the two faces of the cut are displaced by vectors $\mathbf{u}_A, \mathbf{u}_B, \mathbf{u}_C$ and $\mathbf{u}_D$ and that a piece of the same material is subsequently inserted into the hole and welded to the original material of the cylinder. In such a way we would obtain a Somigliana dislocation. In Table 1.1 we show diverse types of defects thus obtained.

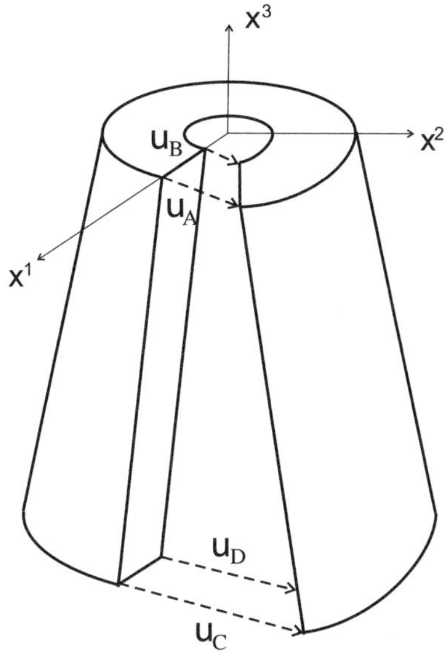

Fig. 1.10. Somigliana dislocations

[7] Extra matter as well as neutron irradiation are two other cases of quasi-plastic nonmetric deformation (cf. [MM-89]). For brevity they are not considered in more detail in subsequent chapters.

Table 1.1. Classification of Somigliana dislocations

Somigliana defect	Displacement components different from zero
Edge x^1-dislocation	$u_A^1 = u_B^1 = u_C^1 = u_D^1 > 0$
Edge x^2-dislocation	$u_A^2 = u_B^2 = u_C^2 = u_D^2 > 0$
Screw x^3-dislocation	$u_A^3 = u_B^3 = u_C^3 = u_D^3 > 0$
Wedge x^1-disclination	$u_C^2 = u_D^2 > 0$
Twist x^2-disclination	$u_C^1 = u_D^1 > 0$
Wedge x^3-disclination	$u_A^2 = u_C^2 > 0$

In a similar way we may imagine that into an ellipsoidal hole a larger ellipsoid is inserted causing residual strain of the inclusion as well as of the matrix (cf. [Esh57, Mur88]).

Following a clear and simple insight of Eshelby we may call the relative strain increment of the ellipsoid $\Delta\mathbf{e}^u$ when it is out of the bulk medium (i.e., matrix) *Eshelby's unconstrained strain*. In the same way, the relative strain increment of the ellipsoid $\Delta\mathbf{e}^c$, when it is inserted into the matrix and welded to it, is called *Eshelby's constrained strain*. They are connected by the famous relationship

$$\Delta\mathbf{e}^c = \mathbb{S} : \Delta\mathbf{e}^u, \tag{1.85}$$

where the 4-tensor $\mathbb{S}$ defined in spatial coordinates of (χ_t) configuration is termed the *Eshelby tensor*.

Denoting volume averaging by means of

$$\langle\bullet\rangle \equiv \frac{1}{V}\int_V (\bullet)dV, \tag{1.86}$$

these two relative strains are represented by (cf. [Mur88])

$$\Delta\mathbf{e}^c = \mathbf{e} - \langle\mathbf{e}\rangle, \qquad \Delta\mathbf{e}^u = \mathbf{e}^u - \langle\mathbf{e}^u\rangle. \tag{1.87}$$

Here following two examples of unconstrained strains are of interest.

- *Unconstrained plastic strains.* Let us consider the special case when all the strains are small. Then in spatial coordinates of (χ_t) we would have (cf. [Mur88]):

$$\begin{aligned}\langle\mathbf{e}\rangle &= \mathbb{D}_\chi^{-1} : \mathbf{T} + \langle\boldsymbol{\Phi}^{-T}\mathbf{e}_P\boldsymbol{\Phi}^{-1}\rangle \\ &\approx \mathbb{D}_\chi^{-1} : \mathbf{T} + \langle\mathbf{e}_P\rangle,\end{aligned} \tag{1.88}$$

  where $\mathbb{D}_\chi$ is Hooke's 4-tensor of material constants of elasticity and $\mathbf{T} \equiv \mathbf{T}_\chi$ is Cauchy stress tensor. By making use of (1.85) and (1.88) we get

$$\mathbf{e} - \langle\mathbf{e}\rangle \approx \mathbb{S} : (\mathbf{e}_P - \langle\mathbf{e}_P\rangle), \tag{1.89}$$

  which is the starting equation for subsequent analysis of polycrystal behaviour in [Mur88] based on Taylor's papers [Tay38]. It should be noted, however, that here a special section is devoted to polycrystal behaviour when plastic strains are large and effective plastic strain is more general than that obtained by simple volume averaging (cf. Chapter 7).

- *Unconstrained quasi-plastic strains.* If the origin of constrained matrix and inclusion strains are thermal unconstrained strains, then a similar analysis might be performed where the role of plastic strain is now taken by pure thermal strains. A detailed analysis of this problem is given in [Lev76] with a more sophisticated way of defining effective unconditional (i.e., thermal) strain and effective coefficients of thermal expansion.

Concerning explicit values of Eshelby's tensor they were derived in [Esh57] for the most special case when the inclusion as well as the matrix are of the same isotropic material. If ratios of semiaxes of unconstrained and constrained ellipsoid are a, b and c, respectively, then for a Cartesian frame situated along these semiaxes, the only nonvanishing components of $\mathbb{S}$ read

$$
\begin{aligned}
S_{aaaa} &= \frac{3\,a^2}{8\,\pi(1-\nu)}\,I_{aa} + \frac{1-2\nu}{8\,\pi(1-\nu)}\,I_a, \\
S_{aabb} &= \frac{3\,b^2}{8\,\pi(1-\nu)}\,I_{ab} - \frac{1-2\nu}{8\,\pi(1-\nu)}\,I_a, \\
S_{abab} &= \frac{3\,(a^2+b^2)}{16\,\pi(1-\nu)}\,I_{ab} + \frac{1-2\nu}{16\,\pi(1-\nu)}\,(I_a+I_b),
\end{aligned}
$$

where

$$
\begin{aligned}
I_a &= 2\pi abc \int_0^\infty \frac{dt}{(a^2+t)\varDelta}, \\
I_{aa} &= 2\pi abc \int_0^\infty \frac{dt}{(a^2+t)^2\varDelta}, \\
I_{ab} &= \frac{2}{3}\pi abc \int_0^\infty \frac{dt}{(a^2+t)(b^2+t)\varDelta}, \\
\varDelta^2 &= (a^2+t)(b^2+t)(c^2+t).
\end{aligned}
$$

The other nonzero components are obtained by permutation of semiaxes indices.[8]

1.5 Residual stresses by the incompatibility approach

Since the configurations (χ_t), (χ_0) and (κ) are embedded into the Euclidean 3-space their curvature tensors (cf. (1.78)) must vanish. In the papers [KS-59, Sto62] this fact has been explored by splitting total strain into

[8] It has to be noted, however, that for real polycrystals the grains are neither ellipsoidal nor isotropic. Therefore, anisotropy must be taken into account and for anisotropic matrix as well as inclusion, the integral expressions in [Mur88, page 116, formula (17.19)] make possible, although tedious and orientation dependent, numerical finding of Eshelby's tensor components.

component parts and introducing stress function. On the other hand, by a somewhat extended procedure an interaction between dislocations and inhomogeneous thermal field was analyzed in the report by [Mic74a]. In all the cases disclinations have not been taken into account.

For simplicity, the procedure is shortly explained for the special case when the body contains only plastic but not quasi-plastic strains. When plastic rotation of natural stress elements is taken to be unity (cf. (1.21)), the total material strain becomes

$$\mathbf{E} = (\mathbf{1} + \mathbf{E}_P)^{1/2}\,\mathbf{E}_E\,(\mathbf{1} + \mathbf{E}_P)^{1/2} + \mathbf{E}_P. \tag{1.90}$$

The curvature tensor expressed in convective holonomic material coordinates of (χ_t) vanishes:

$$\mathbb{R}\big[\mathit{\Gamma}(\chi_t)\big] \equiv \mathbb{R}_\chi\big[\mathit{\Gamma}(\mathbf{G})\big] \equiv \mathbb{R}_\chi\big[\mathcal{G}\big] = 0, \tag{1.91}$$

where coefficients of connection $\mathit{\Gamma}(\mathbf{1}+\mathbf{E})$ are

$$\begin{aligned} \mathit{\Gamma}_{\dot{A}\dot{B}}{}^{C}(\chi_t) &= \mathcal{G}_{\dot{A}\dot{B}}{}^{C} \\ &= \frac{1}{2}G^{CD}(-\partial_D G_{AB} + \partial_B G_{AD} + \partial_A G_{BD}) \end{aligned} \tag{1.92}$$

found from (1.72) replacing the metric tensor $h_{\alpha\beta}$ by $G_{KL} = \delta_{KL} + E_{KL}$ and taking the nonmetricity, the anholonomic object and the torsion all to be equal to zero. In this way a long second-order nonlinear differential equation containing partial derivatives of $\mathbf{E}_E$ and $\mathbf{E}_P$ with respect to material coordinates is obtained (for details see [Mic74a]). If Hooke's law in its inverse form is inserted, then the functions to be determined are stress and plastic strain. However, in order to find how dislocations induce residual stresses we have to explore distant parallelism structure of non-Euclidean space (i.e., $\mathcal{S}_\nu \neq 0, \mathbb{R}_\nu = 0$) of continuously joined natural state elements with respect to the same material coordinates:

$$\begin{aligned} \mathit{\Gamma}_{\dot{A}\dot{B}}{}^{C}(\nu_t) &= \mathcal{H}_{\dot{A}\dot{B}}{}^{C} + \mathcal{T}_{\dot{A}\dot{B}}{}^{C} \\ &= \frac{1}{2}H^{CD}(-\partial_D H_{AB} + \partial_B H_{AD} + \partial_A H_{BD}) + \mathcal{T}_{\dot{A}\dot{B}}{}^{C}. \end{aligned} \tag{1.93}$$

Here the metric tensor in (ν_t) configuration is $\mathbf{H} = \mathbf{1} + \mathbf{E} - \mathbf{E}_P$ whereas the co-torsion term is the same as (1.74) but Greek indices are replaced by capital Latin indices. Since (ν_t) configuration has vanishing curvature

$$\mathbb{R}\big[\mathit{\Gamma}(\nu_t)\big] \equiv \mathbb{R}_\nu\big[\mathit{\Gamma}(\mathbf{G})\big] \equiv \mathbb{R}_\nu\big[\mathcal{H} + \mathcal{T}\big] = 0 \tag{1.94}$$

and the co-torsion term (1.74) contains torsion tensor of (ν_t) configuration, equations (1.91) and (1.94) give us nonlinear differential equations containing the first derivatives of $\mathcal{S}$ as well as first and second derivatives of the stress tensor $\mathbf{T}$.

Suppose now that the first gradients of plastic strain are so small that they may be dropped from (1.91). It is convenient now to introduce into consideration Kröner's incompatibility tensor $\boldsymbol{\eta}$ by means of the second-rank Einstein's curvature tensor $\mathbf{R}$ through

$$\boldsymbol{\eta} = \frac{1}{2}(\mathbf{R} + \mathbf{R}^T), \quad \mathbf{R} = \frac{1}{4}\mathcal{E}_0 : \mathbb{R} : \mathcal{E}_0, \tag{1.95}$$

where $\mathcal{E}_0$ is Ricci's permutation tensor in material coordinates.

2

Crystalline Materials with Thermo-inelastic Memory

The objective of this chapter is to consider a relation between inelastic materials with internal variables and damaged inelastic materials with memory. If an evolution equation for plastic strain rate is given for the first class of materials, then its integration leads to the description represented by integrals whose kernels are responsible for the plastic memory. Here the opposite and more difficult way is taken: to see how functionals appearing in a description of damaged inelastic materials with memory may be transformed into the corresponding evolution equations.

For a correct constitutive theory a geometric description, able to describe properly the most important microstructural changes during an inelastic deformation process, is indispensable. Here only slight amendments to considerations of the first chapter are made, whereas the last section of this chapter is reserved for formulation of the theory.

2.1 A kinematic addendum towards damage-thermo-plastic straining history

For the sake of easier reading we first repeat here briefly some geometrical issues given in the first chapter. Then we extend mainly the notion of quasi-plastic strain of Section 1.3.1 (from page 23 forward).

Consider a crystalline body, $\mathcal{B}$, in a real configuration (χ_t) with defects—such as dislocations, voids, impurities—and an inhomogeneous time-dependent temperature field $\theta(X,t) \equiv T(X,t) - T_0$. Corresponding to (χ_t) there exists an observable global reference configuration (χ_0) (cf. Fig. 1.4) with mentioned defects (but differently distributed) at a homogeneous temperature $\theta_0 = 0$ without surface tractions. Due to inhomogeneously distributed defects such a configuration is not stress-free but contains an equilibrated residual stress (sometimes termed *back stress*).

M.V. Mićunović, *Thermomechanics of Viscoplasticity: Fundamentals and Applications*, Advances in Mechanics and Mathematics 20, DOI: 10.1007/978-0-387-89490-4_2,

© Springer Science + Business Media, LLC 2009

The linear mapping function $\mathbf{G}(.,t) : (\chi_0) \mapsto (\chi_t)$ is compatible second-order *total deformation gradient tensor*. Here t as scalar parameter allows for time developing family of deformed configurations.

According to the approach depicted in Fig. 1.4 on page 11 it is convenient to introduce also a global ideal crystal configuration (κ) at the reference temperature $\theta_0 = 0$ [Bil60]. Then, like in the previous chapter we get an alternative total deformation gradient tensor $\mathbf{F}(.,t) : (\kappa) \mapsto (\chi_t)$.

In accordance with papers dealing with continuum representation of dislocation distributions (χ_t) configuration is imagined to be isothermally cut into small elements denoted by $(\nu_t^{d\theta})$ keeping at each point its temperature unchanged. Then the elastic distortion is defined by $\mathbf{\Phi} : (\nu_t^{d\theta}) \mapsto (\chi_t)$.

Similarly, *pure thermal distortion* is obtained when temperatures of these elements, free of neighbors, are brought to the reference value $\theta_0 = 0$ leading to (ν_t^d), i.e., $\mathbf{\Pi}^\theta : (\nu_t^d) \mapsto (\nu_t^{d\theta})$ (cf. [Mic74b]). According to (1.82) in the special case of thermal isotropy deviatoric part of this tensor vanishes.

The transformation tensor $\mathbf{\Delta}_E(.,t) : (\nu_t^d) \mapsto (\chi_t)$ satisfying the relationship

$$\mathbf{\Delta}_E = \mathbf{\Phi}\mathbf{\Pi}^\theta, \tag{2.1}$$

obtained in such a way, being incompatible as its constituents, should be called the *thermoelastic distortion tensor.*

These local configurations[1] in our case still possess voids. Thus, the term *damaged natural state elements* seems appropriate for them. As a final step an exhaust of voids from these elements leads to (ν_t) configurations which are called *as-received natural states*. The transformation tensor $\mathbf{\Pi}^d(.,t) : (\nu_t) \mapsto (\nu_t^d)$ obtained in such a way is again incompatible and may be termed the *damage distortion tensor.*

Now, the quasi-plastic distortion has two components and reads

$$\mathbf{\Pi}^\omega := \mathbf{\Pi}^\theta\mathbf{\Pi}^d. \tag{2.2}$$

As mentioned in the first chapter, the term *distortion* had been extensively used for incompatible deformations in early papers dealing with continuum theory of dislocations (e.g., [Kro60, Sto62, Bil60]).

In [Bil60] tensor $\mathbf{\Pi}(.,t) : (\kappa) \mapsto (\nu_t)$ is defined as a plastic distortion, where (κ) is a global ideal crystal having the same intrinsic crystalline structure as (ν_t) elements themselves. However, such a distortion is not unique since there are many indistinguishable configurations (κ) with various shapes but the same intrinsic structure. Thus, such a distortion is not replacement invariant as it is seen from Fig. 1.8. However, by means of the relationship $\mathbf{G} = \mathbf{F}\mathbf{F}_0^{-1}$ as well as (1.64) and introducing the replacement invariant plastic distortion $\mathbf{F}_P = \mathbf{\Pi}\mathbf{\Pi}_0^{-1}$ we see that $\mathbf{F}$ and $\mathbf{G}$ differ just by a constant tensor at each material point.

[1] They are commonly referred to as the natural states in classic papers of Kondo and the others.

Due to this equivalence and in order to simplify writing[2] we will use in this chapter the mapping $\mathbf{F}$ with its component distortions. This approach was adopted also in [Mic91a] where thermo-viscoplasticity without damage was analyzed. Nevertheless, such a choice must be properly followed by the corresponding material symmetry treatment.

Alternatively, material symmetry issues are easier if, like in [Mic93], the mapping $\mathbf{G}$ with its component distortions is chosen. In the mentioned paper damage was included but the thermal effects were dropped from the consideration.

Now, with the auxiliary decomposition

$$\boldsymbol{\Delta}_{P\omega} := \boldsymbol{\Pi}^d \boldsymbol{\Pi}^p, \tag{2.3}$$

we arrive at an extended version of Kröner's decomposition rule:

$$\mathbf{F}(.,t) = \boldsymbol{\Phi}(.,t)\boldsymbol{\Pi}(.,t) \equiv \boldsymbol{\Phi}(.,t)\boldsymbol{\Pi}^{\theta}(.,t)\boldsymbol{\Pi}^{d}(.,t)\boldsymbol{\Pi}^{p}(.,t), \tag{2.4}$$

where $\boldsymbol{\Pi}(.,t)$ may be termed the *damage-thermo-plastic distortion.* Let us recall that $\mathrm{curl}\boldsymbol{\Phi}(.,t)^{-T} \neq \mathbf{0}$ and this *incompatibility* is commonly attributed to an asymmetric second rank tensor of dislocation density (cf. (1.41)).

In the sequel the *damage-plastic stretching* $\mathbf{D}_{P\omega}$ and the *damage-plastic spin* tensor $\mathbf{W}_{P\omega}$ introduced, respectively, by means of symmetric and antisymmetric parts of the damage-plastic "velocity gradient" tensor

$$\mathbf{L}_{P\omega} := D_t \boldsymbol{\Delta}_{P\omega} {\boldsymbol{\Delta}_{P\omega}}^{-1} = \mathbf{D}_{P\omega} + \mathbf{W}_{P\omega}, \tag{2.5}$$

will be important for the analysis. Herein D_t stands for material time derivative.

For easier wording, in the rest of this chapter let us call damage-plastic simply as inelastic. Moreover, let us assume that the body is thermally isotropic.

It should be noted, however, that the above distortion tensors are dependent on choice of natural state rotations, i.e., elastic and inelastic rotation tensors. These rotations are defined as usual by the polar decompositions

$$\boldsymbol{\Phi} = \mathbf{R}_E \mathbf{U}_E = \mathbf{V}_E \mathbf{R}_E \quad \text{and} \quad \boldsymbol{\Delta}_{P\omega} = \mathbf{R}_{P\omega} \mathbf{U}_{P\omega} = \mathbf{V}_{P\omega} \mathbf{R}_{P\omega} \tag{2.6}$$

with $\mathbf{R}_E$, $\mathbf{R}_{P\omega}$, being rotation tensors and $\mathbf{U}_E$, $\mathbf{U}_{P\omega}$, $\mathbf{V}_E$, $\mathbf{V}_{P\omega}$, the corresponding symmetric right and left stretch tensors. The following remark gives some existing alternative approaches to possible choices of these rotations.

Remark 2.1. (Inelastic rotations)

Rotations of natural state configurations are one of the most controversial recent issues in plasticity. They might be taken into account at least in three ways:

(a1) Mandel in his work [Man81] proposed isoclinicity of (κ) and (ν_t) which is the most natural choice for single crystals (cf. Remark 1.3, page 13),

[2] The thermal quasi-plastic distortion is here also present.

(a2) Lubarda and Lee in [LL-81] eliminated elastic rotation by assuming $\mathbf{R}_E = \mathbf{1}$ which introduces sensitivity of natural state elements with respect to rotations of an observer of (χ_t) configuration[3] and
(a3) finally $\mathbf{R}_{P\omega} = \mathbf{1}$ (introduced already in [Mic74b] for plasticity without damage) may be accepted either for single crystals or for polycrystals eliminating in such a way arbitrary rotations of (ν_t^d) elements.

Remark 2.2. (Global unloading)

Instead of the complete local unloading by means of $\mathbf{\Phi}^{-1}$ a partial unloading of (χ_t) by removing only surface tractions would lead to a partially unloaded configuration (χ_0) with nonvanishing back stress. In the paper by [Mrk88], the author proposed a fictitious undamaged global configuration (χ_t^{fict}) by closing all voids in (χ_t). Unfortunately, such an attractive approach is not correct for a general inhomogeneous void distribution.

2.2 Materials with thermo-inelastic memory

Let $\mathcal{R}$ denote the set of all real numbers, $\mathcal{R}^+$ the set of all positive numbers, $\mathcal{L}$ the set of all second-order tensors, and $\mathcal{L}^+$ its subset whose elements have positive determinant. Omitting (for simplicity) in the sequel dependence of distortion tensors on particle considered and introducing notations

$$\theta^t(s) := \theta(t) - \theta(t-s), \qquad \mathbf{\Pi}^t(s) := \mathbf{\Pi}(t) - \mathbf{\Pi}(t-s), \tag{2.7}$$

where $s \in [0, \infty)$, $\theta^t : \mathcal{R}^+ \mapsto \mathcal{R}$ is the *relative temperature history* and $\mathbf{\Pi}^t : \mathcal{R}^+ \mapsto \mathcal{L}$ is the *relative inelastic (thermo-damage-plastic) deformation history*, we formulate a constitutive equation for Cauchy stress by the following functional[4]:

$$\mathbf{T}(t) = \mathcal{T}_{s=0}^{\infty}\left(\mathbf{\Phi}(t), \theta(t), \mathbf{\Delta}_{P\omega}(t), \mathbf{\Pi}^t(s)\right) =: \mathcal{T}(\gamma_1, \mathbf{\Pi}^t). \tag{2.8}$$

Here $\mathbf{T}(t)$ is the actual Cauchy stress at the considered time t, whereas $\gamma_1 \equiv (\mathbf{\Phi}(t), \theta(t), \mathbf{\Delta}_{P\omega}(t))$ is the set of *state variables*. For the postulated functional the following properties are being assumed:

Property 2.3. (Elastic range)

$$\mathcal{T}\left(\mathbf{1}, \theta(t), \mathbf{\Delta}_{P\omega}(t), \mathbf{\Pi}^t\right) = \mathbf{0},$$

---

[3] This is essentially also the approach of [Daf84] where the author makes $\mathbf{R}_E$ arbitrary attributing to it subsequently unit value).

[4] Unlike the analysis performed in the paper by [Mic93] here history of temperature is not taken into account. The reason is that its inclusion for materials of differential type leads to first and higher order rates of temperature. If its first time rate appears, then we arrive at extended thermodynamics of Müller's type (cf. Section 4.2.4).

Property 2.4. (Objectivity)

$$\mathbf{Q}(t)\mathbf{T}(t)\mathbf{Q}(t)^T = \mathcal{T}\left(\mathbf{Q}(t)\mathbf{\Phi}(t), \theta(t), \mathbf{\Delta}_{P\omega}(t), \mathbf{\Pi}^t\right)$$
$$\text{with} \quad \mathbf{Q}(t)\mathbf{Q}(t)^T = \mathbf{1}, \ \det\mathbf{Q}(t) = 1,$$

Property 2.5. (Material symmetry)

$$\mathcal{T}_{s=0}^{\infty}\left(\mathbf{\Phi}(t), \theta(t), \mathbf{\Delta}_{P\omega}(t), \mathbf{\Pi}^t(s)\right) =$$
$$\mathcal{T}_{s=0}^{\infty}\big(\mathbf{\Phi}(t)\mathbf{H}^{\omega}(t), \theta(t), \mathbf{H}^{\omega}(t)^{-1}\mathbf{\Delta}_{P\omega}(t)\mathbf{H}^{\kappa}(t),$$
$$\mathbf{H}^{\omega}(s)^{-1}\mathbf{\Pi}^t(s)\mathbf{H}^{\kappa}(s)\big)$$
$$\text{with} \quad \mathbf{H}^{\omega} \in \mathcal{G}^{\omega} \quad \text{and} \quad \mathbf{H}^{\kappa} \in \mathcal{G}^{\kappa}.$$

The property 2.3 states that in the case of unit elastic distortion tensor (corresponding to vanishing elastic strain) stress disappears. Thus, if during tearing process, explained in the previous chapter, at the mapping $\mathbf{\Phi}^{-1} : (\chi_t) \mapsto (\nu_t^{d\theta})$ shapes and dimensions of considered volume elements are unchanged, then it is logical that accompanying stresses disappear. Obviously, this property holds true also when elastic distortion is rotation tensor when $\mathbf{\Phi}^T\mathbf{\Phi} = \mathbf{1}$ holds.

The next property means the principle of *material frame indifference*, i.e., objectivity. Here this is understood in the way that an observer of (χ_t) configuration may rotate while observers of $(\nu_t^{d\theta})$, (ν_t^{d}) and (ν_t) configurations are fixed and connected either to intrinsic crystal structure (Mandel's approach) or to the fixed shape of the configurations considered. The last choice means in fact that the corresponding principal axes of strain ellipsoids are parallel (with unit damage-plastic rotation tensor). Such an objectivity is different than in [Owe70] where for as-received (nondamaged) materials it was accepted that the same (rotating) observer was responsible for both current and natural-state configurations, i.e., the elastic rotation tensor was taken to be unit tensor.

Finally, the last of the above-listed properties establishes material symmetry whose group with respect to global ideal crystal of (κ) is denoted by $\mathcal{G}^{\kappa}$ whereas the group of $(\nu_t^{d\theta})$ configurations is denoted by $\mathcal{G}^{\omega}$. The relationship between these two groups depends on the way in which we treat plastic rotations. In the important special case of thermal isotropy and a spherical damage, when the deviatoric part of $\mathbf{\Pi}^d$ vanishes, symmetry groups $\mathcal{G}^{\omega}$ of $(\nu_t^{d\theta})$ and $\mathcal{G}^{\nu}$ of (ν_t) configurations coincide.

Remark 2.6. (Symmetry and plastic rotation)

It is worth noting that the corresponding symmetry group $\mathcal{G}^{\kappa}$ of global (κ) configurations cannot be related to $\mathcal{G}^{\nu}$ by Noll's rule (cf. its application in [Ogd03, page 49]), i.e.,

$$\mathbf{H}^{\nu} \neq \mathbf{\Pi}^p\mathbf{H}^{\kappa}(\mathbf{\Pi}^p)^{-1} \quad \text{for} \quad \mathbf{H}^{\kappa} \in \mathcal{G}^{\kappa} \quad \text{and} \quad \mathbf{H}^{\nu} \in \mathcal{G}^{\nu}. \tag{2.9}$$

The reason lies in the fact that such a transformation rule does not hold for crystallographic slips preserving lattice but changing material fibres.
(a1) In order to illustrate this let plastic distortion be homogeneous and given by the following simple shear $\Pi^p_{12} = \gamma$ and $\Pi^p_{\alpha\beta} = \delta_{\alpha\beta}$ for all $(\alpha, \beta) \neq (1,2)$ leaving material symmetries of (ν_t) and (κ) configurations to be the same. Calculating now the right-hand side of (2.9) we see that indeed the indicated inequality holds since both groups must be the same.
(a2) There is, however, a small trap in the choice $\mathbf{R}_{P\omega} = \mathbf{1}$ for material symmetries considerations. To show this, let two damage-plastic distortions giving the same response be $\boldsymbol{\Delta}_{P\omega}$ and $\overline{\boldsymbol{\Delta}}_{P\omega} = (\mathbf{H}^\omega)^{-1}\boldsymbol{\Delta}_{P\omega}\mathbf{H}^\kappa$. Then plastic rotations disappear, i.e., $\overline{\boldsymbol{\Delta}}_{P\omega} = \overline{\boldsymbol{\Delta}}^T_{P\omega}$ only if $\mathbf{H}^\omega = \mathbf{H}^\kappa$. Therefore orientations of $(\nu^{d\omega}_t)$ elements have to be adjusted in such a way that proper directions of damage-plastic distortion $\boldsymbol{\Delta}_{P\omega}$ have to be parallel in configurations (κ) and $(\nu^{d\omega}_t)$ in order to meet this requirement.
(a3) If, on the contrary, Mandel's isoclinicity is accepted, then the condition $\mathbf{H}^\omega = \mathbf{H}^\kappa$ is not a must any more. Lattice vectors in (κ) and $(\nu^{d\omega}_t)$ are parallel but groups $\mathcal{G}^\omega$ and $\mathcal{G}^\kappa$, although identical, may be independently applied.[5]

The second property allows an equivalent formulation of the functional (2.8). By the polar decomposition theorem applied to the elastic distortion (2.6) with $\mathbf{C}_E(t) = \boldsymbol{\Phi}(t)^T\boldsymbol{\Phi}(t) \equiv \mathbf{U}_E(t)^2$ and choosing an elastic rotation tensor to be $\mathbf{Q}(t) = \mathbf{R}_E(t)^T$ the functional (2.8) may be transformed into

$$\mathbf{S}(t) = \mathcal{S}\left(\mathbf{C}_E(t), \theta(t), \boldsymbol{\Delta}_{P\omega}(t), \boldsymbol{\Pi}^t\right) =: \mathcal{S}(\gamma, \boldsymbol{\Pi}^t) =: \mathcal{S}(\mathbf{Y}^t). \tag{2.10}$$

Here

$$\mathbf{S}(t) = (\det\mathbf{C}_E(t))^{\frac{1}{2}}\, \boldsymbol{\Phi}(t)^{-T}\mathbf{T}(t)\boldsymbol{\Phi}(t)^{-1} \tag{2.11}$$

is the second Piola–Kirchhoff stress tensor whereas $\gamma \equiv (\mathbf{C}_E(t), \theta(t), \boldsymbol{\Delta}_{P\omega}(t))$ is the set of objective state variables. Denoting the collection of inelastic histories by $\mathcal{J} := \{\boldsymbol{\Pi}^t\}$ this form of the constitutive functional may be presented by the mapping

$$\mathcal{S} : \mathcal{Y} \mapsto \mathcal{L}, \tag{2.12}$$

where $\mathcal{Y} := \mathcal{L}^+ \times \mathcal{R}^+ \times \mathcal{L}^+ \times \mathcal{J}$ is the space of strain histories and $\mathbf{Y}^t \in \mathcal{Y}$.

Remark 2.7. (Plastic history by path)
Consider now a space dual to $\mathcal{Y}$, namely, a space $\mathcal{X}$ with the elements

[5] A definition of material symmetries for inelastic deformation processes is really a subtle subject. In the recent paper by [Riv02] such a famous author gave an approach to symmetry issue by taking observer rotation $\mathbf{Q}$ equal to $\mathbf{H}^\kappa$. Such a "mixture" of rotations could be even useful for fluids but for crystal substances undergoing inelastic strains it is certainly counterproductive. The symmetry of elastic solid bodies is well illustrated by example of a timber cross-section in [Pod00, page 52].

$$\mathbf{X}^t := \left(\mathbf{S}(t), \theta(t), \mathbf{\Delta}_{P\omega}(t), \mathbf{\Pi}^t\right) \in \mathcal{X}, \tag{2.13}$$

called space of stress histories. Moreover, let us define the *inelastic (thermo-damage-plastic) deformation path length* $p(t) \in \mathcal{P} \subset \mathcal{R}^+$ by the following nondecreasing function[6]

$$p(t) = \int_0^t D_\tau p(\tau) d\tau \quad \text{with} \quad \left(D_t p(t)\right)^2 := \operatorname{tr}\{\mathbf{L}_P(t)\mathbf{L}_P(t)^T\}, \tag{2.14}$$

where $\mathbf{L}_P := D_t\mathbf{\Pi}\mathbf{\Pi}^{-1}$ is the inelastic (thermo-damage-plastic) "velocity gradient" tensor.

In most theories of plasticity tensorial plastic deformation history is simply replaced by a plastic deformation path. Such a replacement is convenient but very restrictive since it does not account for multiaxial deformations (cf. [AMM91]).

For a proper understanding of inelastic behaviour of solids it is essential to introduce a frontier between elastic and inelastic behaviour. Herein this is done as follows.

Property 2.8. (Yield surface) Let

$$f(t) = \mathcal{F}(\mathbf{X}^t) \quad \text{or} \mathcal{F} : \mathcal{X} \mapsto \mathcal{R}.$$

By the above definition we introduce a functional which is termed *yield function* whose sign is the indicator of nature of deformation process considered. Namely, if we define sets

$$\mathcal{U}_x := \left\{\mathbf{X}^t \in \mathcal{X} \mid \mathcal{F}(\mathbf{X}^t) < 0, D_t p(t) = 0\right\}, \tag{2.15}$$

$$\mathcal{V}_x := \left\{\mathbf{X}^t \in \mathcal{X} \mid \mathcal{F}(\mathbf{X}^t) > 0, D_t p(t) > 0\right\}, \tag{2.16}$$

$$\mathcal{B}_x := \left\{\mathbf{X}^t \in \mathcal{X} \mid \mathcal{F}(\mathbf{X}^t) = 0, D_t p(t) = 0\right\}, \tag{2.17}$$

then $\mathcal{U}_x$ is the elastic region, $\mathcal{B}_x$ the corresponding elastic–plastic frontier (i.e., *yield surface*) and $\mathcal{V}_x$ the elastic-viscoplastic region. The following *plastic straining indicator function* $<f>$ which is equal to 1 in $\mathcal{V}_x$ and to 0 in $\mathcal{U}_x$ and $\mathcal{B}_x$ serves usually for this purpose.[7] For notational simplicity their obvious dependence on time t being usually called hardening (when $\mathcal{B}_x$ expands) or softening (when $\mathcal{B}_x$ shrinks) is suppressed. Alternatively, the corresponding sets $\mathcal{U}_y, \mathcal{V}_y, \mathcal{B}_y \subset \mathcal{Y}$ may be introduced in the space of strain histories.

[6] In the next sections a slightly different definition of accumulated plastic strain will be used. Namely, multiplying (2.14) by $(2/3)^{1/2}$ we meet its measured value at uniaxial strain. Hereinafter such a modified strain will be called either accumulated or equivalent plastic strain.

[7] Hereinafter in (3.11), p. 50, we will see that such a function is simply the Heaviside function.

Properties of the constitutive functional (2.10) are inevitably connected to notion of continuity and differentiability of the space of inelastic strain histories $\mathcal{J}$.

Suppose that a *forgetfulness measure* (cf. [Tru72]) is introduced on the damage-plastic deformation path $\mu\{\mathcal{P}\}$ such that during an infinitesimal time interval dt we have

$$d\mu = \begin{cases} h(t)\,dt, & D_t p_\omega(t) > 0, \\ 0, & D_t p_\omega(t) = 0, \end{cases} \tag{2.18}$$

where h is the corresponding forgetfulness function in time domain. It is assumed that $\mu\{\mathcal{P}\}$ is positive and given by the corresponding Lebesgue integral with h almost everywhere positive (except in $\mathcal{U}_y$ and $\mathcal{B}_y$ where $h = 0$). With this measure we may introduce an inner product in the space of strain histories $\mathcal{Y}$ by means of

$$\begin{aligned}(\mathbf{Y}_1^t, \mathbf{Y}_2^t) := \operatorname{tr}\{\mathbf{C}_{E1}(t)\mathbf{C}_{E2}(t)\} + \operatorname{tr}\{\boldsymbol{\Delta}_{P\omega1}(t)\boldsymbol{\Delta}_{P\omega2}(t)\} \\ +\theta(t)^2 + \int_0^\infty \operatorname{tr}\{\boldsymbol{\Pi}_1^t(s)\boldsymbol{\Pi}_2^t(s)\}\,d\mu.\end{aligned} \tag{2.19}$$

Then the recollection of all the strain histories $\mathcal{Y}$ with the norm $\|\mathbf{Y}^t\|_h^2 := (\mathbf{Y}^t, \mathbf{Y}^t)$ forms a Hilbert space $\mathcal{H}$.[8]

Let us define the elastic deformation history $\mathbf{Y}^{\#t}$ as the projection of the actual inelastic deformation history $\mathbf{Y}^t$ from $\mathcal{V}_y$ to $\mathcal{B}_y$ by means of

$$\mathbf{Y}^{\#t} = (\mathbf{C}_{E0}(t), \theta(t), \boldsymbol{\Delta}_{P\omega}(t), \mathbf{0}^t) \equiv (\gamma^{\#}, \mathbf{0}^t) \in \mathcal{B}_y, \tag{2.20}$$

where elastic history $\mathbf{0}^t(s) = \mathbf{0}$ holds either for initial or some subsequent elastic region. This history is an elastic "shadow tail" consisting of configurations with the same microstructure as the actual configuration (χ_t) given by the unchanged $\boldsymbol{\Pi}(t)$ and it is necessary in the following definitions. Without loss of generality in the sequel instead of state variables γ the constitutive functional (2.10) will be expressed in terms of $\gamma^{\#}$.

According to the common terminology of viscoplasticity by the following formula:

$$\mathcal{S}(\gamma^{\#}, \boldsymbol{\Pi}^t) = \mathcal{S}(\gamma^{\#}, \mathbf{0}^t) + \mathbf{s}(\gamma^{\#}, \boldsymbol{\Pi}^t) \equiv \mathbf{S}_0(\gamma^{\#}) + \mathbf{s}_\gamma(\boldsymbol{\Pi}^t), \tag{2.21}$$

the notions of *overstress tensor* $\mathbf{s}_\gamma(\boldsymbol{\Pi}^t)$, dynamic stress tensor $\mathcal{S}(\gamma^{\#}, \boldsymbol{\Pi}^t)$ and static stress tensor $\mathbf{S}^{\#}(t) := \mathcal{S}(\gamma^{\#}, \mathbf{0}^t)$ are introduced ([CN-61, CN-60, Hol73]).

Now, the following two properties of *directional (weak) continuity and differentiability* are assumed to hold. For some scalars α , β and some tensors $\mathbf{p}$, $\mathbf{q} \in \mathcal{J}$ satisfying Property 2.8 (cf. also [Tru72])

---

8 An alternative equally legitimate measure of distance between two histories was given by Rivlin as follows: $\langle \boldsymbol{\Pi}_1^t, \boldsymbol{\Pi}_2^t \rangle := \sup \left\{ (\boldsymbol{\Pi}_1^t - \boldsymbol{\Pi}_2^t)^T (\boldsymbol{\Pi}_1^t - \boldsymbol{\Pi}_2^t) \right\}^{1/2}$.

Property 2.9. (Weak continuity)

$$\begin{aligned}
&\forall \varepsilon > 0, \exists \eta > 0, \quad \text{such that}\\
&\forall \mathbf{Y}_1^t, \mathbf{Y}_2^t = \mathbf{Y}_1^t + \beta \mathbf{q} \in \mathcal{V}_d, \quad \beta \in \mathcal{R}^+,\\
&\|\mathbf{Y}_2^t - \mathbf{Y}_1^t\| < \eta \mapsto \|\mathcal{S}(\mathbf{Y}_2^t) - \mathcal{S}(\mathbf{Y}_1^t)\| < \varepsilon,
\end{aligned}$$

Property 2.10. (Weak differentiability)

$$\begin{aligned}
&\mathbf{s}_\gamma(\boldsymbol{\Pi}^t) = \textstyle\sum_{k=1}^n \delta^k \mathbf{s}_\gamma(\boldsymbol{\Pi}^t, \mathbf{p}) + \mathbf{r}(\gamma^\#, \boldsymbol{\Pi}^t), \quad \text{with}\\
&\delta^k \mathbf{s}_\gamma(\boldsymbol{\Pi}^t, \mathbf{p}) = \tfrac{d^k}{d\alpha^k} \mathbf{s}_\gamma(\boldsymbol{\Pi}^t + \alpha \mathbf{p})|_{\alpha=0}, \quad \alpha \in \mathcal{R}^+, \quad \text{and}\\
&\lim_{\|\mathbf{Y}^t\|_h \to 0} \|\mathbf{Y}^t\|_h^{-1} \mathbf{r}(\gamma^\#, \boldsymbol{\Pi}^t) \mapsto 0.
\end{aligned}$$

Such a continuity has here the same meaning as in Coleman and Noll's *principle of fading memory*, i.e., *the constitutive functional for stress $\mathcal{S}$ of a material with inelastic memory is continuous at the elastic deformation history* (cf. [Tru72]) and this statement here may be referred to as the *weak principle of inelastic fading memory.*

On the other hand, Property 2.10 means that $\mathbf{s}$ is directionally differentiable at the zero relative plastic deformation history (i.e., at paths slightly departing from elastic deformation history).

The k-linear homogeneous functional $\delta^k \mathbf{s} : \mathcal{H} \mapsto \mathcal{L}$ appearing in the differentiability property is the kth *variation* or k^{th} *Gateaux differential* of $\mathbf{s}$ at the zero relative inelastic deformation history. In the special case of inelastic fading memory of order 1, according to the Riesz–Fisher theorem of the theory of Hilbert spaces every bounded continuous linear functional may be written as an inner product. Such products may be not only scalars but second-rank tensors as well. Therefore ([TN-65, CN-61])

$$\delta \mathbf{s}(\mathbf{P}_\omega^t) = \int_0^\infty h(\tau) \mathbb{K}(\gamma^\#, \tau) : \boldsymbol{\Pi}^t(\tau)\, d\tau$$

with the kernel being a fourth-rank tensor and, as hereinbefore, $\gamma^\# := (\mathbf{C}_{E0}(t), \theta(t), \boldsymbol{\Delta}_{P\omega}(t))$. In this way by means of the above two properties the representation (2.21) reduces to

$$\mathbf{S}(t) \approx \mathbf{S}_0(\gamma^\#) + \int_0^\infty h(\tau) \mathbb{K}(\gamma^\#, \tau) : \boldsymbol{\Pi}^t(\tau)\, d\tau. \tag{2.22}$$

The above formula is said to describe a *linear viscoplastic material* of integral type. Obviously, a similar procedure could lead to higher order materials of integral type.

Remark 2.11. (Discrete memory)

An alternative approach is to take the discrete memory influence extending $\gamma^{\#}$ to include

$$\delta_{pm} := \max_{\tau} \|\boldsymbol{\varepsilon}_P(\tau)\| \equiv \max_{\tau} \|\mathbf{V}_P(\tau) - \mathbf{1}\|$$

which is especially suitable for cyclic inelastic processes (cf. [MSD73, Mic87b]). In this case $h(t)$ is given by corresponding Dirac delta-functions at temporal points where such maxima appear.

Suppose, now, that the considered plastic motion is slow and $\mathbf{s}$ is k times Gateaux-differentiable. Under such conditions its asymptotic approximation (cf. Theorem 2 in [CN-60]) may be written in the following way:

$$\mathbf{S}(t) = \mathbf{S}_0(\gamma^{\#}) + \sum_{I_k} \mathcal{P}_{m_1 \cdots m_k}(\gamma^{\#})\big(\overset{(m_1)}{\mathbf{\Pi}}(t) \otimes \cdots \otimes \overset{(m_k)}{\mathbf{\Pi}}(t)\big) + o(\alpha^k), \quad (2.23)$$

where I_k is the maximal set of ordered indices $m_1, \ldots, m_k$ whose sum is never larger than k and

$$\overset{(k)}{\mathbf{\Pi}}(t) = D_\tau^k \mathbf{\Pi}(t-\tau)|_{\tau=0},$$

whereas the residual term satisfies the condition

$$\lim_{\alpha \to 0} \alpha^k o(\alpha^k) = 0.$$

A material characterized by (2.23) may be called an *elastic-viscoplastic material of plastic differential type and complexity k.* Its following special classes deserve special attention:

$$\mathbf{S}(t) = \mathbf{S}_0(\gamma^{\#}) + \mathbb{P}_1(\gamma^{\#}) : D_t\mathbf{\Pi}(t), \quad (2.24)$$

for $k = 1$ and

$$\begin{aligned} \mathbf{S}(t) &= \mathbf{S}_0(\gamma^{\#}) + \mathbb{P}_1(\gamma^{\#}) : D_t\mathbf{\Pi}(t) \quad (2.25) \\ &+ \big(\mathfrak{P}_{11}(\gamma^{\#}) : D_t\mathbf{\Pi}(t)\big) : D_t\mathbf{\Pi}(t) + \mathbb{P}_2(\gamma^{\#}) : D_t D_t\mathbf{\Pi}(t), \end{aligned}$$

for $k = 2$. As already mentioned $\Delta\mathbf{S} \equiv \mathbf{S} - \mathbf{S}_0(\gamma^{\#})$ is the overstress tensor where for simplicity the difference between functions and their values is neglected.

A danger of losing contact with real materials forces us, however, to check carefully the meaning of all the coefficients in (2.25) having support in the experimental evidence and thermodynamic considerations (e.g., [Mul71, AMM91, MM-89]).

2.3 Discussion and brief summary of the chapter

Now, let us express thermo-damage-plastic distortion rate in terms of its constituents by making use of the Leibnitz rule, i.e.,

$$D_t\mathbf{\Pi} = D_t\theta\left(\frac{d}{d\theta}\mathbf{\Pi}^\theta\right)\mathbf{\Pi}^d\mathbf{\Pi}^p + \mathbf{\Pi}^\theta D_t\mathbf{\Pi}^d\mathbf{\Pi}^p + \mathbf{\Pi}^\theta\mathbf{\Pi}^d D_t\mathbf{\Pi}^p. \tag{2.26}$$

Then, replacing this formula in (2.24), we acquire that stress is a linear function of temperature, damage distortion and plastic distortion rates. Performing again time differentiation of (2.26) and replacing the result into (2.25) we would obtain a rather complex result including linear terms in second time derivatives of temperature, damage distortion and plastic distortion and second-order products of their first time derivatives.

What is the meaning of all these terms? In order to give an answer let us remember that only existing constitutive theory—according to author's knowledge—which includes temperature time derivatives is given in [Mul71].

Remark 2.12. (Temperature derivatives)

In Müller's extended thermodynamics of thermoelastic bodies (which corresponds to $\mathbf{\Delta}_{P\omega}(t) = \mathbf{1}$ and $D_t\mathbf{\Delta}_{P\omega}(t) = \mathbf{0}$) the elastic stress (i.e., stress during elastic deformation) is assumed to depend on $\gamma^\#$ and $D_t\theta(t)$ but not on $D_t^2\theta(t) \equiv D_tD_t\theta(t)$ as well. It seems logical then to neglect $\mathbb{P}_2(\gamma^\#)$ from (2.25). In this way (2.25) may be expressed by means of

$$\begin{aligned}\mathbf{S}(t) &= \mathbf{S}_0(\gamma^\#) + \mathbb{P}_1(\gamma^\#) : D_t\mathbf{\Pi}(t) \\ &\quad + \left(\mathfrak{P}_{11}(\gamma^\#) : D_t\mathbf{\Pi}(t)\right) : D_t\mathbf{\Pi}(t),\end{aligned} \tag{2.27}$$

where $\mathbb{P}_1\left(\gamma^\#\right)$ is the fourth-rank *tensor of plastic viscosity coefficients* and $\mathfrak{P}_{11}(\gamma^\#)$ the sixth-rank tensor of plastic viscosity coefficients.

Let us assume now that, according to Remark 2.1, we choose $\mathbf{R}_{P\omega} = \mathbf{1}$. Then from (2.5) and (2.6) it follows that

$$\begin{aligned}2\,\mathbf{D}_{P\omega} &= D_t\mathbf{V}_{P\omega}\mathbf{V}_{P\omega}{}^{-1} + \mathbf{V}_{P\omega}{}^{-1}D_t\mathbf{V}_{P\omega}, \\ 2\,\mathbf{W}_{P\omega} &= D_t\mathbf{V}_{P\omega}\mathbf{V}_{P\omega}{}^{-1} - \mathbf{V}_{P\omega}{}^{-1}D_t\mathbf{V}_{P\omega}.\end{aligned}$$

Consider moreover thermally isotropic materials subject to the special case of spherical damage when $\mathbf{\Pi}^d = (1+\omega)\mathbf{1}$. This leads to the specification $\mathbf{V}_{P\omega} = \big(1+\alpha(\theta-\theta_0)\big)(1+\omega)\mathbf{V}_P$. If, moreover, the invertibility of (2.27) is assumed (cf. (2.15)–(2.17) as well as [AMM91]), then from (2.27) there follow two evolution equations:

$$\begin{aligned}D_t\mathbf{V}_P = {<}f{>}\,\big[&d_1(\pi)\mathbf{1} + d_2(\pi)\mathbf{V}_P + d_3(\pi)\mathbf{V}_P^2 \\ &+ d_4(\pi)\mathbf{S} + d_5(\pi)\mathbf{S}^2 + d_6(\pi)(\mathbf{S}\mathbf{V}_P + \mathbf{V}_P\mathbf{S}) \\ &+ d_7(\pi)(\mathbf{S}^2\mathbf{V}_P + \mathbf{V}_P\mathbf{S}^2) + d_8(\pi)(\mathbf{S}\mathbf{V}_P^2\mathbf{V}_P^2\mathbf{S}) \\ &+ d_9(\pi)(\mathbf{S}^2\mathbf{V}_P^2 + \mathbf{V}_P^2\mathbf{S}^2)\big],\end{aligned} \tag{2.28}$$

$$D_t\omega = {<}f{>}\,\hat{\omega}(\pi) \tag{2.29}$$

with invariants formed from the state variables appearing in (2.13), i.e.,

$$\gamma_X \equiv \pi = \{\theta, \omega, \mathrm{tr}\mathbf{S}, \mathrm{tr}\mathbf{S}^2, \mathrm{tr}\mathbf{S}^3, \mathrm{tr}\mathbf{V}_P, \mathrm{tr}\mathbf{V}_P^2, \mathrm{tr}\mathbf{V}_P^3, \\ \mathrm{tr}\mathbf{V}_P\mathbf{S}, \mathrm{tr}\mathbf{V}_P^2\mathbf{S}, \mathrm{tr}\mathbf{V}_P\mathbf{S}^2, \mathrm{tr}\mathbf{V}_P^2\mathbf{S}^2\}.$$

The explanation for the above two evolution equations is simple. Namely, if we apply the tensor generators listed in [Spe71] to the tensor function $D_t\mathbf{V}_P(\mathbf{S}, \mathbf{V}_P)$, then we arrive at (2.28). The scalar coefficients, in general, depend on the scalar invariants π.

Therefore, a necessity to consider a separate evolution equation for plastic spin does not exist. Such a conclusion should hold also if another choice for rotations of natural state configurations is accepted. Indeed, if Mandel's approach (i.e., isoclinicity) is accepted, then it is possible to show that the plastic distortion $\mathbf{\Pi}$ can be realized by six independent crystallographic shears (cf. (1.23) on page 14) and this is the number of components of plastic stretch $\mathbf{V}_P$ or, equivalently, $\mathbf{U}_P$.

Remark 2.13. (Endochronic constitutive equation)

Introducing the mapping $p : (-\infty, t] \mapsto [0, p(t)]$ responsible for the intrinsic elapsed time into (2.22) this becomes *a general endochronic constitutive equation* (e.g., [Faz88]). The inverse transformation, however, is not possible if a sequence of elastic and plastic deformation processes happens. In order to overcome this indeterminacy let us recall that there follows from (2.18) that forgetfulness function $h(t)$ disappears at the segments where plastic strain is constant, i.e., in all the elastic regions of the considered history. Therefore we may introduce the inverse mapping by means of

$$h(t)dt = h_\pi(p)dp, \tag{2.30}$$

which permits a new integral representation for stress

$$\mathbf{S}(t) \approx \mathbf{S}_0(\gamma^\#) + \int_0^{p_f} h_\pi(p')\mathbb{K}_\pi(\gamma^\#, p') : \mathbf{\Pi}^{\mathrm{endo}}(p')\, dp' \tag{2.31}$$

instead of (2.22). Herein

$$\mathbf{\Pi}^{\mathrm{endo}}(p') \equiv \mathbf{\Pi}(p) - \mathbf{\Pi}(p - p')$$

is the corresponding relative inelastic history.

Intrinsic time is introduced by (2.30) as an increasing function of plastic deformation path.[9] If intrinsic time is identified with $p(t)$ itself, then $p(t) \in [0, p_f]$ with p_f being the *intrinsic lifetime*, i.e., the intrinsic time at rupture. Obviously, $p(t)$ may thus be imagined to be an effective age measure (i.e., intrinsic

[9] Hereinafter, in Section 4.3, the intrinsic time will be given more profound thermodynamic meaning connected with dissipation. This concept is used extensively in the next sections.

elapsed time) of material of the considered body accounting for damage as well as plasticity.

Finally, the above transformation (2.30) is not formal but has its deep physical meaning in the sense that (2.31) permits transformation into *endochronic materials of differential type* in the same way as we did passing from (2.22) into (2.23) not only for slow but for fast processes as well. Namely, for a high time rate $D_t\mathcal{A}$ of an arbitrary tensor $\mathcal{A}$ its endo-rate $d\mathcal{A}/dp$ could even be small permitting asymptotic approximation as well. Of course, this extends the field of application of differential type materials.

Remark 2.14. (Infinitesimal memory)

Suppose that only an infinitesimal part of the endochronic inelastic history is relevant for the instant inelastic behaviour. Then the forgetfulness function has the additional restriction that

$$h_\pi(p) = \begin{cases} h_\pi(p), & p < \varepsilon, \\ 0, & p > \varepsilon. \end{cases} \tag{2.32}$$

Then, developing kernel $\mathbb{K}_\pi$ from (2.31) into Taylor series we would get an endochronic material of differential type.

A brief summary of results obtained in this chapter may be stated as follows:

- Rotations of the intermediate local reference configuration are discussed taking into account the principle of objectivity and material symmetries.
- Using the fact that by its very definition elastic deformation is not hereditary a principle of determinism based on damage-plastic deformation history is formulated.
- Viscoplastic materials of integral type and of differential type are obtained from description based on functionals. Both types of materials may be connected to endochronic interpretation.
- A general nonassociate flow rule for materials of differential type is formulated on the basis of tensor function representation.

Another interesting issue is the question of normality, i.e., associativity of flow rule. To this problem, the next chapter of this monograph is devoted.

Acknowledgment. Discussions with Professors Witold Kosinski and Aleksandar Baksa, as well as with late Professor Henryk Zorski, on the subject of this chapter are highly appreciated.

3

Normality Rule? Plastic Work Extremals and Related Topics

The task in this chapter is to reexamine the background of the so-called "normality" principle, i.e., the widely accepted assumption that plastic strain rate tensor is perpendicular to a yield surface in the corresponding stress space. In the papers [Dru52, Ili61, Hil68] this conclusion has been drawn from analysis of the plastic work properties. For instance, in [Dru52], rate-independent plasticity was considered and contemplating that all stress increments were admissible normality was concluded from the assumption of plastic work positiveness for arbitrary stress cycle. In the one-dimensional case this positiveness would mean that

- elastic unloading is possible from every point of stress-plastic strain curve and
- stress σ increases with plastic strain ϵ_P (plastic strain hardening).

Such a dependence is given by a yield function

$$f^{\#} \equiv f(\sigma^{\#}, \epsilon_P),$$

which is zero if plastic deformation takes place and negative in elastic domain of material behaviour (cf. also Property 2.8, page 39). Since the strain rate is extremely low, $\sigma^{\#}$ is called *static stress.* It fulfills the static yield condition $f^{\#} = 0$ when plastic straining takes place.

However, for the rate-sensitive materials we must include the plastic strain rate into the corresponding nonpositive function $F(\sigma, \epsilon_P, D_t\epsilon_P)$ (where, as hereinbefore, D_t stands for the material time derivative). Such a function may be called *dynamic yield function.* Of course,

$$F(\sigma, \epsilon_P, 0) \equiv f(\sigma, \epsilon_P)$$

might have all three signs. It is positive in viscoplastic region, negative in elastic region and vanishes on the static yield surface. Its connection with the static yield function reads

M.V. Mićunović, *Thermomechanics of Viscoplasticity: Fundamentals and Applications*, Advances in Mechanics and Mathematics 20, DOI: 10.1007/978-0-387-89490-4_3,

© Springer Science + Business Media, LLC 2009

$$F(\sigma^{\#}, \epsilon_P, 0) \equiv f(\sigma^{\#}, \epsilon_P) = f^{\#}.$$

Suppose now that an elastic unloading is performed with $D_t\epsilon_P = 0$ and $\epsilon_{PA} =$ const by a sudden jump from a dynamic point σ_A to its static counterpart $\sigma_A^{\#}$. This, however, contradicts the fact that for each $\sigma_1 \in [\sigma_A^{\#}, \sigma_A]$ the inequality $f(\sigma_1, \epsilon_{PA}) > 0$ holds. Thus, during decrease from σ_A to $\sigma_A^{\#}$ plastic strain has to be changed so that such a pure elastic unloading is not possible without an additional plastic straining and this should permit $d\sigma d\epsilon_P < 0$ as well. This does not mean any plastic strain softening but simply unloading from viscoplastic point to its elasto-plastic bounding counterpart. In the sequel the multiaxial viscoplasticity is considered by means of a variational method in order to examine this conclusion in more detail.

3.1 Postulate of minimal plastic work

3.1.1 Preliminary evolution equations by tensor function representation

1. Let us suppose that temperature inhomogeneity is so small that it may be neglected, i.e., $|\theta(X,t)/T_0| \equiv |T(X,t)/T_0 - 1| \ll 1$ and that damage may also be disregarded. Then the quasi-plastic distortion becomes pure rotation (cf. (2.2), page 34). Thus, without loss of generality it may be taken that (see also (2.1) and (2.3))

$$\boldsymbol{\Delta}_E \approx \boldsymbol{\Phi}, \quad \boldsymbol{\Delta}_{P\omega} \approx \boldsymbol{\Pi}^p \equiv \boldsymbol{\Pi}. \tag{3.1}$$

Let us apply again Teodosiu's decomposition:

$$\mathbf{F}_P := \boldsymbol{\Pi}(t)\boldsymbol{\Pi}(t_0)^{-1}, \tag{3.2}$$

which is very convenient for defining the *plastic stretching* $\mathbf{D}_P$ and the *plastic spin* tensor $\mathbf{W}_P$ introduced, respectively, by means of symmetric and anti-symmetric parts of the plastic "velocity gradient" tensor

$$\mathbf{L}_P := (D_t\mathbf{F}_P)\mathbf{F}_P{}^{-1} = (D_t\boldsymbol{\Pi})\boldsymbol{\Pi}^{-1} = \mathbf{D}_P + \mathbf{W}_P. \tag{3.3}$$

These two quantities will be important for the subsequent analysis.

It should be noted, however, that the above tensors are dependent on choice of natural state rotations, i.e., elastic and plastic rotation tensors. They are defined as in previous chapters by the polar decompositions

$$\boldsymbol{\Phi} = \mathbf{R}_E\mathbf{U}_E = \mathbf{V}_E\mathbf{R}_E \quad \text{and} \quad \boldsymbol{\Pi} = \mathbf{R}_P\mathbf{U}_P = \mathbf{V}_P\mathbf{R}_P \tag{3.4}$$

with $\mathbf{R}_E, \mathbf{R}_P$, being rotation tensors and $\mathbf{U}_E, \mathbf{U}_P, \mathbf{V}_E, \mathbf{V}_P$, the corresponding symmetric right and left stretch tensors. These rotations of natural state configurations are explained in detail in Remark 2.1 on rotations of (ν_t) elements on page 35.

2. It is convenient to take in the sequel that $\mathbf{R}_P = \mathbf{1}$. Then from (3.3) and (3.4) it follows that

$$\begin{aligned} 2\,\mathbf{D}_P &= (D_t\mathbf{V}_P)\mathbf{V}_P{}^{-1} + \mathbf{V}_P{}^{-1}D_t\mathbf{V}_P\,, \\ 2\,\mathbf{W}_P &= (D_t\mathbf{V}_P)\mathbf{V}_P{}^{-1} - \mathbf{V}_P{}^{-1}D_t\mathbf{V}_P. \end{aligned} \tag{3.5}$$

Consider moreover isotropic materials with a very general evolution equation for plastic stretching (cf. also [AMM91] as well as (2.28) on page 43)

$$\begin{aligned} D_t\mathbf{V}_P \;=\; <\!f\!> \big[& d_0(\pi)\mathbf{1} + d_1(\pi)\mathbf{V}_P + d_2(\pi)\mathbf{V}_P^2 + d_3(\pi)\mathbf{S} + d_4(\pi)\mathbf{S}^2 \\ & + d_5(\pi)(\mathbf{S}\mathbf{V}_P + \mathbf{V}_P\mathbf{S}) + d_6(\pi)(\mathbf{S}^2\mathbf{V}_P + \mathbf{V}_P\mathbf{S}^2) \\ & + d_7(\pi)(\mathbf{S}\mathbf{V}_P^2 + \mathbf{V}_P^2\mathbf{S}) + d_8(\pi)(\mathbf{S}^2\mathbf{V}_P^2 + \mathbf{V}_P^2\mathbf{S}^2)\big]\,. \end{aligned} \tag{3.6}$$

Here $<\!f\!>$ equals 1 if plastic deformation changes with time and equals 0 otherwise, while invariants

$$\pi \equiv \{\,\mathrm{tr}\mathbf{S}, \mathrm{tr}\mathbf{S}^2, \mathrm{tr}\mathbf{S}^3, \mathrm{tr}\mathbf{V}_P, \mathrm{tr}\mathbf{V}_P^2, \mathrm{tr}\mathbf{V}_P^3, \mathrm{tr}\mathbf{V}_P\mathbf{S}, \mathrm{tr}\mathbf{V}_P^2\mathbf{S}, \mathrm{tr}\mathbf{V}_P\mathbf{S}^2, \mathrm{tr}\mathbf{V}_P^2\mathbf{S}^2\}$$

have already been introduced by (2.28). On the other hand, objectivity requires to take as a variable not the actual Cauchy stress $\mathbf{T}$ but the second Piola–Kirchhoff stress tensor

$$\mathbf{S} = (\det\mathbf{\Phi})\,\mathbf{\Phi}^{-1}\mathbf{T}\mathbf{\Phi}^{-T} \tag{3.7}$$

or convective stress tensor

$$\widetilde{\mathbf{S}} = \mathbf{S}/\det\mathbf{\Phi}, \tag{3.8}$$

both defined with respect to (ν_t) configuration.[1]

For convenience, let us introduce Biot's strain by means of (cf. (1.24) on page 14):

$$\boldsymbol{\varepsilon}_P = \mathbf{V}_P - \mathbf{1}.$$

Then, by means of tensor representation [Spe71] the evolution equation for plastic "velocity gradient" is given by

$$\begin{aligned} \mathbf{L}_P = <\!f\!>\, \big[& l_0(\pi)\mathbf{1} + l_1(\pi)\boldsymbol{\varepsilon}_P + l_2(\pi)\boldsymbol{\varepsilon}_P^2 + l_3(\pi)\mathbf{S} \\ & + l_4(\pi)\mathbf{S}^2 + l_5(\pi)\mathbf{S}\boldsymbol{\varepsilon}_P + l_6(\pi)\boldsymbol{\varepsilon}_P\mathbf{S} \\ & + l_7(\pi)\mathbf{S}^2\boldsymbol{\varepsilon}_P + l_8(\pi)\boldsymbol{\varepsilon}_P\mathbf{S}^2 + l_9(\pi)\mathbf{S}\boldsymbol{\varepsilon}_P^2 \\ & + l_{10}(\pi)\boldsymbol{\varepsilon}_P^2\mathbf{S} + l_{11}(\pi)\mathbf{S}^2\boldsymbol{\varepsilon}_P^2 + l_{12}(\pi)\boldsymbol{\varepsilon}_P^2\mathbf{S}^2)\big]\,. \end{aligned} \tag{3.9}$$

Clearly, thirteen material constants $l_0, \ldots, l_{12}$ are easily obtained from nine coefficients $d_0, \ldots, d_8$ if equation (3.6) multiplied by $\mathbf{V}_P^{-1}$ (being calculated by

[1] Whereas in the chapter devoted to ferromagnetics such a stress is related to convective material coordinates it is more convenient here to define (3.8) with respect to convective structural coordinates (1.70).

Caley–Hamilton theorem) is substituted into (3.5) and made equal to (3.9). Therefore, a necessity to consider a separate evolution equation for plastic spin does not exist. Such a conclusion should hold also if another choice for rotations of natural state configurations is accepted. Indeed, if isoclinicity by Mandel's approach (discussed in Remark 2.1 on page 35) is accepted, then it is possible to show that the plastic distortion $\mathbf{\Pi}$ can be realized by six independent crystallographic shears and this is the number of components of plastic stretch $\mathbf{V}_P$ (or, equivalently, $\mathbf{U}_P$) as well as of plastic stretching $D_t\mathbf{V}_P$ (cf. also (1.23) on page 14).

Moreover, since we assumed that pure thermal as well as pure damage strains vanish, then

$$tr\mathbf{L}_P = 0$$

and this additional equation relates material functions $l_0(\pi), \ldots, l_{12}(\pi)$ with the invariants listed in π (given immediately after (3.6)). On replacing such a relationship into (3.9) all the tensor generators have to be substituted by their deviatoric parts, i.e.,

$$\begin{aligned} \mathbf{L}_P = <f> \operatorname{dev} \{ & l_1(\pi)\boldsymbol{\varepsilon}_P + l_2(\pi)\boldsymbol{\varepsilon}_P^2 + l_3(\pi)\mathbf{S} \\ & + l_4(\pi)\mathbf{S}^2 + l_5(\pi)\mathbf{S}\boldsymbol{\varepsilon}_P + l_6(\pi)\boldsymbol{\varepsilon}_P\mathbf{S} \\ & + l_7(\pi)\mathbf{S}^2\boldsymbol{\varepsilon}_P + l_8(\pi)\boldsymbol{\varepsilon}_P\mathbf{S}^2 + l_9(\pi)\mathbf{S}\boldsymbol{\varepsilon}_P^2 \\ & + l_{10}(\pi)\boldsymbol{\varepsilon}_P^2\mathbf{S} + l_{11}(\pi)\mathbf{S}^2\boldsymbol{\varepsilon}_P^2 + l_{12}(\pi)\boldsymbol{\varepsilon}_P^2\mathbf{S}^2 \} , \end{aligned} \tag{3.10}$$

where dev$\mathbf{A}$ is the deviatoric part of a second-order tensor, i.e., dev$\mathbf{A} \equiv \mathbf{A} - (1/3)\,\mathbf{1}\mathrm{tr}\mathbf{A}$. Notwithstanding, it must be borne in mind that replacement of (3.9) by means of (3.10) when damage evolution appears, i.e., when $D_t\mathbf{\Pi}^d \neq 0$, does not hold any more since in that case $\mathrm{tr}\mathbf{L}_P \neq 0$.

3. According to the analysis given in the introduction of this chapter it is useful to apply the notion of the dynamic yield surface in the sequel. For convenience, let us take in the sequel *dynamic yield function* in the following way:

$$F(\mathbf{S}, \mathbf{V}_P, \mathbf{L}_P) = f(\mathbf{S}, \mathbf{V}_P) - D_t p \leq 0. \tag{3.11}$$

Herein the *plastic strain rate intensity*

$$D_t p := \left(\mathrm{tr}\{\mathbf{L}_P\mathbf{L}_P^T\}\right)^{\frac{1}{2}} \geq 0 \tag{3.12}$$

is positive whenever plastic strain changes with time.

The function (3.11) covers the whole region of viscoplasticity since for larger plastic strain rates $D_t p$ the configuration point $(\mathbf{S}, \mathbf{V}_P)$ is more distant from the *static yield surface*. This surface is the boundary between viscoplastic and elastic region. It is obtained from the dynamics yield function when $D_t p \mapsto 0$. Explicitly we have

$$F(\mathbf{S}^{\#}, \mathbf{V}_P, \mathbf{0}) = f\left(\mathbf{S}^{\#}, \mathbf{V}_P\right) \equiv f^{\#} \leq 0. \tag{3.13}$$

The equality sign holds for points on the static yield surface. Thus the *static yield function* $f^{\#}$ is negative in an elastic region and zero if $D_t p > 0$. As already mentioned in the introduction of this chapter, values of the function $f(\mathbf{S}, \mathbf{V}_P)$ in general could have all three signs (positive in viscoplastic region, negative in elastic region and zero on yield surface) as explained by Property 2.8 (page 39). The difference between dynamic f and static $f^{\#}$ yield function disappears whenever plastic stretching becomes infinitesimal. The simplest way to describe plastic range, elastic range and yield surface is by means of the equation

$$<f> D_t f^{\#} = 0, \tag{3.14}$$

where $<x> \equiv \eta(x)$ equals the Heaviside function being $\eta(x) = 1$ for x positive and $\eta(x) = 0$ otherwise.

For practical determination of the *static yield stress* $\mathbf{S}^{\#}$ it is plausible to assume that it is aligned with the dynamic yield stress, i.e., $\mathbf{S}^{\#} = k^{\#}\mathbf{S}$. Then, this coefficient $k^{\#} < 1$ in viscoplastic region, $k^{\#} > 1$ in elastic region and its value $k^{\#} = 1$ gives a point on the static yield surface.

It should be noted here that in engineering-oriented papers the notion of equivalent plastic strain rate is introduced by means of

$$D_t \varepsilon^p_{eq} = \sqrt{2/3}\, D_t p. \tag{3.15}$$

This quantity has a convenient feature that it is equal to plastic strain rate component along the applied stress at uniaxial tension test. Its integral is called either equivalent plastic strain or accumulated plastic strain (cf. also (2.14)), i.e.,

$$\varepsilon^p_{eq}(t) = \int_0^t D_\tau \varepsilon^p_{eq}(\tau)\, d\tau. \tag{3.16}$$

For the same reason *equivalent stress* being equal to

$$\sigma_{eq} = \sqrt{3/2}\, \|\mathbf{S}\| \equiv \sqrt{3/2}\, \left(\mathrm{tr}\{\mathbf{S}^2\}\right)^{1/2} \tag{3.17}$$

becomes the same as stress during uniaxial tensile experiments.[2]

4. Work per unit volume (of (χ_t) configuration) uncompensated by inertial and body forces is obtained by time integral of the corresponding power as follows

$$W = \int_0^t \mathrm{tr}\{\mathbf{T}(\tau)\mathbf{D}(\tau)\} d\tau, \tag{3.18}$$

where (cf. Fig. 1.4 on page 11)

[2] This will be explained in detail in the fifth chapter.

$$\mathbf{L} := (D_t\mathbf{G})\mathbf{G}^{-1} = (D_t\mathbf{F})\mathbf{F}^{-1}$$

is the velocity gradient tensor and

$$\mathbf{D} = \frac{1}{2}\left(\mathbf{L} + \mathbf{L}^T\right)$$

is the stretching tensor. The corresponding elastic and plastic work per unit volume of (χ_t) configuration read, respectively,

$$W_E = \int_0^t \mathrm{tr}\{\widetilde{\mathbf{S}}(\tau)D_t\mathbf{E}_E(\tau)\}d\tau, \tag{3.19}$$

$$\begin{aligned} W_P &= \int_0^t \mathrm{tr}\{\widetilde{\mathbf{S}}(\tau)\mathbf{C}_E(\tau)\mathbf{L}_P(\tau)\}d\tau \\ &= \int_0^t \mathrm{tr}\{\mathbf{V}_P(\tau)^{-1}\widetilde{\mathbf{S}}(\tau)\mathbf{C}_E(\tau)D_t\mathbf{V}_P(\tau)\}d\tau, \end{aligned} \tag{3.20}$$

where $\mathbf{C}_E := \mathbf{\Phi}^T\mathbf{\Phi} \equiv \mathbf{1} + 2\mathbf{E}_E$ is the elastic deformation tensor and $\widetilde{\mathbf{S}}$ is the convective stress tensor defined hereinabove.

5. Suppose that elastic strain $\mathbf{E}_E$ is small whereas plastic strain $\boldsymbol{\varepsilon}_P \equiv \mathbf{V}_P - \mathbf{1}$ is finite. Then (cf. also [AMM91] as well as (1.88), page 28) Hooke's law holds in the following form:

$$\mathbf{E}_E = \mathbb{D}^{-1} : \mathbf{S}, \tag{3.21}$$

where fourth-rank tensor of elastic "constants" depends in general on temperature and proper invariants of plastic strain.[3] In the special case of isotropic materials it has the following well-known form:

$$\begin{aligned} \mathbb{D} &\equiv \lambda\mathbf{1}\otimes\mathbf{1} + \mu\big(\mathbf{1}\,\square\,\mathbf{1} + \mathbf{1}\,\overset{T}{\square}\,\mathbf{1}\big) \\ &\equiv \lambda\mathbf{1}\otimes\mathbf{1} + \mu\big(\mathbf{1}\,\square\,\mathbf{1} + \mathbf{1}\diamond\mathbf{1}\big) \\ &\equiv \lambda\mathbf{1}\otimes\mathbf{1} + 2\,\mu\mathbb{I} \end{aligned} \tag{3.22}$$

with compact notations

$$(\mathbf{A}\otimes\mathbf{B})_{\alpha\beta\gamma\delta} \equiv A_{\alpha\beta}B_{\gamma\delta},$$

$$(\mathbf{A}\,\square\,\mathbf{B})_{\alpha\beta\gamma\delta} \equiv A_{\alpha\gamma}B_{\beta\delta} \quad \text{and} \quad (\mathbf{A}\,\overset{T}{\square}\,\mathbf{B})_{\alpha\beta\gamma\delta} \equiv A_{\alpha\delta}B_{\beta\gamma}$$

used throughout the following text in order to simplify writing.

Strictly speaking λ and μ must be functions of plastic strain. This means that for isotropic materials they depend on values of the set of invariants

[3] Its best definition is with respect to intermediate (ν_t) configuration. As shown in the previous chapter (Property 2.5 on page 37), in such a configuration the issue of material symmetries is easiest to be analyzed.

$\{\operatorname{tr}\mathbf{V}_P, \operatorname{tr}\mathbf{V}_P^2, \operatorname{tr}\mathbf{V}_P^3\}$. However, experimental results show that such a dependence is so slight that it can be disregarded. On the other hand, their dependence on damage must not be forgotten. However, in this chapter we drop damage from consideration and, consequently, λ and μ are taken as constants.

3.1.2 Formulation of the postulate. Nonassociativity tensor

Consider the following optimization problem in twelve-dimensional stress-plastic strain space:

> **Postulate 3.1.1** *(Minimal plastic work)* *For two arbitrarily chosen points $A_1 \equiv A(\mathbf{S}_1, \mathbf{V}_{P1}, \mathbf{L}_{P1})$ and $A_2 \equiv A(\mathbf{S}_2, \mathbf{V}_{P2}, \mathbf{L}_{P2})$ find the optimal path connecting them such that plastic work is minimal under the unilateral constraint given by the inequality (3.11).*

It is to be noted, however, that in this postulate (formulated for the first time in [Mic92b]) the plastic "velocity gradient" is not an additional independent variable since it is determined by stress and plastic strain through equations (3.6) and (3.9).

Let a path denoted by the triplet $(\mathbf{S}^\star(\tau), \mathbf{V}^\star(\tau), \mathbf{L}_P^\star)(\tau)$, $\tau \in [t_1, t_2]$ with $A(t_1) \equiv A_1$, $A(t_2) \equiv A_2$ be optimal. Then, necessary condition for minimum of the plastic work functional (3.20) reads

$$\delta W_P = \int_{t_1}^{t_2} (\delta \operatorname{tr}\{\mathbf{S}(\tau)\mathbf{C}_E(\tau)\mathbf{L}_P(\tau)\} - \Lambda <f> \delta f)\, d\tau = 0. \tag{3.23}$$

Before proceeding to calculations two notes are worthwhile:

- A yield surface is defined for a fixed intrinsic structure given in our case by $\mathbf{V}_P$. This implies that for each value of stress variation $\delta\mathbf{S} \equiv \mathbf{S} - \mathbf{S}^\star$ we have $\delta\mathbf{V}_P = \mathbf{0}$, i.e., at each real trajectory from A_1 to A_2 plastic strain variation equals zero.[4]
- Another note is connected to lack of commutativity of variation and material time derivative for plastic deformation tensor, i.e., $\delta D_t \mathbf{V}_P - D_t \delta \mathbf{V}_P \neq \mathbf{0}$ which may be easily proved from (3.6).

By making use of (3.11) and performing some simple calculations the above integral transforms into

$$\begin{aligned} \delta W_P &= \int_{t_1}^{t_2} \delta\mathbf{S} : [\,\mathbf{C}_E\mathbf{L}_P + 2\mathbb{D}^{-1} : (\mathbf{L}_P\mathbf{S}) \\ &+ \frac{\Lambda}{D_t p}\mathbb{H} : \mathbf{L}_P - \Lambda <f> \partial_{\mathbf{S}} f + \mathbb{H} : (\mathbf{S}\mathbf{C}_E)\,]dt = 0, \end{aligned} \tag{3.24}$$

[4] This means that we consider strain controlled variations of the direct path from A_1 to A_2.

where

$$\mathbb{H} \equiv \partial_{\mathbf{S}}\mathbf{L}_P = <f>\Big(\mathbf{1}\,\Box\,(l_5\varepsilon_P + l_9\varepsilon_P^2) + (l_6\varepsilon_P + l_{10}\varepsilon_P^2)\,\Box\,\mathbf{1}$$
$$+ (\mathbf{1}\,\Box\,\mathbf{S} + \mathbf{S}\,\Box\,\mathbf{1})(l_7\varepsilon_P + l_{11}\varepsilon_P^2) \tag{3.25}$$
$$+ (l_8\varepsilon_P + l_{12}\varepsilon_P^2)(\mathbf{1}\,\Box\,\mathbf{S} + \mathbf{S}\,\Box\,\mathbf{1})\Big)$$

is a fourth-rank tensor found explicitly by differentiating (3.9) with respect to $\mathbf{L}_P$.

Introducing notations

$$\mathbb{M} \equiv \mathbf{C}_E\,\Box\,\mathbf{1} + 2\mathbb{D}^{-1}\mathbf{S} + \frac{\Lambda}{D_t p}\mathbb{H},$$

remembering the notation $2\text{sym}\mathbf{A} \equiv 2\mathbf{A}_s \equiv \mathbf{A} + \mathbf{A}^T$ and taking into account that the above integral (3.24) disappears for all stress variations we acquire the following constitutive restriction:

$$\text{sym}\,\{\mathbb{M}\;:\;\mathbf{L}_P\} = \Lambda <f>\partial_{\mathbf{S}} f - \text{sym}\,\{\,\mathbb{H} : (\mathbf{S}\mathbf{C}_E)\,\}\,. \tag{3.26}$$

For very small elastic strains $\mathbf{C}_E\,\Box\,\mathbf{1} \approx \mathbf{1}\,\Box\,\mathbf{1}$ approximately holds such that the preceding restriction becomes more transparent, i.e.,

$$\mathbf{D}_P - \Lambda <f>\partial_{\mathbf{S}} f^{\#} = -\text{sym}\Big\{\,\mathbb{H} : \Big(\mathbf{S}\mathbf{C}_E + \frac{\Lambda}{D_t p}\,\mathbf{L}_P\Big)\Big\}. \tag{3.27}$$

This clearly shows that flow rule is in general nonassociative. Moreover, the deflection from associativity is proportional to the tensor $\mathbb{H}$. For this reason this tensor could be termed the *nonassociativity tensor*. It is negligible only for very small plastic strains (which is evident from (3.25)). Accordingly, for very small plastic strains associativity of flow rule could be only approximately accepted.

Indeed, the conclusion about nonassociativity of flow rule has been supported by experiments [AMM91].

3.2 Comparison with Drucker's and Ili'ushin's principles

In the preceding analysis the most fundamental ingredient is that plastic dissipation (i.e., plastic power) is positive and minimal compared to all virtual paths fulfilling yield condition [Mic92b]. In the sequel a comparison of such a postulate with Drucker's and Ili'ushin's postulates is briefly given.

Before writing their formulations let us remember that specific total stress power (per unit volume of natural space local (ν_t) reference configuration) reads

$$W = \mathrm{tr}\{\mathbf{TD}\}\det\mathbf{\Phi} = \mathrm{tr}\big\{\mathbf{S}\big(D_t\mathbf{E}_E + (\mathbf{1}+2\mathbf{E}_E)\mathbf{L}_P\big)\big\} \equiv \mathrm{tr}\{\mathbf{SD}_\nu\} \qquad (3.28)$$

where the work-conjugate of the Piola–Kirchhoff stress $\mathbf{S}$

$$\mathbf{D}_\nu \equiv D_t\mathbf{E}_E + (\mathbf{1}+2\mathbf{E}_E)\mathbf{L}_P \qquad (3.29)$$

is the stretching tensor transformed to (ν_t) configuration[5] (compare also (3.20) as well as (3.23)). Further, as a consequence of absence of damage evolution it is supposed that in Hooke's law (3.21) 4-tensor of elasticity coefficients $\mathbb{D}$ is constant.

More than fifty years ago in his famous paper [Dru52] the author proposed the following postulate.

Postulate 3.2.1 *(Drucker) Excess work in stress space is nonnegative for arbitrary cycle.*

Before analyzing the general three-dimensional case let us depict a typical stress–strain curve for a uniaxial tension test—Fig. 3.1. Let the initial point of the stress cycle be denoted by A and the final point by B.

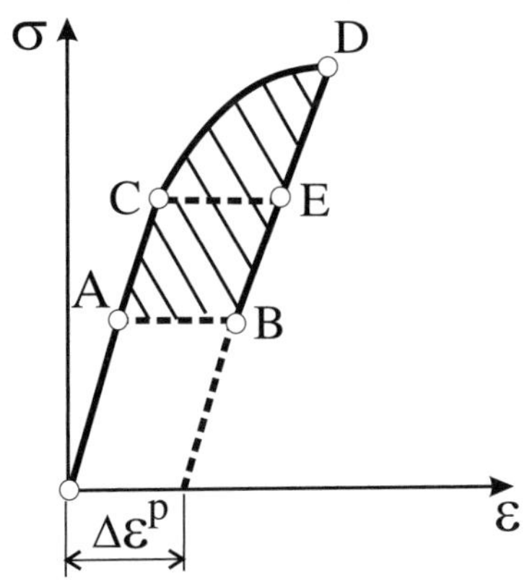

Fig. 3.1. Illustration of Drucker's stress cycle with $\sigma_A = \sigma_B$

Then, this postulate may be formulated as

$$W_{AB} - W_{AB}^{\mathrm{ini}} = \int_{ACDEB} \mathrm{tr}\left\{(\mathbf{S}-\mathbf{S}_A)\mathbf{D}_\nu\right\} dt \geq 0. \qquad (3.30)$$

where segments AC and DEB belong to elastic ranges, i.e., plastic strain is constant in them. Performing the integration and taking into account that $\mathbb{D} = \mathrm{const}$ we arrive at

[5] This is, in fact, the reduced form of the damage-thermo-plastic stretching determined from (2.4) when thermal effects and damage are negligibly small.

$$W_{AB} - W_{AB}^{\text{ini}} = \int_{CD} \text{tr}\,\{(\mathbf{S} - \mathbf{S}_A)(\mathbf{C}_E \mathbf{L}_P)\}\, dt. \tag{3.31}$$

It is worth noting that if points C and D are infinitesimally distant, then (3.30) and (3.31) permit the following inequality:

$$\delta W = \text{tr}\,\{d\mathbf{S}(\mathbf{C}_E \mathbf{L}_P)\}\, dt \geq 0. \tag{3.32}$$

Since elastic strains are small, due to symmetry of the Piola–Kirchhoff stress $\mathbf{S}$ this inequality permits its local representation:

$$\text{tr}\,\{d\mathbf{S}\,\mathbf{D}_P\}\, dt \geq 0. \tag{3.33}$$

This inequality is often referred to as the *fundamental inequality of plasticity.*[6] Drucker's postulate infers that during loading yield surface grows. Therefore, if the yield surface is convex and plastic stretching is perpendicular to the static yield surface $f^{\#} = 0$, then

$$\mathbf{D}_P = \Lambda\, \partial_{\mathbf{S}} f. \tag{3.34}$$

This is the famous associate flow rule, first derived by Drucker and afterwards applied in almost all plasticity papers. It has been criticized above on the basis of minimal plastic work postulate.

On the other hand, in mathematical formulation of Drucker's postulate (3.31) there does not appear change of elastic strain and the integral domain reduces to the parts of the closed stress trajectory where plastic strain changes. It should be noted that such a result holds also if points D and E coincide, i.e., if at some parts of the curve CD there appears a softening with $d\sigma d\varepsilon_P < 0$. If we take that points A and C coincide, we see that for the postulate to hold, positive areas above the line $\overline{AB} \equiv \overline{CD}$ must be larger than negative areas below this line. Nevertheless, we see that multiple maxima are not allowed since for a placement of the point A such that between A and B we have one minimum gives negative value of the excess work (3.30).

Suppose now that a viscoplastic straining takes place and that we have a two-step loading:

- First at a higher stress rate we have that an initial configuration point C is outside the yield surface. It is chosen in such a way that the loading curve CD has the same form as in Fig. 3.1.
- Then, a relaxation path DAE with $\mathbf{D}_\nu = \mathbf{0}$ takes place[7] until the point E reaches the yield surface. In Fig. 3.1 such a line would be parallel to σ-axis and directed downwards.
- Finally, a subsequent loading at a smaller strain rate will take place outside the new yield surface along the corresponding curve EB which is similar to the path CD.

---

[6] More precisely, such a name bears its version written with respect to spatial coordinates of (χ_t) configuration.

[7] At a uniaxial test this relaxation would happen along the line $\varepsilon = \text{const}$.

Now we will apply Drucker's postulate only for the path AEB. In this way it is possible to envisage that a new reference point $A \in \overline{DE}$ is chosen arbitrarily but the point B must fulfill Drucker's condition $\sigma_A = \sigma_B$. Taking into account that along the relaxation path plastic straining also happens and calculating the excess work for the path AEB from (3.31) we obtain that it is now negative, i.e.,

$$W_{AB} - W_{AB}^{\text{ini}} = \int_{AEB} \operatorname{tr}\{(\mathbf{S} - \mathbf{S}_A)\mathbf{D}_\nu\}\, dt < 0. \tag{3.35}$$

Consequently, this postulate does not allow softening as well as viscoplastic behaviour when change of plastic strain appears during plastic unloading.

After this postulate Ili'ushin also made a similar assumption but in strain space ([Ili61]).

> **Postulate 3.2.2** *(Ili'ushin) Excess work in strain space is nonnegative for arbitrary cycle.*

Such a cycle is depicted in Fig. 3.2. The formula which follows this postulate differs from (3.30) only by integral limits:

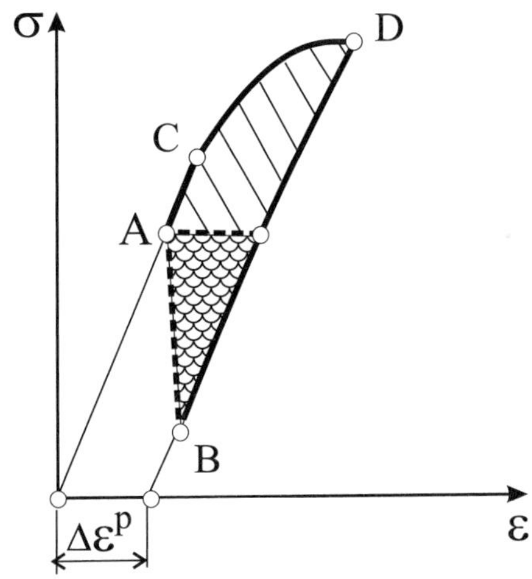

Fig. 3.2. Illustration of Ili'ushin's strain cycle with $\varepsilon_A = \varepsilon_B$

$$W_{AB} - W_{AB}^{\text{ini}} = \int_{ACDB} \operatorname{tr}\{(\mathbf{S} - \mathbf{S}_A)\mathbf{D}_\nu\}\, dt \geq 0. \tag{3.36}$$

After some simple calculations taking into account that $\mathbb{D} = \text{const}$ we acquire the following explicit relationship:

$$W_{AB} - W_{AB}^{\text{ini}} = \int_{CD} \text{tr} \ \{(\mathbf{S} - \mathbf{S}_A)(\mathbf{C}_E \mathbf{L}_P)\} \, dt \\ + \frac{1}{2}[(\mathbf{E}_E)_A - (\mathbf{E}_E)_B] : \mathbb{D} : [(\mathbf{E}_E)_A - (\mathbf{E}_E)_B]. \tag{3.37}$$

Comparing Fig. 3.2 with Fig. 3.1 we see that the excess work according to the Ili'ushin principle is larger than excess work from Drucker's postulate by the area of the shaded triangle.

However, much more important is that this formulation permits softening as well as viscoplastic unloading discussed above. Indeed, such a conclusion may be supported by a sequence of loadings and unloadings similar to the above discussion accompanying Drucker's postulate.

The following postulate [Hil68, Man64] generalizes the above two statements.

Postulate 3.2.3 *(Hill–Mandel) Excess work is positive for an open trajectory either in stress or in strain space.*

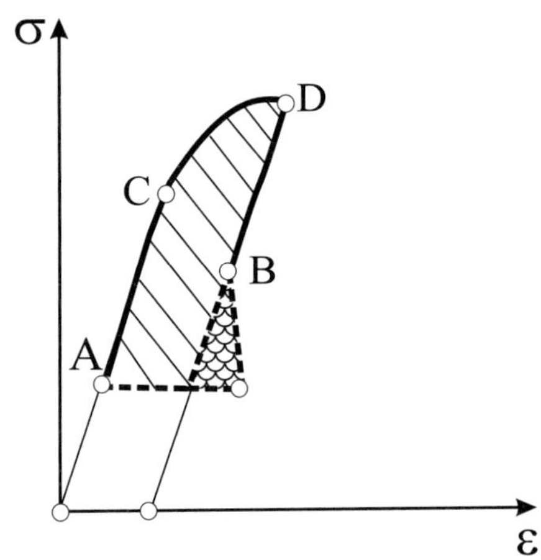

Fig. 3.3. Illustration accompanying Hill–Mandel's postulate with open trajectory

In Fig. 3.2 an open trajectory $ACDB$ is chosen such that the excess work is given by

$$W_{AB} - \mathbf{S}_A : [(\mathbf{E}_E)_B - (\mathbf{E}_E)_A] - \mathbf{S}_A : \int_{CD} (\mathbf{C}_E \mathbf{L}_P) dt \geq 0. \tag{3.38}$$

From this postulate we could obtain as a special case Drucker's postulate for closed stress cycle if condition $\mathbf{S}_A = \mathbf{S}_B$ or, equivalently, $(\mathbf{E}_E)_B = (\mathbf{E}_E)_A$ is fulfilled. On the other hand, Ili'ushin's postulate is obtained from the above formula if $\mathbf{E}_B = \mathbf{E}_A$ holds. All the above comments, of course, are valid and their shortcomings as well as their advantages are included when these special paths are chosen.

3.3 A brief summary of the chapter

The results presented in this chapter may be briefly stated as follows:

- A nonassociate flow rule for viscoplastic materials (of differential type) is formulated and issues concerning plastic spin and Drucker's normality are discussed. It is shown that such a normality cannot be a consequence from postulate that plastic power is minimal on direct trajectory in stress-plastic strain space.
- If the postulate of extremal plastic work is accepted, then the widely accepted associativity of flow rule could be approximately applied only for very small plastic strains. In such a case the nonassociativity tensor may be neglected.
- Validity of all the postulates listed hereinabove in this section must be judged, in the author's opinion, from their agreement or disagreement with thermodynamics as well as experimental results characterizing inelastic behaviour of real materials.

4

Thermodynamics of Inelasticity

This chapter is devoted to the manager of the huge manufactory of thermo-inelastic processes—the second law of thermodynamics [Nem75]. Although a sound definition of the most fundamental items like temperature and entropy is not yet well established for thermodynamic processes we have to deal with them in order to hear the orders of this manager.

The subject is so comprehensive that here the most practical point of view must be taken: *what are the consequences of the second law that must not be forgotten and how to use them in formulating constitutive equations*?

4.1 Balance equations and Clausius–Duhem inequality

Before analyzing the above-stated question and its possible answers let us shortly review balance laws. According to the terminology and language of [Gya70] for an arbitrary tensor field $\mathcal{A}(X,t)$ the local equation of balance reads

$$\varrho D_t \mathcal{A} + \mathrm{div} \mathcal{J}_{\mathcal{A}} = \varrho \sigma_{\mathcal{A}} \tag{4.1}$$

where $D_t(\bullet) \equiv \frac{\partial(\bullet)}{\partial t} + \mathbf{v}\,\mathrm{grad}(\bullet)$ denotes material time derivative ("substantial" in terms of [Gya70]) of the considered tensor $\mathcal{A}(X,t)$, whereas $\mathcal{J}_{\mathcal{A}}$ and $\sigma_{\mathcal{A}}$ are its flux and source density, respectively, and ϱ is mass density.

In Table 4.1, ingredients of such a general balance equation are demonstrated for momentum $\varrho\mathbf{v}$, internal energy u, heat q and nonequilibrium entropy density s.

It must be noted that in this table, the ratio of heat increase and time differential is not total differential, i.e., it depends on the path of thermo-mechanical history. It is usually called *nonmechanical power*. To accentuate this fact the specific notation $\mathfrak{D}_t q$ different from $D_t q$ is employed in Table 4.1.

The balance equations are to be completed either by standard mass conservation equation

M.V. Mićunović, *Thermomechanics of Viscoplasticity: Fundamentals and Applications*, Advances in Mechanics and Mathematics 20, DOI: 10.1007/978-0-387-89490-4_4,
© Springer Science + Business Media, LLC 2009

Table 4.1. Fluxes and sources for balance equations

Tensor $\mathcal{A}$	Rate of $\mathcal{A}$	Flux $\mathcal{J}_{\mathcal{A}}$	Source density $\sigma_{\mathcal{A}}$
$\mathbf{v}$	$D_t\mathbf{v}$	$-\mathbf{T}$	$\mathbf{b}$
u	$D_t u$	$\mathbf{J}_u \equiv \mathbf{q}$	$\sigma_u = (1/\varrho)\,\mathrm{tr}\{\mathbf{TD}\}$
q	$\mathfrak{D}_t q$	$\mathbf{J}_q \equiv \mathbf{q}$	σ_q
s	$D_t s$	$\mathbf{J}_s$	$-\sigma_s$

$$D_t\varrho + \varrho\,\mathrm{div}\,\mathbf{v} = 0, \tag{4.2}$$

or by the corresponding balance equation in the case of volumetric growth (cf. [EM-00, GFG03]) when a mass source must be inserted into the above equation.

In the above balance equations $\mathbf{D}$ and $\mathbf{T}$ are total stretching and Cauchy stress (cf. (1.35) and (1.88)), while $\mathbf{q}$ and $\mathbf{J}_s$ are heat flux and entropy flux vectors. Sometimes, in the balance equation for energy (i.e., the first law of thermodynamics) the free energy density is introduced by means of

$$g := u - Ts. \tag{4.3}$$

Of course, for such a definition temperature T and entropy s must be well defined. Whether introduction of nonequilibrium temperature and entropy is justified will be treated in the sequel.

Our main objective—the *second law* is now locally written (e.g., [Gya70, Mul73]) by means of the condition that entropy source must be nonnegative for all processes:

$$\varrho\sigma_s = \varrho D_t s + \mathrm{div}\mathbf{J}_s \geq 0, \tag{4.4}$$

where irreversibility is emphasized by the inequality sign. The above inequality is known as the Clausius–Duhem inequality.

Now, in the so-called *rational thermodynamics* (*RTIP*) and *classical thermodynamics of irreversible processes* (*CTIP*) entropy flux is connected to the heat flux by the following relationship:

$$\mathbf{J}_s = \frac{1}{T}\mathbf{q}. \tag{4.5}$$

Moreover, in both of these approaches to thermodynamics, entropy source density is related to heat source density by means of

$$\sigma_s = \frac{1}{T}\sigma_q. \tag{4.6}$$

Of course, many authors are against introducing these sources into the theory (e.g., [Woo82]), but if used they are connected by (4.6).[1]

[1] It is interesting to underline the same structure of (4.5) and (4.6).

On the other hand, in diverse versions of *extended thermodynamics of irreversible processes* (ETIP) the above two equalities do not hold in general (cf. [Mul85, CLJ01]).

The above issues have been discussed in [Hut77], where the author gave a comprehensive review. He preferred to use $\varrho\gamma_s \equiv \varrho D_t s + \mathrm{div}\mathbf{J}_s - \varrho\sigma_s \geq 0$ as the second law instead of (4.4), calling γ_s entropy production. Moreover, in this paper the second law is given also in its integral form at a fixed time instant but for an arbitrary part of the considered body $\mathcal{P} \subseteq \mathcal{B}$, i.e.,

$$\int_{\mathcal{P}} \gamma_s dm \equiv D_t \int_{\mathcal{P}} s\, dm + \oint_{\partial\mathcal{P}} \mathbf{nJ}_s - \int_{\mathcal{P}} \sigma_s dm \geq 0 \tag{4.7}$$

with $dm \equiv \varrho dV$.

After these introductory remarks, a more detailed analysis is given in the following sections.

4.2 Classical and rational versus extended thermodynamics

4.2.1 Highlights of some classical views

Some classical contributions to irreversible thermodynamics like [Car09], [Bor49] are very precious and deserve careful attention. Let us recall them very briefly in the following notes.

Remark 4.1. (Caratheodory)

In his paper [Car09] the author stated the famous theorem on integrating factor of Pfaff's equations and gave two axioms, namely,
(a1) With every phase of the considered system it is possible to associate a function of generalized coordinates, which is proportional to the volume of the phase and which is called its internal energy.
(a2) In every arbitrarily close neighborhood of a given initial state there exist states that cannot be approached arbitrarily closely by adiabatic processes.

The proposition that thermodynamics can be justified without recourse to any hypothesis that cannot be verified experimentally—as one of its most noteworthy results—requires that chosen generalized coordinates must be either measurable or uniquely defined by some measurable features. Moreover, we often do not know whether the chosen coordinates are mutually independent or not.

From the first law of thermodynamics and the first axiom internal energy increase is found by the work for adiabatic processes. For the other processes heat enters the first law. The phases here may be understood as entering either a polycrystal or some (solid or fluid) mixture.

The second axiom was formulated on the basis of Max Planck's formulation of the second law of thermodynamics. It shows that real thermodynamic processes take place either on diverse reversible narrow islands which are mutually connected at isolated crossovers or are completely immersed into the sea of irreversibility. This is of paramount importance when an analysis of the second law of thermodynamics is performed—*we must be very careful choosing system coordinates and proclaiming that some coordinates of the system are independent.* To this issue we will return in the sequel.

Remark 4.2. (Born)
This author ([Bor49]) commented on Caratheodory's second axiom, known also as Caratheodory's principle, and defined adiabatic as well as diathermanous walls.
(a1) He gave a precise approach to irreversible processes stating that the reversal of time

$$t \mapsto -t \tag{4.8}$$

is not allowed at irreversible processes.
(a2) Moreover, in the customary definition of absolute temperature and entropy as an integrating factor and corresponding total differential whose product equals the heat increment he showed that either such integrating factor does not exist or if it exists that it is not unique.

The simplest and still very illustrative example for Born's irreversibility criterion is found at the balance equation for heat when the Fourier law of heat conduction holds. Then we get a linear partial differential equation for the absolute temperature with its first time derivative and Laplacian. Such an equation is not invariant to the time reversal (4.8) and, consequently, describes irreversible behaviour. On the contrary, elastic wave equation contains the second-order time derivative of displacement as well as some of its spatial derivatives (like Laplacian)—thus it describes a reversible process.

From the second explanation it follows that a temperature scale may be chosen arbitrarily. However, if an absolute temperature scale (like Kelvin's scale) is chosen, then for equilibrium states there is no arbitrariness in the definition of entropy.

4.2.2 Classical thermodynamics of irreversible processes

Remark 4.3. (Prigogine)
In his book [Pri55] the author first classified systems as (a) isolated without exchange with exterior, (b) closed with exchange of energy and (c) open with exchange of both energy and mass. Then he split the entropy into two parts: $d_e s$ due to interaction with the exterior and the other part denoted by $d_i s$ and caused only by changes inside the system. Then,
(a1) he formulated the second law as the inequality

$$d_i s \equiv \sum_k J_k X_k dt \geq 0 \tag{4.9}$$

with rates (generalized fluxes) X_k and affinities (generalized forces) J_k where the inequality sign holds for all irreversible processes;
(a2) accepted validity of Onsager relations

$$J_k = \sum_m L_{km} X_k \tag{4.10}$$

for processes sufficiently close to equilibrium and
(a3) postulated and applied the principle of minimum entropy production.

For a continuous medium such a formulation of the second law must be replaced by a formulation with entropy flux and entropy production. This will be discussed in the sequel.

In general, Onsager relations must be replaced by some more general evolution equations for irreversible thermodynamic processes far from equilibrium (cf. [Zie63]). The symmetry or antisymmetry of the so-called reciprocity coefficients L_{km} is derived from microscopic considerations and discussed in [Gya70].

The postulate of minimum entropy production implies that irreversible processes operate in such a way to reduce the value of entropy production. Finally, it has been shown in [Gya70] by means of a variational principle that Prigogine's principle of minimum entropy production is equivalent to the *principle of least energy dissipation.* It is worth noting that the postulate of minimal plastic work ([Mic92b]) stated on page 53 of the previous section has the same meaning as these two equivalent principles.

Remark 4.4. (Bataille and Kestin)

These authors, extending the approach of [Mei76], gave in their paper [BK-79] some clear advantages and shortcomings of rational (RTIP) and classical thermodynamics of irreversible processes (CTIP):
(a1) RTIP operating with measurable histories of deformation gradient and temperature has remarkable beauty and simplicity.
(a2) Taking as granted existence of nonequilibrium entropy and temperature, it postulates validity of the Clausius–Duhem inequality for nonequilibrium processes and
(a3) it leaves the most important question of practical determination of response functionals outside the theory.
Their new postulates in [BK-79] are the following:
(b1) To each nonequilibrium state there exists a neighboring *accompanying* state, being in constrained equilibrium, rendered by finite number of appropriate internal variables.
(b2) Entropy and temperature of accompanying state fulfill the Clausius–Duhem inequality.

(b3) According to [Lam77], there exists an integration factor $(\Lambda)^{-1}$ of the form $\sum_k J_k X_k dt$ such that a dissipation potential $\psi(u, \mathbf{F}, X_k)$ leads to the expression for generalized forces

$$J_k = \Lambda \frac{\partial \psi(u, \mathbf{F}, X_m)}{\partial X_k} \tag{4.11}$$

generalizing in such a way Onsager's relations. It should be noted here that in the paper by [Zie63] the author has proposed the following value for this integration factor:

$$\Lambda = \psi(u, \mathbf{F}, X_m) \left[\sum_k X_k \partial \psi(u, \mathbf{F}, X_m)/\partial X_k \right]^{-1} . \tag{4.12}$$

The corresponding value of X_k is named by this author *the least irreversible force*, while (4.11) and (4.12) express *Ziegler's principle of extremum entropy production.*
Their main conclusions read:
(c1) Assuming that a nonequilibrium entropy exists and depends only on actual values of deformation gradient and internal energy they have shown that Clausius–Duhem inequality in general is questionable. Thus, the list of arguments must be extended either by the histories of these arguments or by some appropriate internal variables.
(c2) In the accompanying state heat flux is zero whereas density and internal energy are the same as in actual nonequilibrium state.
(c3) Following [Mei76] they have shown that a nonequilibrium entropy is not unique and that it depends on numbers of constrained and unconstrained internal variables.
(c4) Generalized forces could contain also a *nonpotential term* (analogous to gyroscopic forces) J_k^{gyr} which does not contribute to dissipation, i.e., $\sum_k J_k^{\mathrm{gyr}} X_k dt = 0$. Then Onsager's reciprocal relations do not hold. Such a proposition is intimately connected to our discussion on nonassociativity (nonnormality) given in the previous chapter (cf. (3.27)). Moreover, applying such a reasoning to (3.27) permits experimental verification whether such a nonpotential term contributes to dissipation or not.

4.2.3 Rational thermodynamics of irreversible processes

Concerning advantages and shortcomings of the so-called rational thermodynamics of irreversible processes (RTIP), it is probably the simplest and the most effective way to analyze briefly the famous paper [CN-63] which has been mostly glorified and criticized.

Let us shortly review their analysis of the second law applied to linear viscoelastic materials. They employed (4.5) as well as $\sigma_s = \sigma_q/T$ for entropy flux and entropy source in terms of heat flux $\mathbf{J}_q \equiv \mathbf{q}$ and heat source σ_q. Then in their version the Clausius–Duhem inequality in local form (cf. its global form (4.7)) reads

$$\varrho\gamma_s \equiv \varrho D_t s + \operatorname{div}\left(\frac{1}{T}\mathbf{q}\right) - \frac{\varrho}{T}\sigma_q \geq 0. \tag{4.13}$$

For the following analysis it is useful to take second-order material of differential type (cf. (2.25), page 42) instead of the linear viscoelastic material given in the original paper [CN-63]. In such a case the following set of constitutive equations:

$$\begin{gathered} u = \hat{u}(\mathbf{F},T) + \left(\mathbb{A}_u : \mathbf{L}\right) : \mathbf{L}, \quad s = \hat{s}(\mathbf{F},T) + \left(\mathbb{A}_s : \mathbf{L}\right) : \mathbf{L}, \\ \mathbf{q} = \hat{\mathbf{q}}(\mathbf{F},T,\mathrm{grad}T), \\ \mathbf{T} = \hat{\mathbf{T}}(\mathbf{F},T) + \mathbb{L}_1(\mathbf{F},T) : \mathbf{L} + \left(\mathfrak{L}_{11}(\mathbf{F},T) : \mathbf{L}\right) : \mathbf{L} + \mathbb{L}_2(\mathbf{F},T) : D_t\mathbf{L} \end{gathered} \tag{4.14}$$

is appropriate, where $\mathbf{L} := (D_t\mathbf{F})\,\mathbf{F}^{-1}$ is the velocity gradient tensor defined in (1.35) whereas $\mathbb{L}_1(\mathbf{F},T)$ and $\mathfrak{L}_{11}(\mathbf{F},T)$ are, respectively, the first-order 4-tensor of linear and the first-order 6-tensor of quadratic viscosity coefficients while $\mathbb{L}_2(\mathbf{F},T)$ is the second-order 4-tensor of linear viscosity coefficients. The above form for the stress is more general than in [CN-63] where only linear terms are kept, i.e., they assumed that $\mathfrak{L}_{11} = 0$ as well as $\mathbb{L}_2 = 0$.

This choice of independent variables is convenient when the free energy (4.3) is used instead of internal energy, i.e.,

$$g = u - Ts = \hat{g}(\mathbf{F},T) + \left(\mathbb{A}_g : \mathbf{L}\right) : \mathbf{L}, \tag{4.15}$$

where $\hat{g} = \hat{u} - T\hat{s}$ and $\mathbb{A}_g = \mathbb{A}_u - T\mathbb{A}_s$.

Suppose now that $\mathbf{L}$ and $D_t\mathbf{L}$ are replaced by $\alpha\mathbf{L}$ and $\beta D_t\mathbf{L}$ where α and β are some scalars. Then, neglecting heat (and entropy) source, using energy balance equation and choosing free energy instead of internal energy, the second law is transformed into the differential inequality:

$$\begin{gathered} -\varrho\big(\hat{s} + \partial_T\hat{g}(\mathbf{F},T)\big)D_tT - \frac{1}{T}\hat{\mathbf{q}}(\mathbf{F},T,\mathrm{grad}T)\,\mathrm{grad}T \\ +\alpha\,\beta\left(\left(\mathbb{L}_2(\mathbf{F},T) - 2\varrho\,\mathbb{A}_g(\mathbf{F},T)\right) : \mathbf{L}\right) : D_t\mathbf{L} \\ +\alpha\left(\hat{\mathbf{T}}(\mathbf{F},T) - \varrho\mathbf{F}\partial_{\mathbf{F}}\hat{g}(\mathbf{F},T)\right) : \mathbf{L} \\ +\alpha^2\left(\mathbb{L}_1(\mathbf{F},T) : \mathbf{L}\right) : \mathbf{L} \\ +\alpha^3\left(\left(\left(\mathfrak{L}_{11}(\mathbf{F},T) - \varrho\,\mathbf{F}\,\partial_{\mathbf{F}}\mathbb{A}_g(\mathbf{F},T)\right) : \mathbf{L}\right) : \mathbf{L}\right) : \mathbf{L} \geq 0. \end{gathered} \tag{4.16}$$

In the subsequent analysis it is assumed that $D_tT, \mathrm{grad}T, \mathbf{L}$ and $D_t\mathbf{L}$ are independent. Then the following outcomes are possible.

- Suppose that $\mathbf{L} = \mathbf{0}$, $D_t\mathbf{L} = \mathbf{0}$ and that $\mathrm{grad}T = \mathbf{0}$. Then

$$\hat{s} = -\partial_T \hat{g}(\mathbf{F}, T). \tag{4.17}$$

- In the second special case let $D_tT = 0$, $\mathrm{grad}T = \mathbf{0}$ and $D_t\mathbf{L} = \mathbf{0}$. For us this situation is most pregnant. For arbitrary (either positive or negative) α we would have the conclusions:

$$\begin{aligned} \hat{\mathbf{T}}(\mathbf{F}, T) &= \varrho\, \mathbf{F}\, \partial_{\mathbf{F}} \hat{g}(\mathbf{F}, T), \\ \mathfrak{L}_{11}(\mathbf{F}, T) &= \varrho\, \mathbf{F}\, \partial_{\mathbf{F}} \mathbb{A}_g(\mathbf{F}, T). \end{aligned} \tag{4.18}$$

  The reduced dissipation inequality in this case becomes

$$\big(\mathbb{L}_1(\mathbf{F}, T) : \mathbf{L}\big) : \mathbf{L} \geq 0. \tag{4.19}$$

  It is worth noting that the first of the equations (4.18) and the above reduced inequality (4.19) have been obtained in [CN-63].
- Let us assume that again $D_tT = 0$, $\mathrm{grad}T = \mathbf{0}$ but the velocity gradient $\mathbf{L}$ has some fixed value. Then, taking β to be arbitrary (either positive or negative) the second law permits

$$\mathbb{L}_2(\mathbf{F}, T) = 2\varrho\, \mathbb{A}_g(\mathbf{F}, T). \tag{4.20}$$

- Finally, consider the case of an inhomogeneous constant temperature field and a homogeneous constant deformation field, i.e., $\mathbf{L} = \mathbf{0}$, $D_t\mathbf{L} = \mathbf{0}$ and $D_tT = 0$. Then the reduced dissipation inequality takes the familiar form

$$\hat{\mathbf{q}}(\mathbf{F}, s, \mathrm{grad}T)\, \mathrm{grad}T \leq 0, \tag{4.21}$$

  which is the very well-known reduced dissipation inequality showing that heat flows to colder particle.

At first sight everything seems perfect. The reasoning of RTIP has given us all the restrictions needed. Even more, we have got a class of nonlinear viscoelastic materials, more general than in [CN-63], that satisfies the second law of thermodynamics. Nevertheless, let us remark that an embarrassing relationship is obtained between the first- and the second-order viscosity coefficients, namely,

$$2\mathfrak{L}_{11}(\mathbf{F}, T) = \varrho\, \mathbf{F}\, \partial_{\mathbf{F}} \mathbb{L}_2(\mathbf{F}, T). \tag{4.22}$$

If we look carefully at the last of (4.14), we will see that (4.22) connects the second time rate of deformation gradient $D_tD_t\mathbf{F}$ with products of their time rates $D_t\mathbf{F} \otimes D_t\mathbf{F}$. For the special case when constitutive equations (4.14) are linear when

$$\mathfrak{L}_{11} = 0, \mathbb{A}_g = 0,$$

like in [CN-63], we may forget about all these difficulties. However, in our opinion the real materials must permit nonlinearity in the first time derivatives in spite of the second time derivatives not appearing in evolution equations. As easily checked, this is not permitted if the above procedure is strictly obeyed. Stated explicitly, there is no reason why $\mathfrak{L}_{11}$ must vanish whenever $\mathbb{L}_2$ disappears.[2]

Anyway, the appealing mathematical beauty of the RTIP procedure is really attractive.

Remark 4.5. (Woods' criticism of RTIP)

Perhaps the strongest reaction against rational thermodynamics of irreversible processes has been given in [Woo82]. In this paper the author derived the second law in the form (4.4) from purely statistical arguments by making use of a "collision integral" defined to account for collisional scattering of molecules. The main objections are the following:

(a1) The *principle of local action* used extensively in RTIP states for simple materials that when devising constitutive equations the second-order spatial derivatives are ignored while the second-order and higher time derivatives are permitted. On the contrary to such an approach, Woods claims that entropy should depend on as many variables, including higher spatial derivatives, as necessary to get better information about the process and material of the considered body. Indeed the above-presented analysis of RTIP of nonlinear viscoelastic materials (cf. (4.14)–(4.22)) supports this criticism.

(a2) The so-called *principle of equipresence* is disliked by many researchers showing our ignorance about the real nature of the irreversible process considered.[3]

(a3) There is no justification for assumed independence of time derivatives of temperature, deformation gradient and the other constitutive variables.

In our opinion the last objection is most evident in inelastic deformation processes where changes of temperature, plastic strain and some inelastic variables like residual magnetization certainly are not mutually independent.

4.2.4 Müller's version of extended thermodynamics

There are at least two versions of extended thermodynamics of irreversible processes. Both are mentioned in the introduction to this chapter. Let us see how one of them works in the case of viscoplasticity. The brief analysis given in the sequel is mainly based on the paper by [Mic86a]. Following [Mul71],

[2] Perhaps this is the reason why these authors have restricted their attention to linear viscoelasticity.

[3] In its partial defense we might say that a synergetic interaction among diverse physical fields should be very useful in problems treating coupled fields and especially for sophisticated treatment of experiments on inelasticity.

let us first introduce an *empirical temperature* ϑ, being a scalar measured by some device.

For simplicity we drop source terms, consider generalized Fourier material, small elastic strains and assume that a linear relationship between plastic stretching and stress holds.

If plastic rotation tensor is taken to be unity (for details see Remark 1.3 on page 13), then chosen constitutive variables are shortly denoted by $\gamma_M \equiv \{\vartheta, D_t\vartheta, \mathbf{E}_E, \mathbf{V}_P\}$. The corresponding set of viscoplastic constitutive equations appropriate to this approach is the following (cf. also (2.25) and (2.28) as well as (2.27)):

$$\mathbf{S} = \mathbf{S}^{\#}(\gamma_M) + \mathbb{P}(\gamma_M) : D_t\mathbf{V}_P, \tag{4.23}$$

$$\mathbf{J}_q^\nu = \mathbf{K}_q(\gamma_M)\,\mathrm{grad}_\nu\vartheta, \tag{4.24}$$

$$\mathbf{J}_s^\nu = \mathbf{K}_s(\gamma_M)\,\mathrm{grad}_\nu\vartheta, \tag{4.25}$$

$$g = g(\gamma_M), \tag{4.26}$$

$$s = s(\gamma_M). \tag{4.27}$$

Herein dynamic and static stress, heat $\mathbf{J}_q^\nu \equiv \mathbf{\Phi}\mathbf{J}_q$ and entropy flux $\mathbf{J}_s^\nu \equiv \mathbf{\Phi}\mathbf{J}_s$ as well as free energy and entropy density are specified, respectively. We use the notation $\mathrm{grad}_\nu\vartheta := \mathbf{\Phi}^{-1}\,\mathrm{grad}\vartheta$ for "natural space" gradient of empirical temperature (used already in (1.71), page 24). By making use of linearity of (4.24) and (4.25), the tensor representation (cf. [Spe71]) gives us the explicit forms of the constitutive *heat conductivity tensor* $\mathbf{K}_q(\gamma_M)$ and its entropy counterpart $\mathbf{K}_s(\gamma_M)$ as follows:

$$\mathbf{K}_\beta = \sum_{k=0}^{11} I_\beta^k\, \mathbf{\Xi}^k \qquad (\beta \in \{q, s\}). \tag{4.28}$$

The above scalar coefficients $I_q^k, I_s^k, k \in \{0, 1, \ldots, 11\}$ are functions of empirical temperature and its material time rate as well as of proper and mixed invariants of elastic strain and plastic left stretch tensors $\big\{\mathrm{tr}\mathbf{V}_P, \mathrm{tr}\mathbf{V}_P^2, \mathrm{tr}\mathbf{V}_P^3,$ $\mathrm{tr}\mathbf{E}_E, \mathrm{tr}\{\mathbf{E}_E\mathbf{V}_P\}, \mathrm{tr}\{\mathbf{E}_E\mathbf{V}_P^2\}\big\}$. On the other hand, the twelve tensor generators comprise $\{\mathbf{\Xi}^0, \ldots, \mathbf{\Xi}^{11}\} := \{\mathbf{1}, \mathbf{V}_P, \mathbf{V}_P^2, \mathbf{E}_E, \mathbf{E}_E\mathbf{V}_P, \mathbf{V}_P\mathbf{E}_E, \mathbf{E}_E\mathbf{V}_P^2,$ $\mathbf{V}_P^2\mathbf{E}_E, \mathbf{V}_P\mathbf{E}_E\mathbf{V}_P, \mathbf{V}_P\mathbf{E}_E\mathbf{V}_P^2, \mathbf{V}_P^2\mathbf{E}_E\mathbf{V}_P, \mathbf{V}_P^2\mathbf{E}_E\mathbf{V}_P^2\}$. It is worth noting here that if the conditions

$$I_\beta^4 = I_\beta^5,\; I_\beta^6 = I_\beta^7,\; I_\beta^9 = I_\beta^{10} \qquad (\beta \in \{q, s\}) \tag{4.29}$$

are satisfied, then skew-symmetric parts of the above two conductivity tensors vanish.

These equations must obey the second law of thermodynamics in the form (4.4). Moreover, the elastic and plastic strain ranges are separated by the yield surface (cf. Property 2.8, page 39). For the subsequent calculation it is useful to write all the balance laws with respect to the intermediate local reference (ν_t) configuration (cf. page 10).

Let us recall the plastic straining indicator function (cf. (2.15)–(2.17)) $<f>$ which equals unity whenever plastic straining takes place and vanishes if elastic straining happens. It is useful in description of spatial and temporal properties of elastic and viscoplastic regions. Choosing the dynamic yield function in the form (3.11), i.e.,

$$\begin{aligned} f(\mathbf{S}, \mathbf{V}_P, \mathbf{L}_P) &\approx f\left(\mathbf{S}^{\#}, \mathbf{V}_P, \mathbf{0}\right) + \varDelta f \\ &\equiv f^{\#} + \phi\left(\mathbf{S}, \mathbf{V}_P, D_t p\right), \end{aligned} \tag{3.11$'$}$$

we cover these properties by means of the equation

$$<f> D_t f^{\#} = 0. \tag{3.14$'$}$$

If damage is negligible and only plastic straining takes place ($\mathbf{V}_{P\omega} = \mathbf{V}_P$ from (2.6)), then it would be necessary to add two analogous relationships to the above constraining equations, which follow from spatial and temporal differentiation of the condition $\det \mathbf{V}_P = 1$. We consider the more general case when this condition does not hold.

Taking the equation (3.14$'$) together with internal energy balance equation and applying Liu's procedure with Lagrange's multipliers ([Liu72]) we arrive at the extended entropy inequality in the following form:

$$\begin{aligned} \varrho_\nu \sigma_s &= \varrho_\nu D_t s + \operatorname{div}_\nu \mathbf{J}_s^\nu \\ &\quad + \varLambda^u \left(\mathbf{S} : (D_t \mathbf{E}_E + \mathbf{C}_E\, \mathbf{L}_P) - \operatorname{div}_\nu \mathbf{J}_s^\nu - \varrho_\nu D_t u\right) \\ &\quad - \varLambda^u \varLambda^f <f> D_t f^{\#} + \varLambda^\pi \mathbf{V}_P^{-1}\, D_t \mathbf{V}_P \geq 0. \end{aligned} \tag{4.30}$$

In the above inequality the expression for total power has been used in the form derived in (3.19) and (3.20).

Suppose now, following Müller, that the multiplier $\varLambda^u$—named by Müller *coldness function*—depends only on temperature and its rate and that it connects entropy and heat fluxes by Müller's relationship:

$$\mathbf{J}_s^\nu = \varLambda^u(\vartheta,\, D_t \vartheta)\, \mathbf{J}_q^\nu. \tag{4.31}$$

This function was found in [Mul71] to be universal for isotropic thermoelastic solids and heat conducting fluids. The author defined ideal walls as boundaries between different substances at which normal component of heat as well as entropy flux vectors are not changed.

According to the procedure of Coleman and Noll in rational thermodynamics of irreversible processes as explained in the previous subsection coefficients multiplying the highest order derivatives $D_t \mathbf{C}_E$, $D_t D_t \vartheta$, $D_t \operatorname{grad}_\nu \vartheta$, $\operatorname{grad}_\nu \mathbf{C}_E$, $\operatorname{grad}_\nu \mathbf{V}_P$, $\operatorname{grad}_\nu \operatorname{grad}_\nu \vartheta$ must vanish.[4] It should be noted here that

[4] This analysis is akin to the example on nonlinear viscoelasticity treated in Section 4.2.3.

coefficient with $D_t\mathbf{V}_P$ does not vanish since this rate has been included into constitutive variables into the relation for stress (4.23) In his approach to thermodynamics Müller also accepts the reasoning that coefficients with the highest degree derivatives vanish. This gives rise to the following constitutive restrictions:

$$\varrho_\nu\big(\partial_{\mathbf{C}_E}s - \Lambda^u\partial_{\mathbf{C}_E}u\big) + \frac{1}{2}\Lambda^u\big(\mathbf{S}^\# + \mathbb{P} : D_t\mathbf{V}_P\big)$$

$$+\Lambda^f <f> \partial_{\mathbf{C}_E}f^\# = 0, \tag{4.32}$$

$$\varrho_\nu\big(\partial_{D_t\vartheta}s - \Lambda^u\partial_{D_t\vartheta}u\big) + \Lambda^f <f> \partial_{D_t\vartheta}f^\# = 0, \tag{4.33}$$

$$\varrho_\nu\big(\partial_{\mathrm{grad}_\nu\vartheta}s - \Lambda^u\partial_{\mathrm{grad}_\nu\vartheta}u\big) + \Lambda^f <f> \partial_{\mathrm{grad}_\nu\vartheta}f^\#$$

$$+\partial_{D_t\vartheta}\mathbf{J}_s^\nu - \Lambda^u\,\partial_{D_t\vartheta}\mathbf{J}_q^\nu = \mathbf{0}, \tag{4.34}$$

$$\partial_{\mathbf{C}_E}\mathbf{J}_s^\nu - \Lambda^u\partial_{\mathbf{C}_E}\mathbf{J}_q^\nu = 0, \tag{4.35}$$

$$\partial_{\mathbf{V}_P}\mathbf{J}_s^\nu - \Lambda^u\partial_{\mathbf{V}_P}\mathbf{J}_q^\nu = 0, \tag{4.36}$$

$$\mathrm{sym}\,\big(\partial_{\mathrm{grad}_\nu\vartheta}\mathbf{J}_s^\nu - \Lambda^u\partial_{\mathrm{grad}_\nu\vartheta}\mathbf{J}_q^\nu\big) = 0, \tag{4.37}$$

where by sym$\mathbf{A}$ is denoted the symmetric part of the second-order tensor $\mathbf{A}$. When all these terms are extracted from the dissipation inequality it becomes reduced as follows

$$\varrho_\nu\sigma_s^{res}\big(\vartheta, D_t\vartheta, \mathbf{V}_P, \mathbf{C}_E\big) \geq 0. \tag{4.38}$$

Now, from the last of the above restricting equations, i.e., (4.37), the relationship between fluxes reads

$$\mathbf{J}_s^\nu = \Lambda^u\mathbf{J}_q^\nu + \mathrm{skw}\big(\mathbf{K}_s + \Lambda^u\,\mathbf{K}_q\big)\mathrm{grad}_\nu\vartheta. \tag{4.39}$$

Herein skw$\mathbf{A}$ is the skew-symmetric part of the second-order tensor $\mathbf{A}$. Now, from (4.28) there does not follow, in general, that skew-symmetric parts of $\mathbf{K}_s$ and $\mathbf{K}_q$ vanish. It is essential that both of these tensors depend on the two time–dependent 2-tensors $\mathbf{V}_P$ and $\mathbf{E}_E$ such that we do not have any justification to assume that (4.29) are valid in the general case. Therefore, it seems that the relationship (4.31) which holds for (virgin) elastic materials and heat conducting fluids should be extended with the new term $\big(\mathrm{skw}\mathbf{K}_s + \Lambda^u\,\mathrm{skw}\mathbf{K}_q\big)\mathrm{grad}_\nu\vartheta$ for viscoplastic materials.

The conclusion drawn from (4.39) seems quite satisfactory but there appears another problem. Namely, from (4.32)–(4.37) we should conclude that plastic viscosity tensor $\mathbb{P}$ must vanish if RTIP approach is strictly applied but this contradicts the viscoplasticity approach with overstress concept.

In order to overcome this inconsistency let us suppose that free energy density is a linear function of plastic left stretch rate, i.e.,

$$g = g^\#(\gamma_M) + \mathbf{A}_g(\gamma_M) : D_t\mathbf{V}_P. \tag{4.40}$$

If we repeat the same analysis as hereinabove, then aside from the above-mentioned highest derivatives we have the term

$$-\varrho_\nu \Lambda^u \mathbf{A}_g(\gamma_M) : D_t D_t \mathbf{V}_P.$$

Accepting the RTIP approach we are forced, although unwillingly, to conclude that $\mathbf{A}_g$ must disappear. If we assume that free energy is not linear but quadratic function in $D_t\mathbf{V}_P$ like (4.15) (cf. (2.27) on page 43), i.e.,

$$g = g^{\#}(\gamma_M) + \left(\mathfrak{P}_{11}(\gamma_M) : D_t\mathbf{V}_P\right) : D_t\mathbf{V}_P,$$

then the same analysis would yield that the 6-tensor of plastic viscosity coefficients $\mathfrak{P}_{11}$ must vanish.

Concluding this subsection we may state that the very promising and history including Müller's approach to thermodynamics has the mentioned drawbacks of RTIP obeying its method of analysis. Anyway, this analysis will be the most useful when we will apply an alternative method of extended thermodynamics (cf. [CLJ01]). At this point it is necessary to stress that such an approach has evolution equation for heat flux rate and the simplified version of connection between entropy flux and heat flux vectors. It will be used in the section devoted to viscoplasticity of ferromagnetics as one of the approaches for thermodynamic analysis.

4.2.5 Synergetics issues. Importance for experiments

For all people dealing with multiaxial experiments it is often a nightmare when they try to identify when plasticity commences, i.e., points on yield surface in stress space. When a tensile experiment is performed, it is well known that in elastic region we have cooling [BCN82, BBC86] of the specimen. On the other hand, there appear heating during compression and no temperature change during shear as should be expected according to the thermoelastic effect.

When plasticity begins, dislocations start moving relatively with respect to bulk material and temperature rises in all three cases [Mic92a]. This is described well for slow thermomechanical histories. Therefore, for slow tension test, such characteristic minimum point in the diagram $(\mathrm{tr}\mathbf{T}, T)$, i.e., temperature as a function of stress tensor trace is precious for a genuine detection of plasticity triggering.

As we will see in the chapter devoted to experimental issues it is never possible to measure all the fields relevant for constitutive equations, especially for very fast processes when there is a delay between a measured history and its record. Then, like in a detective story approach, we have to collect all the information possible and try to reconstruct what is going on. Although often disliked—due to abundant structure of the principle of equipresence—the idea of synergetics [Hak87] would be very useful either in fast inelastic dynamic experiments or in cyclic inelastic experiments. In fact a cyclic inelastic strain of ferromagnetic specimens (e.g., [EGD97]) would be a very good example for such a synergetic approach where magnetic field should replace temperature as a nonmechanical device for detection of characteristic points in inelastic mechanical behaviour.

4.3 Endochronic thermodynamics

The main idea in the so-called *endochronic thermodynamics* (cf. [Vak70, Val71]) is to replace actual time by means of some nondecreasing scalar function of inelastic strain history responsible for aging whose ultimate value leads finally to rupture of the body. Vakulenko called this function *thermodynamic time* ([Vak70]). In such a concept purely elastic strain does not contribute to the thermodynamic time.

Such a time was introduced in [Vak70] by means of the inelastic entropy source accumulation (cf. Table 4.1) as follows. First, divergence of entropy flux is split into pure thermal and inelastic parts:

$$\operatorname{div}\left(\frac{\mathbf{q}}{T}\right) = \frac{1}{T}\operatorname{div}\mathbf{q} + \mathbf{q}\operatorname{grad}\left(\frac{1}{T}\right). \tag{4.41}$$

The second term disappears for a homogeneous temperature field inside the considered body. Then, the entropy source is split into its thermal and inelastic parts by means of

$$\varrho\sigma_s^P = \varrho\sigma_s - \mathbf{q}\operatorname{grad}\left(\frac{1}{T}\right) \equiv \varrho(\sigma_s - \sigma_s^T). \tag{4.42}$$

In the case of thermoelasticity, Clausius–Duhem inequality reduces to $\varrho\sigma_s^T = \mathbf{q}\operatorname{grad}(\frac{1}{T}) \geq 0$ showing that heat "flows" to the particle with lower temperature.[5] In addition, Vakulenko assumes that $\varrho\sigma_s^P \geq 0$ and defines the thermodynamic time by means of (cf. also [Fom75])

$$\zeta(t) := \int_0^t \varrho\sigma_s^P(t')\,dt'. \tag{4.43}$$

The function $\zeta(t)$ is piecewise continuous and nondecreasing in that $D_t\zeta(t) = 0$ within elastic ranges and $D_t\zeta(t) > 0$ when plastic deformation takes place.

Assuming that at each time instant the considered state can be obtained either by *instantaneous loading or by unloading*[6] (cf. Fig. 4.1) and splitting the whole time history into a sequence of infinitesimal segments Vakulenko claimed that a superposition and causality exists—extending in such a way Boltzmann's and Volterra's superposition to nonlinear inelastic phenomena.

[5] According to [Kro86] this is Clausius' formulation of the second law.

[6] This assumption should be dropped if viscoplastic unloading discussed in the previous chapter takes place. Even without it his theory works well.

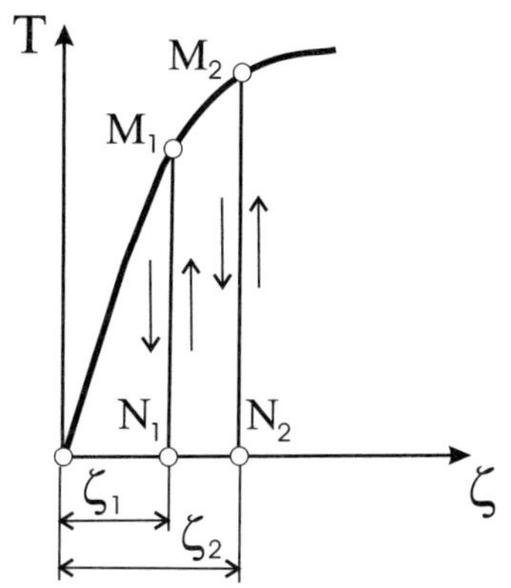

Fig. 4.1. Vakulenko's instantaneous loading and unloading

Proceeding with accumulation of infinitesimal memory he obtained an integral relationship between Eulerian plastic strain deviator (1.8) and stress deviator history (2.10) as follows:

$$\mathbf{e}_{Pd}(\zeta) = \int_0^{\zeta} \mathbf{\Phi}\Big(\zeta - \xi, \mathbf{S}(\xi), \frac{d}{d\xi}\mathbf{S}(\xi)\Big)\, d\xi. \tag{4.44}$$

Therefore, in this setting the plastic strain tensor is obtained as a functional of stress and stress rate history. It should be noted that Vakulenko's original formulation has been adapted here to the geometry given in the first chapter.

In the book by [Fom75] the author applied Vakulenko's approach to diverse media splitting the free energy into a part dependent on temperature and elastic strain and a part which contains as arguments inelastic internal parameters and temperature. In such a way elastic constants appearing in Hooke's law do not depend on such inelastic internal parameters. A similar treatment will be applied also in the sequel to inelastic ferromagnetics and polycrystals and compared to some other thermodynamic approaches.

If we compare the hereinabove function of thermodynamic time to accumulated plastic strain discussed in Remark 2.13 on page 44, then we see that ζ defined by (4.43) is more general and able to account for nonlinear evolution equations in a simple way.

In conclusion we may say that the approach by the endochronic thermodynamics is less general than the others applied by ETIP or RTIP but, as we will see in the next chapters, is much more convenient and able to cover a variety of phenomena. On the other hand, a lack of excessive generality could be an advantage if clear understanding of the considered process is possible.

4.4 Statistical approach to thermodynamics and dislocation distributions

From all the above considerations it is possible to draw a conclusion that a lot of issues in thermodynamics are open. One of the fundamental questions is whether statistical thermodynamics can help us clarify the shadows of classical, rational and extended thermodynamics. In the paper by [Kro86] the author gave an impressive view of this problem. His leitmotif is that continuum thermodynamics must follow from statistical thermodynamics.

- Question: "How general is thermodynamics" must be completed with another question: "to which physical systems" do we want to apply it? Our physical system of interest here is inelasticity.
- Essential problem: a good selection of a few most representative internal variables. Kröner claims that good variables are elastic strain and infinite number of dislocation measures. If correlation functions are not taken into account, then such a statistical theory is not much complicated. The fact that necessary number of internal variables is not finite but unlimited is of utmost importance.
  In order to make these internal variables explicit, let us divide a typical volume element—later in this text, in Chapter 7, called a representative volume element (RVE)—to n cells. Then through each Λ-th cell passes a number of dislocation segments $\mathbf{t}_{\Lambda_\alpha}(\mathbf{x}_\Lambda), \Lambda \in \{1, N_g\}, \alpha \in \{1, \ldots, N_s\}$ with Burgers vectors $\mathbf{b}_\alpha, \alpha \in \{1, \ldots, N_s\}$ (cf. (1.39)). Then, *Kröner's elementary dyadic* ([Kro70]) is defined by

$$\mathbf{k}_{\Lambda_\alpha}(\mathbf{x}_\Lambda) := \mathbf{t}_{\Lambda_\alpha}(\mathbf{x}_\Lambda) \otimes \mathbf{b}_\alpha, \tag{4.45}$$

  whereas *Kröner's m-th order dislocation correlation function* is calculated by the following ensemble average:

$$\begin{aligned}\mathcal{K}^{(m)}(\mathbf{x}_1, \mathbf{x}_2, \ldots, \mathbf{x}_m) &:= \\ &\langle \mathbf{k}_{\Lambda_\alpha}(\mathbf{x}_1) \otimes \mathbf{k}_{\Lambda_\alpha}(\mathbf{x}_2) \otimes \ldots \otimes \mathbf{k}_{\Lambda_\alpha}(\mathbf{x}_m)\rangle,\end{aligned} \tag{4.46}$$

  where the summation is taken over all N_g cells (i.e., grains—cf. page 122) and all Burgers vectors (i.e., N_s-slip systems) assuming that the *ergodic hypothesis* holds ([Kro86]) stating that

$$\textit{volume average over RVE} = \textit{ensemble average}. \tag{4.47}$$

- As Kröner remarked in his CISM lectures [Kro71] there is no known statistical solution without ergodic hypothesis being employed. It is important to know size ΔV of RVE and elementary time increment Δt where this hypothesis should be applied. According to [Kro86] allowable size of RVE could be 1 μm^3 which for solids contains roughly 10^{13} atoms,[7] whereas Δt must be small compared to the time required for the body to arrive at equilibrium when left alone but much larger than lattice vibration times.

- Internal energy is found from the average of Hamiltonian and distribution function for such an estimation is of supreme importance. It depends on extensive state variables and fulfills Gibbs fundamental equation. Its derivatives give us intensive conjugate generalized forces in equilibrium.
- Important conclusion: entropy and temperature are well defined if ergodic hypothesis holds and for each nonequilibrium state there exists a close state with constrained equilibrium.[8] This reminds us about the accompanying state of Bataille and Kestin (cf. Remark 4.4 on page 65). Again the fundamental assumption is that these two states are very close. Then the entropy differential is defined by the ratio of heat increase at transition from one constrained state to the other and absolute temperature.
- Without going much into further details we quote here the final conclusion that derivation of the second law from a statistical theory by means of Liouville's theorem is still an open problem.

4.5 A discussion with some comments

According to the author's taste, the issues of thermodynamics of inelasticity worthy to be emphasized are the following:

- Bataille and Kestin derived in the aforementioned paper a procedure for deriving memory approach from internal variables approach as follows. First, they assumed that the given set of evolution equations for internal variables is necessary and sufficient for their integration and that they vanish at $t \mapsto -\infty$. As arguments, deformation gradient and its rate, gradient of temperature (of accompanying state), internal energy and a list of internal variables are chosen. By means of integration they obtained stress, heat flux and entropy of accompanying state as functionals of histories of deformation gradient, temperature of accompanying state and its gradient.
  Compared to the approach where plastic memory is taken, this procedure is reminiscent of a black box. Namely, if viscoplasticity is combined either with magnetostriction or with neutron irradiated damage, then at constant deformation gradient and temperature we may envisage that change of these fields induces change of plastic strain as well as elastic strain causing stress change as well. Therefore, *inelastic memory cannot be replaced by total memory approach.* The latter does not give complete information about the actual state of the body.

[7] This issue has been extensively analyzed also in [Ost02].
[8] Such an equilibrium is defined in [Lmb74].

- Related to the former conclusion, consulting with statistical thermodynamics seems worthwhile. Indeed, Kröner insisted that the number of internal variables must be infinite, i.e., $m \mapsto \infty$. Suppose that we have all the necessary evolution equations governing temporal behaviour of Kröner's dislocation correlation functions in the form of Bataille and Kestin:

$$D_t \mathcal{K}^{(m)} = \Upsilon^m \left(\mathcal{K}^{(m)}, \mathbf{F}, D_t \mathbf{F}, \mathrm{grad} T, u \right) \qquad (m = 1, 2, \ldots) \tag{4.48}$$

  and that only thermomechanical process without nonmechanical effects is considered. Then, although for a reasonable application we keep $\max(m) = 2$ the general transition suggested in the previous item is not possible.
- Another approach to statistical issues at inelastic deformation processes was given in [Zor68]. This author gave a very general theory of infinitesimal dislocation loops moving in accordance with the laws of dynamics of discrete dislocations applying then statistical hydrodynamics of [IK-50] to the so-called *dislocation fluid.* The resulting compound medium is a mixture that contains material and nonmaterial constituent.[9] Under the assumption that the theory is linear (case of small strains) the field equations are obtained by statistical averaging. Parallel to material forces and laws it was assumed that a generalized force (*kinetic stress on dislocation lines*) responsible for relative motion of dislocation loops is governed by adiabatic motion of a barotropic dislocation fluid. It is the simplest constitutive assumption for this fluid. Such a constitutive assumption is indispensable since a set of phenomenological equations derived by means of statistical methods is never closed.
- A special two-dimensional distribution of long parallel screw dislocations was afterwards analyzed in papers [Mic77, Mic79] applying Zorski's approach and the mentioned statistical analysis explained in [IK-50]. The derived field equations consist of
  - relationship between dislocation density and elastic distortion (like (1.42)),
  - a connection between elastic distortion rate and dislocation flux [Teo70],
  - a conservation law for dislocation density,
  - a balance equation for dislocation momentum disturbed by the kinetic stress and
  - a constitutive equation for *kinetic stress* similar to [Zor68].

  In the second of these papers the kinetic stress[10]—necessary to complete the set of field equations—was derived in a more general way assuming that dislocation lines have distribution of van der Waals gas and applying virial expansion of statistical physics. For the case when kinetic energy

---

[9] In his recent paper [Aer02] this author also introduces a similar compound medium.

[10] It belongs to a wide class of "configurational forces" which are generalized forces responding to diverse types of structural defects. In the review paper [Mau95] they are called *material forces.*

(i.e., their mass) is negligible, "configurational" entropy satisfies "grand canonical" distribution for dislocation lines. Although such a treatment seems idealized, in the sense that dislocation lines are assumed to be infinite, it could model individual grains in a polycrystal. Let us recall that also the approach in [Kro70] was to analyze dislocation segments inside individual grains ("cells"). Of course, the same approach could be applied not only to screw but to edge as well as mixed dislocations and their networks but calculations become much more tedious.

- A main conclusion from the mentioned statistical papers concerning thermodynamics of inelasticity is that a statistical approach does not help very much to understanding the existence of nonequilibrium entropy since it is configurational entropy and again linked to equilibrium states. We could only hope that actual states are sufficiently close to such equilibrium states.
- By making use of the standard procedure of rational thermodynamics of irreversible processes applied to nonlinear viscoelastic materials it is possible to conclude that such a procedure must be cautiously examined for each considered material. Especially doubtful is the assumption of independence of spatial and time derivatives of relevant physical quantities. For instance, in the case of plastic deforming the yield function restricts possible directions of plastic stretching. In view of this fact, the principle of local action is convenient for its simplicity but the conclusion that simple materials do not include as arguments higher gradients of deformation (total or plastic) is only a convenient simplifying assumption.
- Concerning the extended thermodynamics of irreversible processes (ETIP) we may conclude that it gives more freedom in consideration of memory effects. In Müller's approach coldness function has a role of taking into account infinitesimal memory of temperature. The assumption of ideal walls helps consideration of thermal contact between two different material bodies. It gives rise to finite speed of heat propagation. However, drawbacks of RTIP imposed by analysis proposed by Coleman and Noll make its application to viscoplasticity difficult.
- The principal features of endochronic thermodynamics introduced by Vakulenko could be stated very shortly that it has smaller generality but great applicability to inelastic material behaviour especially when coupled fields are present. As an illustration of its ability we will see that inelasticity of ferromagnetics is very well described introducing thermodynamic time into consideration.
- There are many standpoints in the literature with arguments that entropy and temperature cannot be defined for nonequilibrium states. From the practical point of view, either entropy and temperature may be defined and assumed to exist like in RTIP or we can choose one of the two following solutions. Namely, we could either assume that the real nonequilibrium process is always close to equilibrium [BK-79] or imagine that a fast unloading as well as reloading do exist and are known as has been done in [Vak70].

5

Some Multiaxial Viscoplastic Experiments: Relation to Tensor Functions

The aim of the chapter on experiments is to find as simple a relationship as possible between fundamental invariants of stress, plastic strain and plastic strain rate (usually called equivalent plastic strain, equivalent plastic strain rate and equivalent von Mises stress). It would be desirable that the results lead perpetually to the so-called unique flow curve connecting equivalent stress and equivalent strain for a fixed strain rate and diverse stress trajectories. Such curves have been used extensively for numerical purposes like in FEM codes. In order to be objective as much as possible, no constitutive assumption concerning flow rule is introduced throughout the first part of this chapter.

Then the two constitutive models are calibrated comparing plastic stretching tensors, calculated from their evolution equations, with their experimentally found values. This would clarify admissibility of evolution equations in FEM codes. Experiments were performed in the dynamic testing laboratory of JRC-Ispra, Italy, with specimens made of austenitic stainless steel AISI 316.

5.1 Preliminaries

Theoretical consideration of viscoplasticity has become an important item for finite element codes which pretend to perform calculations of complex structures with a high precision. In a majority of them (like ABAQUS, ADINA, MARC, etc.) evolution equation for plastic strain rate is of associate type, i.e., it is perpendicular to yield surface in stress space. Such a surface most commonly is based on the above-mentioned unique flow curve. It should be noted that usually yield function is detected from tension tests and then applied to calculation during arbitrary stress–strain histories appearing in real structures. This procedure could produce significant errors destroying geometrical accuracy which FEM codes offer.

The best check for a theory is to compare it to experiments. We have access to the results of experiments performed in the dynamic testing laboratory of JRC-Ispra, Italy, with specimens made of austenitic stainless steel AISI 316 in

M.V. Mićunović, *Thermomechanics of Viscoplasticity: Fundamentals and Applications*, Advances in Mechanics and Mathematics 20, DOI: 10.1007/978-0-387-89490-4_5,
© Springer Science + Business Media, LLC 2009

the range of small, medium and high strain rates from 0.001 s^{-1} to 1000 s^{-1}. The testing has been done mainly at room temperature.[1]

However, before stating theories and comparing them with experiments, clear stress and strain measures are necessary. It is assumed that damage development is negligible during the analyzed stress and strain histories (cf. (1.66), page 23). Then $\mathbf{\Pi}^{\omega} = \mathbf{1}$ and quasiplastic strain vanishes. From the polar decomposition (1.21) we have $\mathbf{\Pi} = \mathbf{R}_P\mathbf{U}_P = \mathbf{V}_P\mathbf{R}_P$ (with $\mathbf{R}_P^{-1} = \mathbf{R}_P^T$). If as tensor of plastic strain we choose the logarithmic measure

$$\boldsymbol{\varepsilon}_P = \log \mathbf{V}_P = \frac{1}{2}\log(\mathbf{\Pi}\mathbf{\Pi}^T), \tag{5.1}$$

whose main advantage is that under the above assumptions it is traceless:

$$\pi_1 = \mathrm{tr}\boldsymbol{\varepsilon}_P = 0, \quad \pi_2 = \mathrm{tr}\boldsymbol{\varepsilon}_P^2 \neq 0, \quad \pi_3 = \mathrm{tr}\boldsymbol{\varepsilon}_P^3 \neq 0, \tag{5.2}$$

we see that it is a deviatoric tensor. This will be very convenient in the sequel.

In the evolution equations we shall use the *plastic stretching* as symmetric part of the plastic "velocity gradient" tensor

$$\mathbf{D}_P = \mathrm{sym}\mathbf{L}_P \equiv (\mathbf{L}_P)_s, \qquad \mathbf{L}_P := (D_t\mathbf{\Pi})\mathbf{\Pi}^{-1}, \tag{5.3}$$

where D_t stands for material derivative.

Employing subscript d to denote deviatoric part of a second-rank tensor the following set of invariants will be used throughout this chapter[2]:

$$\gamma := \{s_1, s_2, s_3, \pi_2, \pi_3, \mu_1, \mu_2, \mu_3, \mu_4\}, \tag{5.4}$$

where

$$\begin{aligned} &s_1 = \mathrm{tr}\ \mathbf{S}, \quad s_2 = \mathrm{tr}\mathbf{S}_d^2, \quad s_3 = \mathrm{tr}\mathbf{S}_d^3, \quad \mu_1 = \mathrm{tr}\{\mathbf{S}_d\boldsymbol{\varepsilon}_P\}, \\ &\mu_2 = \mathrm{tr}\{\mathbf{S}_d\boldsymbol{\varepsilon}_P^2\}, \quad \mu_3 = \mathrm{tr}\{\mathbf{S}_d^2\boldsymbol{\varepsilon}_P\}, \quad \mu_4 = \mathrm{tr}\{\mathbf{S}_d^2\boldsymbol{\varepsilon}_P^2\}. \end{aligned}$$

Let us recall Hooke's law (3.21) written in terms of tensors expressed in structural coordinates of the intermediate reference (ν_t) configuration

$$\mathbf{E}_E = \mathbb{D}^{-1} : \mathbf{S} \tag{3.21}$$

as well as the invariants (cf. (3.17), (3.15))

$$\sigma_{eq} \equiv \left(\frac{3}{2}s_2\right)^{1/2}, \quad D_t\varepsilon_{eq}^p \equiv \left(\frac{2}{3}\mathrm{tr}\{\mathbf{D}_P^2\}\right)^{1/2}, \tag{5.5}$$

commonly used by experimentalists and called equivalent stress and equivalent plastic strain rate, respectively. They are used throughout this chapter.

---

[1] The author has taken part in these testing programmes.

[2] The choice of the deviatoric stress is motivated by the traditional approach in the existing plasticity papers. This issue is treated in more detail in Section 5.6. Of course, everything derived in this chapter holds if instead of dev$\mathbf{S} \equiv \mathbf{S}_d$ we use the invariants formed by means of $\mathbf{S}$.

5.2 Uni-tensile and bi-tensile experiments

The tensile tests were performed using the standard tensile specimen depicted at the top of Fig. 5.1. Axial forces were applied causing low, medium and high strain rate. The details of the procedure and experimental setup are available in [AM-79, AMM91]. As already mentioned, it is assumed that in the considered range of strains plastic incompressibility holds. Then we introduce conventional notations (cf. Fig. 5.1)

$$\varepsilon^{\text{eng}} = \frac{\Delta l}{l_0}, \ \sigma^{\text{eng}} = \frac{F}{A_0}$$

for the so-called engineering stress and engineering strain. Here l_0 is the initial gage length while A_0 is the initial cross-section area of the undeformed specimen. Neglecting plastic shears gives rise to deformation gradient tensor and plastic distortion tensor, respectively:

$$\mathbf{F} = \begin{Bmatrix} 1+\varepsilon^{eng} & 0 & 0 \\ 0 & 1-\varepsilon^{T} & 0 \\ 0 & 0 & 1-\varepsilon^{T} \end{Bmatrix}, \tag{5.6}$$

$$\mathbf{\Pi} = \begin{Bmatrix} 1+p & 0 & 0 \\ 0 & (1+p)^{-1/2} & 0 \\ 0 & 0 & (1+p)^{-1/2} \end{Bmatrix}. \tag{5.7}$$

Logarithmic plastic strain tensor and plastic stretching are now obtained as follows:

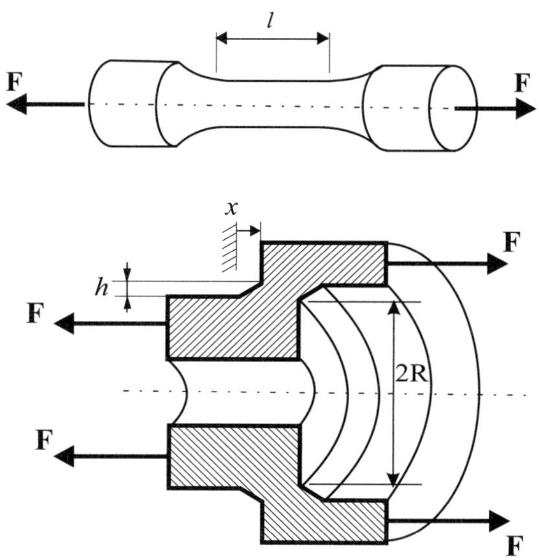

Fig. 5.1. Standard tension and "bicchierino"-shear specimen

$$
\begin{aligned}
\boldsymbol{\varepsilon}_P &= \frac{1}{2}\ln(1+p)\begin{Bmatrix} 2 & 0 & 0 \\ 0 & -1 & 0 \\ 0 & 0 & -1 \end{Bmatrix} \equiv \sqrt{3/2}\ln(1+p)\ \mathbf{N}_T, \\
\mathbf{D}_P &= \operatorname{sym}\left\{D_t\boldsymbol{\Pi}\,\boldsymbol{\Pi}^{-1}\right\} = \sqrt{3/2}\frac{D_t p}{1+p}\mathbf{N}_T.
\end{aligned}
\tag{5.8}
$$

Herein a unit tensor

$$
\mathbf{N}_T \equiv \sqrt{1/6}\begin{Bmatrix} 2 & 0 & 0 \\ 0 & -1 & 0 \\ 0 & 0 & -1 \end{Bmatrix}
\tag{5.9}
$$

(with the property $\mathrm{tr}\{\mathbf{N}_T^2\} \equiv \mathbf{N}_T : \mathbf{N}_T = 1$) is introduced to illustrate the corresponding stress and strain direction. The subscript "T" designates that type of straining is tension. It is clear from the above two formulae that the simple relationship $\mathbf{D}_P = D_t\boldsymbol{\varepsilon}_P$ holds in this special case of specimen geometry and loading.

Now, from the definition of the second Piola–Kirchhoff stress (3.7) and (3.17) we have for the uniaxial tension at Fig. 5.1:

$$
\mathbf{T} = \sigma\begin{Bmatrix} 1 & 0 & 0 \\ 0 & 0 & 0 \\ 0 & 0 & 0 \end{Bmatrix}, \qquad \mathbf{S} = \frac{E}{E+2\sigma}\mathbf{T}.
$$

Under the assumption that the elastic strain is much smaller than total and plastic strain we acquire approximately for isotropic materials

$$
\sigma \approx \sigma^{\mathrm{eng}}\left(1+\varepsilon^{\mathrm{eng}}\right),
$$

$$
p \approx \varepsilon^{\mathrm{eng}} - \frac{\sigma^{\mathrm{eng}}}{E}\left(1+\varepsilon^{\mathrm{eng}}\right)^2,
$$

where E is Young modulus and ν is Poisson's ratio.

In the dynamic testing laboratory of JRC-Ispra there have been a lot of tests with low, medium and high strain rates mainly done by making use of specimens made of the austenitic stainless steel AISI 316H. If we try to establish relationships among $D_t\varepsilon_P^{eq}$, ε_P^{eq}, σ^{eq} from all these data, then it is possible to see that a small increase in equivalent stress causes rapid increase in equivalent plastic strain rate having a few orders of magnitude. Thus a relationship is hard to find. The situation is quite different if we depict $\ln(D_t\sigma_{eq})$ versus $\ln\left(D_t\varepsilon_{eq}^p\right)$. If this is done, then an approximately linear relationship between these two quantities is found which permits the following simple representation ([Mic97]):

$$
D_t\varepsilon_{eq}^p = \exp\left(-\mathcal{M}_T\right)\left(D_t\sigma_{eq}\right)^{\mathcal{A}_T}.
$$

By a standard best fit procedure the above constants are found to be

$$
\mathcal{A}_T = 1.0015, \quad \mathcal{M}_T = 7.3109.
$$

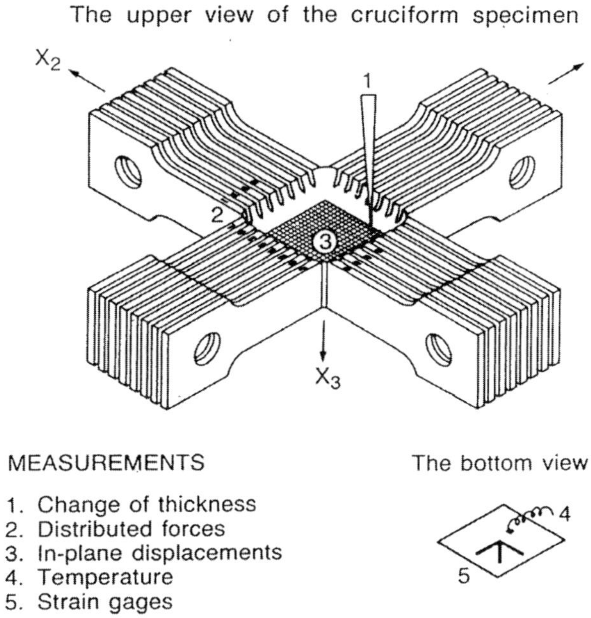

Fig. 5.2. Cruciform specimen of dynamic testing laboratory in JRC-Ispra

Such a logarithmic linearity indeed could be detected since it is observed that the curves $(\varepsilon^p_{eq}, \sigma_{eq})$ differ almost only within translation.

At the end of this section let us note that in the dynamic testing laboratory of JRC-Ispra there has been developed an original specimen in the shape of the cross shown in Fig. 5.2. Its main advantage is the possibility of changing stress direction during bi-tension and homogeneous state of strain until advanced strain values which permits a very precise check of associativity of flow rule. It is explained in detail in [AMM91]. Here only briefly some experimental results achieved by making use of this specimen are discussed.

5.3 Longitudinal axisymmetric shear experiments

The shear tests were performed using the "bicchierino" specimens shown at the bottom of Fig. 5.1, where the gage part consists of a thin circular crown built by means of a slight difference between the outer diameter of the smaller cylindrical part and the inner diameter of the larger cylindrical part. Its shear is given by the ratio $\gamma = x/h$. The detailed development of this specimen has been described in [AMP91]. It has been constructed with a short gage length, which allows a homogeneous stress distribution along the gage length even in the case of very high strain rate. Orders of magnitude for the three ranges of strain rate are here similar to those of tension but low strain rate is much smaller.

Although the maximum ratio $\gamma_{\max} \equiv (x/h)_{\max}$ obtained during longitudinal shear of the "bicchierino" specimen has a value larger than 3 (cf. [AMP91]), only values less than 0.4 could be taken as simple shears ([AM-92]) while at larger displacements of "rigid" parts of the specimen shear is accompanied by tension. Under such a restriction the deformation gradient tensor and both distortion tensors take simple forms:

$$\mathbf{F} = \begin{Bmatrix} 1 & \gamma & 0 \\ 0 & 1 & 0 \\ 0 & 0 & 1 \end{Bmatrix}, \quad \mathbf{\Pi} = \begin{Bmatrix} 1 & \gamma_P & 0 \\ 0 & 1 & 0 \\ 0 & 0 & 1 \end{Bmatrix}, \quad \mathbf{\Phi} = \begin{Bmatrix} 1 & \gamma_E & 0 \\ 0 & 1 & 0 \\ 0 & 0 & 1 \end{Bmatrix}, \tag{5.10}$$

such that Kröner's decomposition formula (1.16) (page 12) leads to the very simple relation $\gamma_P = \gamma - \gamma_E$. Moreover, the engineering shear stress

$$\tau^{eng} = \frac{F}{A_0} = \frac{F}{2R\pi h},$$

detected during the experiments, serves with the above geometric connections to find, respectively, Cauchy stress, the nominal (engineering) stress and the second Piola–Kirchhoff stress tensors (cf. (2.11)) as follows:

$$\begin{aligned} \mathbf{T} &= \mathbf{\Phi}\,\mathbf{S}\mathbf{\Phi}^T/\det\mathbf{\Phi} \approx \sqrt{2}\,\tau^{eng}\,\mathbf{N}_S, \\ \mathbf{T}_R \equiv \mathbf{T}^{eng} &= (\det\mathbf{F})\,\mathbf{T}\,\mathbf{F}^{-T} = \tau^{eng} \begin{Bmatrix} 0 & 1 & 0 \\ 1 & -\gamma & 0 \\ 0 & 0 & 0 \end{Bmatrix}, \\ \mathbf{S} &= \sqrt{2}\,S_{12}\,\mathbf{N}_S, \qquad \text{where } \mathbf{N}_S \equiv \sqrt{1/2} \begin{Bmatrix} 0 & 1 & 0 \\ 1 & 0 & 0 \\ 0 & 0 & 0 \end{Bmatrix}. \end{aligned} \tag{5.11}$$

Hooke's law (3.21) tells us that in this case $S_{12} = G\,\gamma_E$. On the other hand, by the notation $\mathbf{N}_S$ we understand the unit tensor showing straining direction of shear with the property $\mathrm{tr}\{\mathbf{N}_S^2\} \equiv \mathbf{N}_S : \mathbf{N}_S = 1$. It should be noted that in the above expressions for stresses second and higher order terms in γ_E are neglected according to the assumption that elastic strains are much smaller than plastic strains. Finally, we express plastic shear by means of measurable quantities:

$$\gamma_P = \gamma - \frac{\tau^{eng}}{G}.$$

Plastic stretching is easily calculated from

$$\mathbf{D}_P = \frac{1}{2}\left(D_t\mathbf{F}_P\,\mathbf{F}_P^{-1} + \mathbf{F}_P^{-T}\,D_t\mathbf{F}_P^T\right) = \sqrt{1/2}\,D_t\,\gamma_P\,\mathbf{N}_S, \tag{5.12}$$

such that equivalent plastic strain rate amounts to

$$D_t\varepsilon_P^{eq} = \sqrt{2/3}\,\|\mathbf{D}_P\|\,. \tag{5.13}$$

On the other hand, the logarithmic plastic strain

$$\varepsilon_P = \frac{1}{2}\ln\left(\mathbf{\Pi}\,\mathbf{\Pi}^T\right) = \frac{1}{2}\ln\left\{\begin{matrix} 1+\gamma_P^2 & \gamma_P & 0 \\ \gamma_P & 1 & 0 \\ 0 & 0 & 1 \end{matrix}\right\} \tag{5.14}$$

must be calculated after finding proper directions and eigenvalues of the tensor $\mathbf{V}_P$. Although its first invariant $\mathrm{tr}\,\varepsilon_P$ disappears, the inequality

$$\mathbf{D}_P \neq D_t\varepsilon_P$$

is evident. Clearly, in this case it is much easier to find the plastic stretching than to differentiate plastic strain tensor.

Let us turn to the results of shearing experiments. We again have the linearity in the diagram $\ln(D_t\sigma_{eq})$ versus $\ln(D_t\varepsilon_P)$ like in the case of tension. Such a linear logarithmic relationship permits the representation

$$D_t\varepsilon^p_{eq} = \exp\left(-\mathcal{M}_S\right)\,\left(D_t\sigma_{eq}\right)^{\mathcal{A}_S}$$

such that best fit gives values for the above constants as follows:

$$\mathcal{A}_S = 0.9689,\ \mathcal{M}_S = 7.535.$$

The agreement of these values with the corresponding values obtained in the case of tension is considerable, i.e., discrepancy between $\mathcal{A}_T$ and $\mathcal{A}_S$ is about 3 percent and the corresponding discrepancy $\mathcal{M}_T$ and $\mathcal{M}_S$ is even less. This encourages us to claim that $\mathcal{A}$ and $\mathcal{M}$ are material constants. Furthermore, a thermodynamic argumentation suggests that the exponent should be taken to be unity, i.e.,

$$\mathcal{A}_T \approx \mathcal{A}_S \approx \mathcal{A} = 1$$

such that

$$D_t\varepsilon^p_{eq} = \exp(-\mathcal{M})\,D_t\sigma_{eq} \tag{5.15}$$

giving finally the *universal material constant* $\mathcal{M}$ connecting linearly time rates of Mises equivalent stress and equivalent plastic strain rate

$$\mathcal{M}_T \approx \mathcal{M}_S \approx \mathcal{M} = 7.2593.$$

It is worth noting that these final values of $\mathcal{A}$ and $\mathcal{M}$ are obtained from the simultaneous fit of all data for both types of testing in all three regimes of strain rates.

5.4 Time delay of plastic yielding

It has been known by experimentalists for a long time that initial yield stress depends on strain rate or on stress rate. Explicitly stated at higher stress rates initial yield stress is larger. On the other hand, in [Rab80] it has been suggested that there exists the phenomenon of *delayed yielding* inherent with some metals and alloys, i.e., it means that stress exceeds its static value for a certain time called delay time. Following Rabotnov, we postulate an integral equation ([Mic97])

$$\varepsilon^p_{eq}(t) \equiv \int_0^t J\,(t-\tau)\; D_\tau \sigma_{eq}(\tau)\, d\tau \equiv \int_0^t \varphi(t,\tau)\, d\tau,$$

accounting for this yield delay. Let plastic deformation commence at time t^* such that

$$\sigma_{eq}\,(t^*) \equiv Y\big(D_t\sigma_{eq}(t)\,|_{t=t^*}\big), \tag{5.16}$$

i.e., the *initial equivalent dynamic yield stress* depends on the initial time rate of Mises equivalent stress. For very slow tests when $D_t\sigma_{eq} \mapsto 0$ its counterpart Y_0 is the *initial equivalent static yield stress.* Taking into account (5.16) this leads to the following form of the kernel in the above integral equation:

$$J(t-\tau) = \begin{cases} 0, & \tau \geq t^*\,, \\ \exp\big(-\mathcal{M}\big), & \tau < t^*\,, \end{cases} \tag{5.17}$$

based on tension and shear experiments described in the previous two sections. Applying this expression for kernel to the above integral equation we acquire the following representation:

$$D_t\varepsilon^p_{eq}(t) = J\,(0)\;D_t\sigma_{eq}(t) = \begin{cases} 0, & t < t^*, \\ \exp\,(-\mathcal{M})\;D_t\sigma_{eq}(t), & t \geq t^*. \end{cases}$$

The integral appearing above is the Riemann integral because it is uniformly bounded on the interval $[0,t]$ with the least upper bound

$$\sup\,\varphi\,(t,\tau) = \exp\,(-\mathcal{M})\;\max_{\tau\in[0,t]}\,D_\tau\sigma_{eq}(\tau).$$

The kernel and the plastic stretching magnitude can be derived from the two parts of Fig. 5.3, letting $h \to 0$.

5.5 Modified Perzyna–Chaboche–Rabotnov model

In order to incorporate approximately a linear relationship between equivalent plastic strain rate and equivalent stress rate we assume that

$$\mathbf{D}_P = \sqrt{3/2}\,D_t\varepsilon^p_{eq}\,\mathbf{N}_a, \quad \mathrm{dev}\mathbf{S} = \sqrt{2/3}\,\sigma_{eq}\,\mathbf{N}_a, \quad \mathrm{dev}\mathbf{B} = \sqrt{2/3}\,\beta_{eq}\,\mathbf{N}_a.$$

where $\mathbf{N}_a$, $(a \in \{T,S\})$ are unit tensors describing directions of tension and shear (with the property $\mathrm{tr}\mathbf{N}_a^2 = 1$) given already by (5.9) and the last of (5.11). According to evolution equations presented in [Per71, CR-83] and elsewhere, we extend them to cover high strain rates as follows:

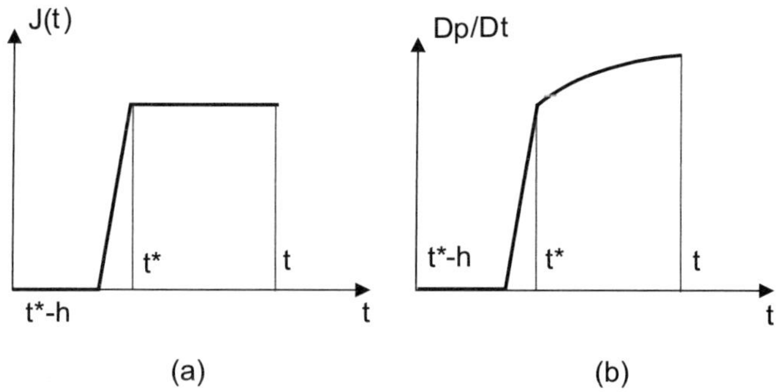

Fig. 5.3. Kernel connecting equivalent plastic strain rate and equivalent stress rate

$$
\begin{aligned}
\mathbf{N}_a \frac{D_t \varepsilon^p_{eq}}{D_t \sigma_{eq}} &= \left\lceil \frac{|\sigma_{eq} - \beta_{eq}| - \sigma^{\#}_{eq}}{K_1} \right\rceil^{K_2} \exp(-\mathcal{M}) \operatorname{sgn}(\sigma_{eq} - \beta_{eq}) \, \mathbf{N}_a, \\
\mathbf{N}_a \frac{D_t \beta_{eq}}{D_t \sigma_{eq}} &= K_4 \, (K_3 - \beta_{eq}) \frac{D_t \varepsilon^p_{eq}}{D_t \sigma_{eq}} \mathbf{N}_a, \qquad a \in \{T, S\}, \qquad (5.18) \\
\frac{D_t \sigma^{\#}_{eq}}{D_t \sigma_{eq}} &= K_6 \left(K_5 - \sigma^{\#}_{eq} \right) \frac{D_t \varepsilon^p_{eq}}{D_t \sigma_{eq}}.
\end{aligned}
$$

Hereinabove dev**B** is deviatoric part of *back stress* (i.e., residual stress), β_{eq} is equivalent back stress while $\sigma^{\#}_{eq}$ denotes equivalent stress at plasticity onset. In traditional terms it is called equivalent yield stress. The Mc Auley bracket is used to show the function $\lceil x \rceil = x$ for $x > 0$ and 0 otherwise.

Essential modifications by this so-called PCR model with respect to its standard version ([Per71, CR-83]) are:

a) replacement of time as independent variable by Vakulenko's thermodynamic time[3] and

b) stress rate sensitivity is taken into account by means of (5.16) such that initial value of $\sigma^{\#}_{eq}$, i.e., $\sigma^{\#}_{0eq}$, depends on logarithm of equivalent stress rate by means of a quadratic function. Thus, at initial yielding $\sigma_{eq} = \sigma^{\#}_{eq}$.

Indeed such evolution equations are endochronic permitting scaling of plastic strain rate. Six constants have been best fitted from all tension and shear tests and results are plotted in Fig. 5.4.

5.6 MAM model with tensor representation

According to [Ric71] the increment of plastic strain tensor is perpendicular to a loading surface $\Omega = \text{const}$ where Ω depends on stress, temperature and *Pattern of Internal Rearrangement* (PIR). Translating this statement into the

[3] Such a replacement is correct due to linearity of (5.15).

language of the previous section an evolution equation for plastic stretching should hold in the following form ([Ric71]):

$$\mathbf{D}_P = \partial_{\mathbf{S}} \Omega(\mathbf{S}, T, PIR). \tag{5.19}$$

Here PIR is described by anholonomic internal variables representing crystal slips over active slip systems. Due to its significance, the whole first chapter of this monograph is devoted to the related geometric issues.

The plastic distortion tensor $\mathbf{\Pi}$ is incompatible, represents also slips and may reflect transformation of anholonomic coordinates. Thus, taking into account that plastic rotation tensor may be either fixed or taken to be unity, it was assumed in [Mic97] that in the above equation PIR may be represented by the plastic strain tensor (5.1). Moreover, we extend the above evolution equation inserting in it a scalar function Λ which must account for the linear connection between rates of Mises equivalent stress and equivalent plastic strain rate (5.15). The structure of Λ is discussed at the end of this section. Therefore,[4]

$$\mathbf{D}_P(D_t\mathbf{S}, \mathbf{S}, \varepsilon_P, T) \ = \ \Lambda\, \partial_{\mathbf{S}} \Omega(\mathbf{S}, \varepsilon_P, T), \tag{5.20}$$

where Rice's loading function depends on temperature and invariants (5.4), i.e.,

$$\Omega \ = \ \Omega(\gamma, T) \ \equiv \ \Omega(s_1, s_2, s_3, \pi_2, \pi_3, \mu_1, \mu_2, \mu_3, \mu_4, T).$$

Let us approximate now Ω by a fourth-order polynomial with respect to $\mathbf{S}$ and first order in ε_P. With such an approximation we would have

$$\begin{aligned}\Omega &\approx a_1^0\mu_1 + a_2^0 s_2 + a_3^0\mu_3 + a_4^0 s_3 + a_5^0 s_1 s_2 + a_6^0 s_1\mu_3 + a_7^0\mu_1 s_2\\ &+ a_8^0 s_1^3 + a_9^0 s_1^2\mu_1 + a_{10}^0 s_2^2 + a_{11}^0 s_1 s_3 + a_{12}^0 s_1^2 s_2 + a_{13}^0 s_1^4\\ &+ a_{14}^0\mu_1 s_3 + a_{15}^0\mu_1 s_1 s_2 + a_{16}^0\mu_1 s_1^3 + a_{17}^0\mu_3 s_2 + a_{18}^0\mu_3 s_1^2.\end{aligned} \tag{5.21}$$

It is clear now why it is not taken here that the loading function includes also higher order terms in ε_P. Namely, the number of material constants to be determined from experiments would be too large and not practical.

Accepting that plastic stretching vanishes with stress, i.e., that $\mathbf{S} = \mathbf{0} \mapsto \mathbf{D}_P(D_t\mathbf{S}, \mathbf{0}, \varepsilon_P, T) = \mathbf{0}$ we get that $a_1^0 = 0$. Moreover it has been assumed that damage development is negligible. This has as a consequence that the spherical part of (5.20) must vanish. Such a condition simplifies greatly the above expression of loading function. A lot of constants a_k^0, $k \in \{1, \ldots, 18\}$ must vanish while the others are connected by linear relationships. As a result only five independent constants are needed giving rise to the very reduced form of Ω as follows:

$$\begin{aligned}2\,\Omega &= a_1 s_2 + (a_2 + a_4\mu_1)(s_1 s_2 - s_3)\\ &+ \frac{1}{2} a_3 s_2^2 + \frac{1}{3} a_5 (3\mu_3 s_2 - 2\mu_1 s_3).\end{aligned} \tag{5.22}$$

[4] The form of this evolution equation may be related to Ziegler's principle of least irreversible force (4.12), page 66.

Therefore, tensor representation of (5.20) (according to [BS-77, Mrk79, MM-89, AMM91]) allows the next equation:

$$\mathbf{D}_P = \Lambda \sum_{\alpha=1}^{4} \Gamma_\alpha(\gamma)\, \mathbf{H}_\alpha \tag{5.23}$$

with tensor generators[5]

$$\begin{aligned} Y_0\, \mathbf{H}_1 &= \mathbf{S} - \frac{1}{3}\mathbf{1}\,\mathrm{tr}\mathbf{S} \equiv \mathrm{dev}\mathbf{S} \equiv \mathbf{S}_d, & \mathbf{H}_3 &= \boldsymbol{\varepsilon}_P, \\ Y_0^2\, \mathbf{H}_2 &= \mathrm{dev}(\mathbf{S}_d^2), & Y_0\mathbf{H}_4 &= \mathrm{dev}(\mathbf{S}_d\boldsymbol{\varepsilon}_P + \boldsymbol{\varepsilon}_P\mathbf{S}_d) \end{aligned} \tag{5.23a}$$

and corresponding scalar coefficients depending on the listed invariants in the following manner:

$$\begin{aligned} \Gamma_1 &= a_1 + a_2 s_1 + a_3 s_2 + a_4\mu_1 + a_5\mu_3, \\ \Gamma_2 &= -\frac{3}{2}(a_2 + a_4\mu_1) - 2a_5\mu_1, \\ \Gamma_3 &= \frac{1}{2}a_4(s_1 s_2 - s_3) - \frac{2}{3}a_5 s_3, \\ \Gamma_4 &= a_5 s_2. \end{aligned} \tag{5.23b}$$

At this point let us turn our attention to the scalar coefficient Λ appearing in (5.20). If we want to cover the effect of Rabotnov's yielding delay, the best way is to propose this scalar in the form

$$\Lambda = \eta(\sigma_{eq} - Y)\left(\frac{\sigma_{eq}}{Y_0} - 1\right)^{\lambda} D_t\sigma_{eq}\, \exp(-\mathcal{M}). \tag{5.24}$$

Here Y is the dynamic initial equivalent yield stress, Y_0 is its static counterpart, $\eta(x)$ is Heaviside's function, λ is a material constant and $\mathcal{M}$ is the above-introduced and determined material constant. It is worth noting that inserting of (5.24) into (5.20) leads to an evolution equation of incremental form seemingly characteristic for rate-independent materials. At first sight the evolution equation for plastic stretching looks rate independent since it can be transformed into an incremental equation if it is multiplied by an infinitesimal time increment. However, the rate dependence appears in stress-rate-dependent value of the initial yield stress Y which has a triggering role for inelasticity onset. The model could be termed *quasi rate independent* ([Mic97]).

Two special cases of the loading function leading to reduced forms of the evolution equation (5.23) have remarkable simplicity.

- If $a_4 = 0$ and $a_5 = 0$, then the plastic stretching is of third-order power of stress. The loading function becomes[6]

[5] Scaling by the initial yield stress Y_0 leads to nondimensional tensor generators. This is very convenient when calibration of evolution equations is made.

[6] In the sequel this will be called the MAM reduced model.

$$2\,\Omega = a_1 s_2 + a_2(s_1 s_2 - s_3) + \frac{1}{2} a_3 s_2^2. \tag{5.25}$$

- For the second-order stress-dependent plastic stretching the loading function is specified with only two material constants

$$2\,\Omega = a_1 s_2 + a_2(s_1 s_2 - s_3). \tag{5.26}$$

5.7 Calibration of the models and comments

Theories presented in the previous two sections are confronted with experimental data achieved in the dynamic testing laboratory of JRC CEC, Ispra ([AMP91, AM-92]). Specimens depicted in Figs. 5.1 and 5.2 made of austenitic stainless steel AISI 316H were tested at strain rates in the range $D_t \varepsilon^p_{eq} \in [10^{-3}, 10^3]\,\mathrm{s}^{-1}$. Large plastic strains in the range $\varepsilon^p_{eq} \in [0, 0.5]$ are analyzed here.

Let us denote a set of material constants by $\mathcal{A}$. Then for the MAM model with $\mathbf{D}_P$ of third order in $\mathbf{S}$ and linear in $\boldsymbol{\varepsilon}_P$ we have

$$\mathcal{A}_{MAM} = \{\lambda, a_1, \ldots, a_5\}. \tag{5.27}$$

Its reduced version

$$\mathcal{A}_{MAM} = \{\lambda, a_1, \ldots, a_3\} \tag{5.27$'$}$$

has only four material constants to be found from experiments.

On the other hand, the modified PCR model has six material constants

$$\mathcal{A}_{PCR} = \{K_1, \ldots, K_6\}. \tag{5.28}$$

For each model we have used also the universal material constant $\mathcal{M}$.[7]

Now, material constants are determined by standard best-fit procedure, i.e., by minimization of χ^2-functional in $\mathcal{A}$-space. In other words, if measured plastic stretching is denoted by $\mathbf{D}_P^{\mathrm{exp}}$, then such a minimization reads

$$\chi^2 = \|\mathbf{D}_P^{\mathrm{exp}} - \sum_\alpha \Gamma_\alpha(\gamma, \mathcal{A})\mathbf{H}_\alpha\| \to \min. \tag{5.29}$$

In order to show calibration results on single diagrams for all the tests considered let us discuss which components of plastic stretching are different from zero in our tests:

$$\mathbf{D}_P = \begin{Bmatrix} D_{P11} & D_{P12} & 0 \\ & D_{P22} & 0 \\ \mathrm{sym} & & D_{P33} \end{Bmatrix}. \tag{5.30}$$

[7] Of course, it is possible to forget about $\mathcal{M}$. In such a case, however, great difficulties appear to cover low, medium and high strain rates by a single evolution equation. This is, in fact, the main reason for PCR modification.

For tension and shear specimens $D_{P11}, D_{P22}, D_{P33}$ are its axial, radial and circumferential physical component, respectively. On the other hand, for the cruciform specimen such components are self-evident on the basis of Fig. 5.2. On the other hand, within the considered strain ranges only at shear by the bicchierino specimen (bottom of Fig. 5.2), there appear the nondiagonal components.

Comparison of experimental data and corresponding analytical models has been given in Figs. 5.4–5.7. In order to show all the tests and tensorial nature of the plastic stretching all the diagrams have been split into four sections—each for one component of $\mathbf{D}_P$. The upper part of a figure is reserved for experimental plastic stretching histories whereas the corresponding bottom part shows the analytical model. Then each section is composed of the following temporal sequence of tests: shear at low, medium and high strain rate, then uniaxial tension at low, medium and high strain rate and finally biaxial tension (with cruciform specimen) at low strain rate but at four different stress directions S_{11}/S_{22}. It is worth noting that each history is normalized dividing the corresponding plastic stretching component by the right-hand side of (5.15), i.e., at the subdiagrams we have $\mathbf{D}_P \exp(\mathcal{M})/D_t\sigma_{eq}$. In such a way a very good visualization is obtained for processes with very diverse orders of magnitude.

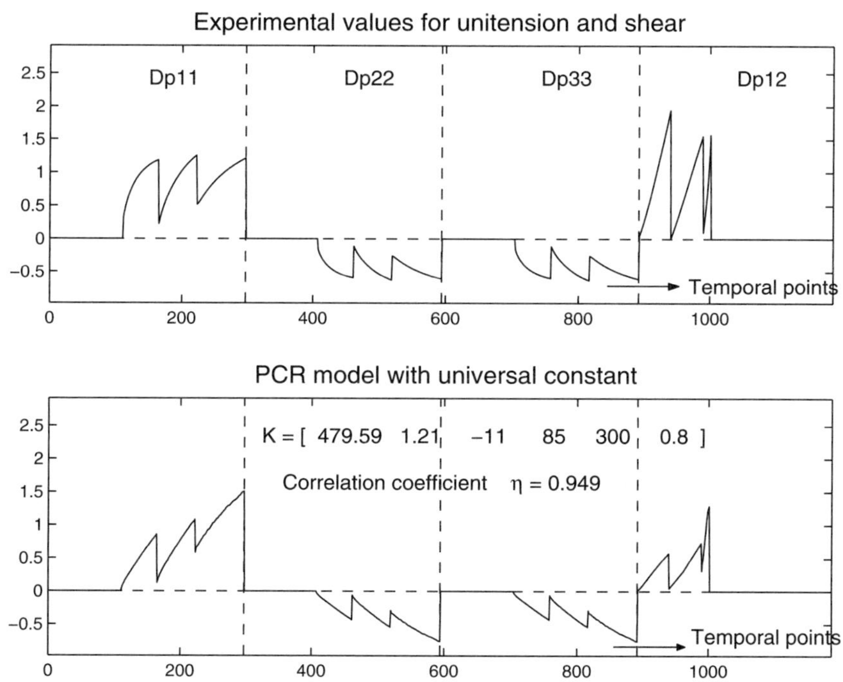

Fig. 5.4. Calibration of modified PCR model

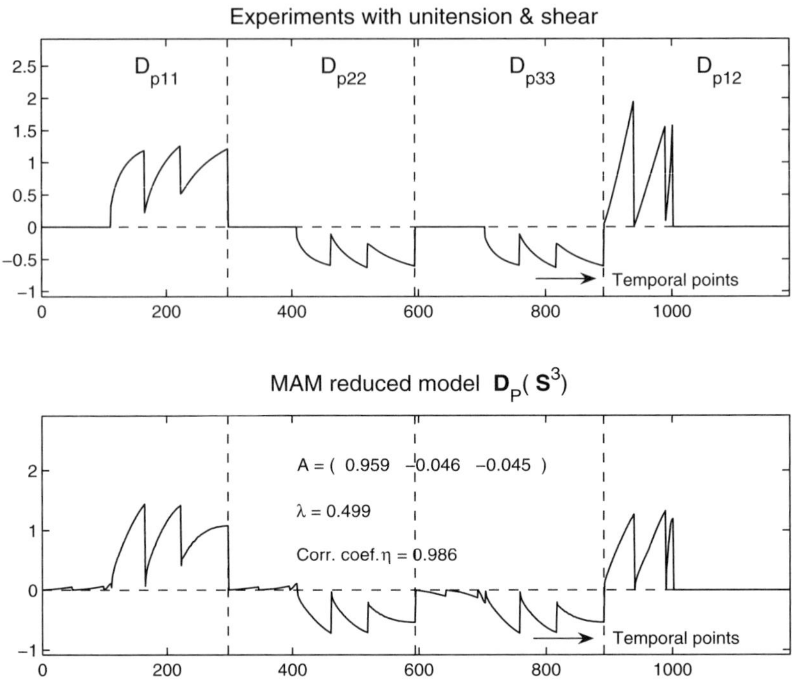

Fig. 5.5. Calibration of MAM reduced model where $\mathbf{D}_P$ depends on $\mathbf{S}^3$

The agreement is judged by the commonly defined correlation coefficient $\eta = X \cdot Y/(|X||Y|)$ for two sets of random variables. These sets in our case correspond to $\mathbf{D}_P^{\rm exp}$ and $\mathbf{D}_P^{\rm anal}$ being either $\Lambda \sum_\alpha \Gamma_\alpha \mathbf{H}_\alpha$ for the MAM model or the corresponding expression (5.18_1) for the PCR model at discretized time instants $t_1, \ldots, t_n$ of the test considered.

Now some important conclusions may be drawn:

- ◇ Introducing the universal viscoplastic constant into a standard viscoplastic evolution equation, where the plastic stretching depends only on stress and, eventually, plastic strain is very important. The scaling makes possible an easy way to account for strain rates from static until almost impact values.
- ◇ At first sight the PCR model is able to account for directionality of diverse tests like shear and tension. Its correlation coefficient (cf. Fig. 5.4) is moderately good. However, the fitting has been obtained by the constrained optimization method keeping values of the statical yield stress in the experimentally found range whereas the back stress was assumed to be very small since reversed tests have not been available. Namely, if we integrate $(5.18)_2$ and $(5.18)_3$, then we get the relations

$$\beta_{eq} = K_3 \left(1 - \exp(-K_4 \, \varepsilon^p_{eq})\right), \qquad (5.18_2')$$

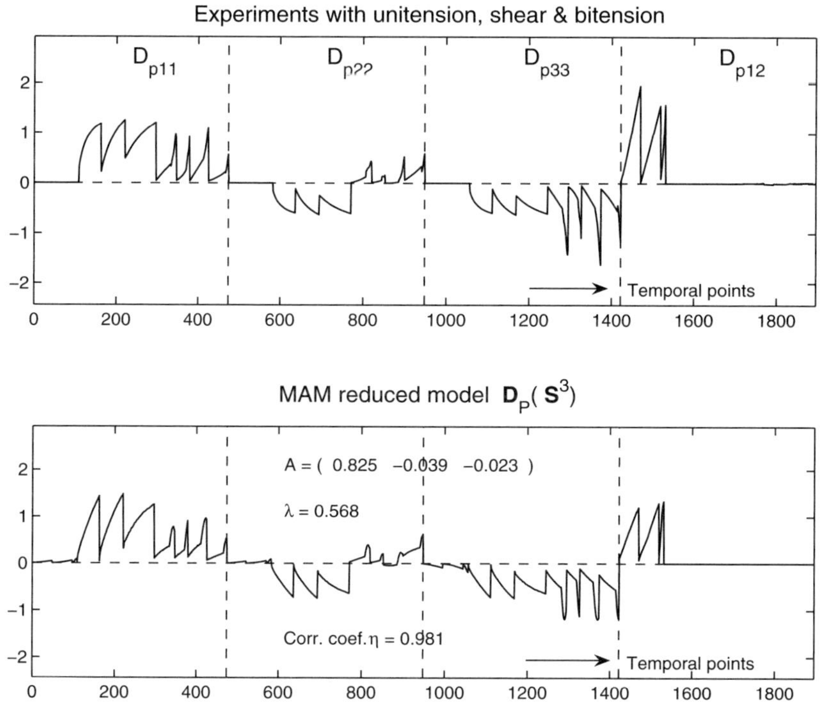

Fig. 5.6. Calibration of MAM reduced model where $\mathbf{D}_P$ depends on $\mathbf{S}^3$ with cruciform specimen included

$$\sigma^{\#}_{eq} = K_5 - (K_5 - Y_0)\exp(-K_6\, \varepsilon^p_{eq}), \tag{5.18$'_3$}$$

which show the meaning and the bounds of the constants K_3 and K_4. On the other hand, it is not clear how to find the direction tensor $\mathbf{N}$ for cruciform specimen (Fig. 5.2). Thus, the PCR model, although considerably improved by introducing thermodynamic time, has two serious drawbacks.

◇ Representation of internal variables (i.e., PIR) by logarithmic plastic strain and using *tensor function representation* for evolution equation for plastic stretching proposed gave very good agreement with multiaxial stress–strain histories. It is remarkable that such obtained MAM reduced model with only four material constants $\lambda, a_1, \ldots, a_3$ gave very good agreement with uni-tension, shear and bi-tension tests with only four material constants. The correlation coefficient is very high, the shape of $\mathbf{D}_P$ histories is the same as experimental and the difference between the MAM reduced model (depicted in Fig. 5.6) and the MAM model with six constants $\lambda, a_1, \ldots, a_5$ (Fig. 5.6) is very small showing small sensitivity of the theory.

◇ For even higher precision of results it is necessary to repeat the same tests at lower strain rates finding in such a way improved "static" strain rate

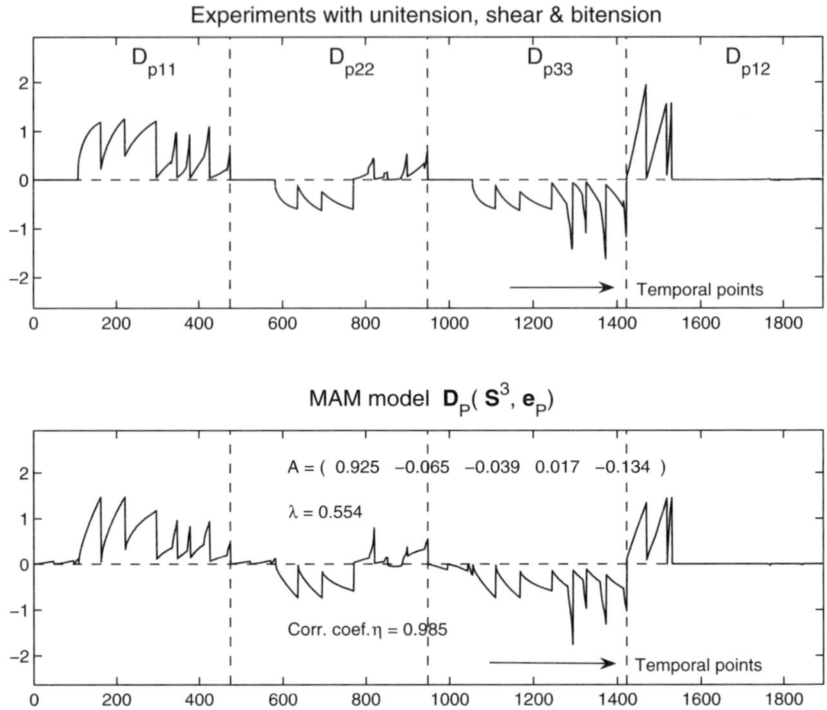

Fig. 5.7. Calibration of MAM model where $\mathbf{D}_P$ depends on $\mathbf{S}^3$ and ε_P with cruciform specimen included

results and to abandon associativity of flow rule as suggested in the third chapter.[8] However, the number of material constants would then be larger.

◇ It must be mentioned that the so-called *universal flow curve*

$$f_{univ}(\sigma_{eq}, \varepsilon_p^{eq}, D_t\varepsilon_p^{eq}) = 0,$$

even when corrected by kinematic hardening, is not capable to cover multiaxial stress–strain histories. Some authors like [Bod87] have tried to account for complex inelastic stress–strain histories treating only equivalent stress and strain and introducing kinematic variables with unclear meaning, obtaining 15 and more material constants. A modification of their theories to include the thermodynamic time, as explained in Section 5.5, would certainly improve results leading to smaller number of material constants. Without the concept of thermodynamic time the ability to describe low, medium and high strain rates within a single evolution equation is very low.

[8] Introducing the nonassociativity tensor given by (3.25), p. 54 is in progress for a proper description of localization at metal forming process.

Acknowledgment. Discussions and long-term joint work with Dr. C. Albertini and the late Dr. M. Montagnani on the subject are highly appreciated. Analysis of the experimental data has been made possible through the courtesy of Dr. C. Albertini.

Part II

Some General Problems

6

Viscoplasticity of Ferromagnetics

The principal objective of this chapter is to give a rational thermodynamic approach to inelasticity of ferromagnetic materials in a simplified version which should serve primarily for subsequent nondestructive electromagnetic examination of inelastic behaviour of reactor steels (cf. [MSC87, ZH-88, Ruu88]). Due to limited space, however, some second-order effects have to be omitted from the consideration.

In this chapter, like in [Mic86a, Mic87a, MM-89], the associativity of flow rule (the normality of the plastic stretching tensor onto a yield surface) has not been taken as granted even if such an approach is accepted in the majority of the papers dealing with the subject. Such normality is seriously questioned not only by the theoretical but by experimental results as well.[1] For these reasons the normality is at first abandoned and instead of such an assumption, evolution equations (exposed in the second section of this chapter) are based on the appropriate geometry of deformation and tensor representation. This geometry is founded on the continuum theory of dislocations (compare to [Kro60, Mic74b, FFP91]) and is shortly reviewed in the next section. As already mentioned in the chapter devoted to thermodynamics a very attractive approach to the extended thermodynamics has been proposed in [Mul71] with a rational analysis of thermodynamic processes leading to the desired thermodynamic restrictions of general constitutive equations. This approach with Liu's theorem ([Liu72]) was applied to viscoplastic materials in [Mic86a] and to inelastic composite materials in [Mic87a]. However, despite its originality an inherent coldness function (which is not quite clear from the experimental point of view) is inevitable. The method with its advantages and drawbacks was discussed in detail in Section 4.2.4.

[1] In the paper by [GHA81] a comparison between tension and torsion was one of the first signs of such a discrepancy. The subject has been discussed in detail in Chapters 3 and 5 of this monograph as well as in [Alb89], where also experiments dealing with cruciform specimens are included.

M.V. Mićunović, *Thermomechanics of Viscoplasticity: Fundamentals and Applications*, Advances in Mechanics and Mathematics 20, DOI: 10.1007/978-0-387-89490-4_6,

© Springer Science + Business Media, LLC 2009

Herein an alternative approach to extended thermodynamics following [CLJ01] (already applied to thermoplasticity in [MM-89] and viscoplasticity of irradiated materials in [Mic91b]) is chosen instead.

As the starting point we have to extend the geometry of deformation presented in the first chapter in order to include magneto-mechanical interaction. This is done in the easiest way following Kröner's *incompatibility method* which in the paper [FFP91] was applied to magnetostriction.[2] The authors assumed that total incompatibility (1.95) composed of purely elastic and magnetostrictive strain is zero while its constituents are not. They assumed magnetization vectors in the natural state elements (ν_t) are isoclinic and inhomogeneous in instant deformed configuration (χ_t) the inhomogeneity being responsible for magnetostrictive strains. Such an assumption is very much in accord with our geometrical approach and is accepted in the sequel.

Extending their reasoning to the more general case of thermo-elasto-magneto-plastic strain we rewrite the formula

$$\mathbf{F} = \mathbf{\Phi}\, \mathbf{\Pi}^{\omega}\, \mathbf{\Pi}^{p} \tag{1.66}$$

derived in Section 1.3.1 (page 23) as follows:

$$\mathbf{\Pi}^{\omega} = \mathbf{\Pi}^{\theta}\, \mathbf{\Pi}^{\mathrm{mag}}. \tag{6.1}$$

This is the *thermo-magnetic quasi-plastic distortion tensor* (cf. also Fig. 1.9, page 24). Suppose now that thermal as well as magnetostrictive strain is much smaller than plastic strain. Then for *quasi-plastic thermo-magnetic strain* linear decomposition approximately holds as follows:

$$\mathbf{E}_P^{\omega} = \mathbf{E}_P^{\theta} + \mathbf{E}_P^{\mathrm{mag}}. \tag{6.2}$$

Here the second part is called *magnetostrictive strain*. It is a bilinear function of magnetization vector. Its explicit form is given below in Section 6.2.

6.1 Evolution and constitutive equations of ferromagnetics

Ignoring ferroelectric effects the next set of objective and Galilei-invariant state variables ([MSC87, NY-81]) should be introduced in general:

$$\Gamma := \{\mathbf{E}, \mathbf{E}_P, A, T, \mathrm{GRAD}\, T, Q_f, \boldsymbol{\mathcal{Q}}, \boldsymbol{\mathcal{J}}, \boldsymbol{\mathcal{P}}, \boldsymbol{\mathcal{M}}, \boldsymbol{\mathcal{M}}_R\}, \quad \Gamma \in \mathcal{G}. \tag{6.3}$$

Unlike the choice accepted in the other chapters the tensorial quantities are chosen here in such a way to be connected to the convective material X-

[2] Due to brevity of exposition their other method called "balance of forces" is omitted here.

coordinates[3] in the deformed instant (χ_t) configuration (cf. Fig. 1.4, page 11).

Herein

$2\,\mathbf{E} = \mathbf{F}^T\mathbf{F} - \mathbf{1} \equiv \mathbf{C} - \mathbf{1}$ is the Lagrangian total strain tensor,

$2\mathbf{E}_P = \mathbf{\Pi}^T\mathbf{\Pi} - \mathbf{1} \equiv \mathbf{C}_P - \mathbf{1}$ - Lagrangian plastic strain tensor,

$\mathrm{GRAD}\,T \equiv \mathbf{F}^{-1}\mathrm{grad}\,T$ - temperature gradient,

$Q_f \equiv Jq_f$ - the electric charge ($J^2 \equiv \det\mathbf{C}$),

$\mathcal{Q} \equiv J\mathbf{F}^{-1}\mathbf{q}$ - the heat flux vector,

$\mathcal{J} = J\mathbf{F}^{-1}(\mathbf{j} - \mathbf{v}q_f) \equiv \mathbf{J} - \mathbf{V}Q_f$ - the electric current vector,

$\mathcal{P} = J\mathbf{F}^{-1}\mathbf{p} \equiv \mathbf{P}$ - the polarization vector,

$\mathcal{M} = J\mathbf{F}^{-1}(\mathbf{m} + \mathbf{v}\times\mathbf{p}/c) \equiv \mathbf{M} + \mathbf{V}\times\mathcal{P}/c$ - the magnetization vector,

$\mathcal{M}_R$ - the corresponding irreversible (residual) magnetization vector,

A - the volume-defined dislocation density (number of dislocation lines per unit volume).

Capital letters are reserved for such a convective representation. Differential operator GRAD $\equiv \mathbf{C}^K\partial_K\otimes$ is referred to such coordinate frame, whereas grad $\equiv \mathbf{g}^a\partial_a\otimes$ is used to denote the same operator in spatial (possibly Cartesian) coordinate frame of (χ_t) configuration. Accordingly, $\mathrm{GRAD}\,T, \mathcal{Q}, \mathcal{J}$, $\mathbf{P}, \mathcal{M}, \mathcal{M}_R$ have convective material either covariant or contravariant components. The above set Γ may be otherwise understood as a point belonging to the *extended configuration (deformation-temperature-electromagnetic) space* $\mathcal{G}$. Its subset $\{\mathbf{E}_P, A, \mathcal{M}_R\}$ collects internal variables responsible for irreversible behaviour.

To this configuration point there corresponds a reaction point represented by the set

$$\Delta_1 := \{\mathbf{T}_\kappa, u, s, \mathbf{S}, \mathcal{E}, \mathcal{B}\}, \quad \Delta_1 \in \mathcal{D}, \tag{6.4}$$

where[4]

[3] Another choice is to accept the convective structural coordinates employing (1.70). However, this is more difficult here since it is necessary to account for non-Euclidean expressions of differential operators.

[4] Our convective electromagnetic vectors coincide with those in the comprehensive reference [Mau88, page 169] with the exception of magnetic induction and magnetization where $\mathcal{M} = J\,\mathbf{C}^{-1}\mathcal{M}_{\mathrm{Maugin}}$ and $\mathcal{B} = J^{-1}\,\mathbf{C}\mathcal{B}_{\mathrm{Maugin}}$ holds. The convective velocity here reads $\mathbf{V} = \mathbf{F}^{-1}\mathbf{v}$. The advantage of the material convective frame is that $\mathrm{GRAD}\mathbf{C} = \mathbf{0}$ which simplifies subsequent equations. Unlike the relationship between $\mathbf{E}$ and $\mathbf{B}$, looking at polarization and magnetization we observe the symmetry breaking (cf. [Mau88, page 47]).

$\mathbf{T}_\kappa = J\mathbf{F}^{-1}\mathbf{T}_\chi\mathbf{F}^{-T}$ is the symmetric Piola–Kirchhoff stress tensor of the second kind related to the material convective coordinates of (χ_t) configuration wherein $\mathbf{T}_\chi$ is the Cauchy stress,

u and s - the internal energy and the entropy densities,

$\mathbf{S} \equiv J\mathbf{F}^{-1}\mathbf{s}$ - the entropy flux vector,

$\boldsymbol{\mathcal{E}} = (\mathbf{e} + \mathbf{v} \times \mathbf{b}/c)\mathbf{F} \equiv \mathbf{E} + \mathbf{V} \times \mathbf{B}/c$ - the electromotive intensity vector and

$\boldsymbol{\mathcal{B}} = (\mathbf{b} - \mathbf{v} \times \mathbf{e}/c)\mathbf{F} \equiv \mathbf{B} - \mathbf{V} \times \mathbf{E}/c$ - the magnetic induction vector.

By means of $\mathcal{D}$ the *extended stress space* is denoted, whose objective and Galilei-invariant elements are listed above in (6.4). Here the constitutive equations are simply stated by the bijective mapping:

$$\varDelta_1 = \mathcal{R}(\varGamma) \equiv \varDelta_1(\varGamma) \qquad \text{or} \qquad \mathcal{R} : \mathcal{G} \to \mathcal{D}, \tag{6.5}$$

which is too general so that the thermodynamic analysis presented henceforth is aimed to supply restrictions concordant with the second law of thermodynamics.

The evolution functions are proposed here in such a way to be compatible with (6.3)–(6.5) and are collected into the set

$$\varDelta_2 := \{\boldsymbol{\mathcal{Q}}^*, \boldsymbol{\mathcal{J}}^*, \mathbf{E}^*, \boldsymbol{\mathcal{M}}^*, A^*\}, \quad \varDelta_2 \in \mathcal{D}, \tag{6.6}$$

such that objective evolution equations simply read

$$D_t\boldsymbol{\mathcal{Q}} = \boldsymbol{\mathcal{Q}}^*(\varGamma), \tag{6.7}$$

$$D_t\boldsymbol{\mathcal{J}} = \boldsymbol{\mathcal{J}}^*(\varGamma), \tag{6.8}$$

$$D_t\mathbf{E}_P = \mathbf{E}^*(\varGamma), \tag{6.9}$$

$$D_t\boldsymbol{\mathcal{M}}_R = \boldsymbol{\mathcal{M}}^*(\varGamma), \tag{6.10}$$

$$D_t\, A = A^*(\varGamma), \tag{6.11}$$

where the material time derivative is designated by D_t. The simplicity of left-hand sides of (6.7)–(6.11) owing to the absence of corotational time derivatives has the origin in the accepted material convective description of constitutive functions and variables listed in (6.3)–(6.4).[5]

A thermodynamic process occurring in the considered body is described by the evolution equations and by the following balance laws which are equivalent but slightly modified with regard to those of [Mau88]:

---

[5] It is also worth noting that the right-hand sides of (6.7)–(6.11) do not include material time derivatives of internal variables—elements of the set $\varDelta_1$ in (6.4). In other words, quasi-rate-independent ferromagnetics are not covered by these evolution equations (6.7)–(6.11) In the last section of this chapter we will extend the approach to include such materials as well.

$$\rho D_t u - \Big(\mathbf{T}_\kappa + \big(\mathbf{P}\otimes\boldsymbol{\mathcal{E}} - \mathbf{B}\otimes\boldsymbol{\mathcal{M}} + (\mathbf{B}\boldsymbol{\mathcal{M}})\,\mathbf{1}\big)\mathbf{C}^{-1}\Big) : D_t\mathbf{E} \\ - \boldsymbol{\mathcal{J}}\boldsymbol{\mathcal{E}} - \boldsymbol{\mathcal{E}}D_t\mathbf{P} - \mathbf{B}D_t\boldsymbol{\mathcal{M}} + \mathrm{DIV}\boldsymbol{\mathcal{Q}} - \rho_0 h = 0, \tag{6.12}$$

$$\rho_0 - \rho J = 0, \tag{6.13}$$

$$\rho D_t\mathbf{v} - \mathbf{f} - \mathbf{f}^{em} - \frac{1}{J}\,\mathrm{DIV}\,(\mathbf{T}_\kappa\mathbf{F}^T) = \mathbf{0} \tag{6.14}$$

(ρ_0 and ρ are mass densities in (κ) and (χ_t) while $\mathbf{v}$ is the velocity of the particle) with conventional notation:

$$J\,\mathbf{f}^{em}\,\mathbf{F} \equiv \frac{1}{c}(\boldsymbol{\mathcal{J}} + D_t\mathbf{P}) \times \mathbf{B} + \mathbf{P}\mathrm{GRAD}(\boldsymbol{\mathcal{E}}\mathbf{F}^{-1})\mathbf{F} \\ + Q_f\,\boldsymbol{\mathcal{E}} + J\,\mathrm{GRAD}(\frac{1}{J}\mathbf{B}\mathbf{F}^T)\mathbf{F}^{-T}\boldsymbol{\mathcal{M}}, \tag{6.15}$$

$$\mathrm{skw}\mathbf{T}_\kappa = \mathrm{skw}\Big(\mathbf{C}^{-1}(\mathbf{B}\otimes\boldsymbol{\mathcal{M}} - \mathbf{P}\otimes\boldsymbol{\mathcal{E}})\Big), \tag{6.16}$$

$$D_tQ_f + \mathrm{DIV}\boldsymbol{\mathcal{J}} = 0, \tag{6.17}$$

$$Q_f - \mathrm{DIV}\,(\mathbf{P} + J\,\mathbf{C}^{-1}\mathbf{E}) = 0, \tag{6.18}$$

$$\mathrm{CURL}\,\boldsymbol{\mathcal{E}} + \frac{1}{c}\,D_t\mathbf{B} = 0, \tag{6.19}$$

$$\mathrm{DIV}\,\mathbf{B} = 0, \tag{6.20}$$

$$\mathrm{CURL}\,(\boldsymbol{\mathcal{B}} - J^{-1}\mathbf{C}\boldsymbol{\mathcal{M}}) - \frac{1}{c}\,D_t\big(\mathbf{P} + J\mathbf{C}^{-1}\mathbf{E}\big) = \frac{1}{c}\,\boldsymbol{\mathcal{J}}, \tag{6.21}$$

which are, respectively, the equation of energy balance, the mass conservation law, the equation of balance of momentum, the equation of balance of angular momentum, the balance of electric charge[6] and Maxwell equations.

Let us restrict more precisely the scope of this section by the assumptions:

Assumption 6.1.1 *Ferroelectric and ferrimagnetic effects, intrinsic spin, exchange forces and gyromagnetic effects are ignored with negligible precessional velocity of magnetization (cf. [Mau88]).*

Assumption 6.1.2 *The considered process is slow enough such that the term $\mathbf{V}/c$ is negligible.*

The consequence of Assumption 6.1.1 is simplification of the set of internal variables losing from it the gradient of the magnetization vector assuming in such a way that the balance law for magnetization ([Nae77]) (i.e., balance of angular momentum of spin continuum in the wording of [Mau88]) is identically

---

6 As explained in [Mau88, page 154] the balance of electric charge (6.17) is obtained from the first and last of Maxwell equations, i.e., (6.18) and (6.21), in the obvious way.

satisfied. One consequence of Assumption 6.1.2 is that all the terms of the form $(\mathbf{V}/c) \times (\cdot)$ in electromagnetic vector fields listed in (6.3)–(6.4) may be disregarded.

The above-listed balance laws imply constraints on the elements of the set $\{\Gamma\} \cup \{D_t\Gamma\}$ causing breaking of their independence, which is the essence of Liu's theorem (given in [Liu72]).

There is still another constraint on these elements in the case of inelastic deformation process: the essential notion of yield surface which divides sharply two regions of material behaviour. In other words, the elastic and plastic strain ranges are separated by the yield surface (cf. Property 2.8, page 39). The dynamic and static scalar yield functions are here defined in the same way as in (3.11)[7]:

$$f = f(\mathbf{T}_\kappa, T, \mathbf{E}_P, \boldsymbol{\mathcal{M}}_R) \equiv h(\Gamma), \tag{6.22}$$

$$f^{\#} = f(\mathbf{T}^{\#}_\kappa, T, \mathbf{E}_P, \boldsymbol{\mathcal{M}}_R) \equiv h_0(\Gamma), \tag{6.23}$$

where $\mathbf{T}^{\#}_\kappa$ is the static stress corresponding to the dynamic viscoplastic stress $\mathbf{T}_\kappa$ (cf. also (2.21)). Their difference is usually termed the *overstress tensor* and in the simplest case it may be represented by a linear function of $D_t\mathbf{E}_P$ as follows:

$$\Delta\mathbf{T}_\kappa := \mathbf{T}_\kappa - \mathbf{T}^{\#}_\kappa = \mathbb{P}(\Gamma) : D_t\mathbf{E}_P, \tag{6.24}$$

with $\mathbb{P}(\Gamma)$ being the fourth-rank *tensor of plastic viscosity coefficients* introduced already in (2.27). Introducing the *plastic strain rate intensity* by

$$D_t p := (D_t\mathbf{E}_P : D_t\mathbf{E}_P)^{1/2} \equiv \|D_t\mathbf{E}_P\| \geq 0, \tag{6.25}$$

the classification:

$f > 0,\ f^{\#} = 0,\ Dp > 0$ - viscoplastic behaviour;
$f = f^{\#} = 0,\ Dp = 0$ - elastoplastic frontier;
$f = f^{\#} < 0,\ Dp = 0$ - elastic behaviour;

and the kinematic constraint:

$$<\!f\!> D_t f^{\#} = 0, \tag{6.26}$$

has already been given by (3.14), page 51.

All thermodynamic processes must obey the master law of nature, i.e., the second law of thermodynamics which in material convective coordinate frame of (χ_t) configuration reads

$$\rho D_t s + \mathrm{DIV}\mathbf{S} - \rho\frac{r}{T} = 0, \tag{6.27}$$

where r/T is the entropy source. Precisely defined a thermodynamic process is a solution of evolution and balance equations which obeys (6.27). The analysis

[7] Let us recall that in the definition (3.11) we have $F := f - D_t p \leq 0$. Then whenever plastic deformation takes place the function F vanishes whereas the dynamic yield function f is positive. It should be noted that the function f does not depend on plastic stretching.

of the above entropy inequality (6.27) by Liu's theorem may be described as follows. Replacing $s^*(\Gamma)$ and $\mathbf{S}^*(\Gamma)$ into (6.27) this becomes a differential inequality linear with respect to the elements of the set $\{D_t\Gamma\} \cup \{\mathrm{GRAD}\Gamma\}$, namely:

$$
\begin{aligned}
&\rho_0 D_t s + \mathrm{DIV}\mathbf{S} - \rho \frac{r}{T} \\
&\quad - \Lambda^u \Big[\rho_0 D_t u - \boldsymbol{\mathcal{J}}\boldsymbol{\mathcal{E}} - \boldsymbol{\mathcal{E}} D_t \mathbf{P} - \mathbf{B} D_t \boldsymbol{\mathcal{M}} + \mathrm{DIV}\boldsymbol{\mathcal{Q}} - \rho_0 h \\
&\quad - \Big(\mathbf{T}_\kappa + (\mathbf{P} \otimes \boldsymbol{\mathcal{E}} - \mathbf{B} \otimes \boldsymbol{\mathcal{M}} + (\mathbf{B}\boldsymbol{\mathcal{M}})\mathbf{1})\mathbf{C}^{-1} \Big) : D_t \mathbf{E} \Big] \\
&\quad - \Lambda^\ell : \mathrm{skw}\Big\{ \mathbf{T}_\kappa - \Big(\mathbf{C}^{-1} (\mathbf{B} \otimes \boldsymbol{\mathcal{M}} - \mathbf{P} \otimes \boldsymbol{\mathcal{E}}) \Big) \mathcal{M} \otimes \mathbf{B}) \Big\} \\
&\quad - \boldsymbol{\Lambda}^v \Big(\rho D_t \mathbf{v} - \mathbf{f} - \mathbf{f}^{em} - J^{-1} \mathrm{DIV}(\mathbf{T}_\kappa \mathbf{F}^T) \Big) \\
&\quad - \Lambda^E : \Big(D_t \mathbf{E}_P - \mathbf{E}^*(\Gamma) \Big) - \boldsymbol{\Lambda}^Q (D_t \boldsymbol{\mathcal{Q}} - \mathbf{Q}^*(\Gamma)) \\
&\quad - \boldsymbol{\Lambda}^J (D_t \boldsymbol{\mathcal{J}} - \mathbf{J}^*(\Gamma)) - \boldsymbol{\Lambda}^M (D_t \boldsymbol{\mathcal{M}}_R - \mathbf{M}^*(\Gamma)) \\
&\quad - \Lambda^A (D_t A - A^*(\Gamma)) - \Lambda^q \left(D_t Q_f + \mathrm{DIV}\boldsymbol{\mathcal{J}} \right) \\
&\quad - \Lambda^f <f> D_t f^\# - \Lambda^1 \Big(Q_f - \mathrm{DIV}\left(\mathbf{P} + J\,\mathbf{C}^{-1}\mathbf{E}\right) \Big) \\
&\quad - \boldsymbol{\Lambda}^2 \left(\mathrm{CURL}\boldsymbol{\mathcal{E}} + \frac{1}{c} D_t \mathbf{B} \right) - \Lambda^3\, \mathrm{DIV}\mathbf{B} \\
&\quad - \boldsymbol{\Lambda}^4 \Big(\mathrm{CURL}\left(\boldsymbol{\mathcal{B}} - J^{-1}\mathbf{C}\boldsymbol{\mathcal{M}}\right) - \frac{1}{c} D_t (\mathbf{P} + J\mathbf{C}^{-1}\mathbf{E}) - \frac{\boldsymbol{\mathcal{J}}}{c} \Big) \geq 0.
\end{aligned}
\tag{6.28}
$$

By introducing Lagrange multipliers all the elements of the set $\{D_t\Gamma\} \cup \{\mathrm{GRAD}\Gamma\}$ except $\mathrm{GRAD}\,T$ (which is already included in Γ) become mutually independent. Hence, in the above-extended inequality all the coefficients with the elements of the set $\{D_t\Gamma\} \cup \{\mathrm{GRAD}\Gamma\}$ must vanish. This gives rise to the following constitutive restrictions (cf. [MM-89]):

$$
\mathbf{S} = \Lambda^u(T)\boldsymbol{\mathcal{Q}} + \Lambda^q(T)\boldsymbol{\mathcal{J}} \equiv (\boldsymbol{\mathcal{Q}} + \lambda^q \boldsymbol{\mathcal{J}})/T, \tag{6.29}
$$

$$
\begin{aligned}
\mathbf{T}_\kappa = \mathbf{C}^{-1} \Big(&\boldsymbol{\mathcal{E}} \otimes \mathbf{P} - \boldsymbol{\mathcal{M}} \otimes \mathbf{B} + (\mathbf{B}\boldsymbol{\mathcal{M}})\mathbf{1} \\
&+ \rho_0\, \partial_{\mathbf{E}} g + <f> T \Lambda^f \partial_{\mathbf{E}} f^\# \Big),
\end{aligned}
\tag{6.30}
$$

$$
s = \partial_T g + <f> T \rho^{-1} \Lambda^f \partial_T f^\#, \tag{6.31}
$$

$$
\boldsymbol{\mathcal{E}} = -\rho_0 \partial_{\mathbf{P}} g - <f> T \Lambda^f \partial_{\mathbf{P}} f^\#, \tag{6.32}
$$

$$
\mathbf{B} = -\rho_0 \partial_{\boldsymbol{\mathcal{M}}} g - <f> T \Lambda^f \partial_{\boldsymbol{\mathcal{M}}} f^\#, \tag{6.33}
$$

$$
\mathbf{0} = \partial_{\mathrm{GRAD}\,T}\, g + <f> T \rho^{-1} \Lambda^f \partial_{\mathrm{GRAD}\,T} f^\#, \tag{6.34}
$$

as well as to the residual dissipation inequality

$$\begin{aligned} &\boldsymbol{\Lambda}^Q \boldsymbol{\mathcal{Q}}^*(\Gamma) + \boldsymbol{\Lambda}^J \boldsymbol{\mathcal{J}}^*(\Gamma) + \boldsymbol{\Lambda}^M \boldsymbol{\mathcal{M}}^*(\Gamma) + \Lambda^E : \mathbf{E}^*(\Gamma) \\ &+ \Lambda^A A^*(\Gamma) - T^{-2}(\boldsymbol{\mathcal{Q}} + \lambda^q \boldsymbol{\mathcal{J}})\mathrm{GRAD}\, T + T^{-1}\boldsymbol{\mathcal{J}}\boldsymbol{\mathcal{E}} \geq 0, \end{aligned} \tag{6.35}$$

where $g := u - s(\Lambda^u)^{-1} \equiv u - Ts$ is the free energy density (already defined in (4.3), p.62). If the thermodynamic process is very near to equilibrium (cf. [MM-89]), then the above residual inequality permits the direct application of Onsager–Casimir reciprocity relations. The above Lagrange multipliers are explicitly given by

$$\begin{aligned} \boldsymbol{\Lambda}^Q &= -\rho_0 T^{-1}\partial_{\mathbf{Q}} g - \langle f\rangle \Lambda^f \partial_{\mathbf{Q}} f^{\#}, \\ \boldsymbol{\Lambda}^J &= -\rho_0 T^{-1}\partial_{\mathbf{J}} g - \langle f\rangle \Lambda^f \partial_{\mathbf{J}} f^{\#}, \\ \boldsymbol{\Lambda}^M &= -\rho_0 T^{-1}\partial_{\mathbf{M}} g - \langle f\rangle \Lambda^f \partial_{\mathbf{M}} f^{\#}, \\ \Lambda^E &= -\rho_0 T^{-1}\partial_{\mathbf{E}} g - \langle f\rangle \Lambda^f \partial_{\mathbf{E}} f^{\#}, \\ \Lambda^A &= -\rho_0 T^{-1}\partial_{A} g - \langle f\rangle \Lambda^f \partial_{A} f^{\#}, \end{aligned}$$

while the others vanish:

$$\boldsymbol{\Lambda}^v = 0,\ \Lambda^1 = 0,\ \boldsymbol{\Lambda}^2 = 0,\ \Lambda^3 = 0,\ \boldsymbol{\Lambda}^4 = 0,\ \Lambda^\ell = 0.$$

The details of the above-given procedure are presented in the reference [MM-89] where plasticity of neutron-irradiated steels was considered.

6.2 Small magnetoelastic strains of isotropic plastically deformed insulators

In order to illustrate the above derived constitutive and evolution equations we accept in this section the following very simplifying assumptions for an isotropic body:

> **Assumption 6.2.1** *Elastic strain, reversible and irreversible magnetization are small of the same order but plastic strain itself is finite (cf. also [MM-89]).*

> **Assumption 6.2.2** *Thermal and electric effects are neglected.*

Such assumptions correspond to the so-called piezomagnetism processes when magnetization is generated by straining processes (e.g., [EGD97], [MA-95]).

Let us take into account that by its very nature the mechanical stress disappears when pure elastic strain vanishes and, similarly, the local magnetic field equals zero if the reversible magnetization vanishes. Then, according to [Ant70], [FFP91] it is reasonable to introduce *magnetostrictive quasi-plastic strain* by means of

$$\mathbf{E}_P^{\mathrm{mag}} = \mathbb{L} : (\boldsymbol{\mathcal{M}}_0 \otimes \boldsymbol{\mathcal{M}}_\mathbf{0}), \tag{6.36}$$

where $\mathbb{L}$ is the fourth-rank tensor of magnetostriction constants symmetric only in indices of the first as well as the second pair whereas the notation $\boldsymbol{\mathcal{M}}_0$ stands for the unit vector of the magnetization vector $\boldsymbol{\mathcal{M}}$.

For convenience, the elastic strain tensor expressed in material convective coordinates, accepted in the previous section, then reads

$$\begin{aligned} \widetilde{\mathbf{E}}_e &:= \frac{1}{2}\mathbf{\Pi}^{pT}(\mathbf{\Pi}^{\mathrm{mag}})^T(\mathbf{\Phi}^T\mathbf{\Phi} - \mathbf{1})\mathbf{\Pi}^{\mathrm{mag}}\mathbf{\Pi}^p \\ &\equiv \mathbf{E} - \mathbf{E}_P - \mathbf{\Pi}^{pT}\mathbf{E}_P^{\mathrm{mag}}\mathbf{\Pi}^p. \end{aligned} \tag{6.37}$$

As already discussed in detail in the first chapter, the constituents of the *Lagrangian strain tensor*, namely, $\widetilde{\mathbf{E}}_e$, $\mathbf{E}_P^{\mathrm{mag}}$ as well as $\mathbf{E}_P$, are incompatible.

With these facts taken into account and the above Assumptions 6.2.1 as well as 6.2.2, mechanical and magnetic constitutive equations are presented in the sequel. First, mechanical part of the stress tensor must be linear in elastic strain (6.37):

$$\begin{aligned} \mathbf{T}_\kappa &= (c_1\mathbf{1} + c_2\mathbf{E}_P + c_3\mathbf{E}_P^2)\, tr\widetilde{\mathbf{E}}_e + 2c_4\widetilde{\mathbf{E}}_e \\ &+ c_5\,(\mathbf{E}_P\widetilde{\mathbf{E}}_e + \widetilde{\mathbf{E}}_e\mathbf{E}_P) + c_6(\mathbf{E}_P^2\widetilde{\mathbf{E}}_e + \widetilde{\mathbf{E}}_e\mathbf{E}_P^2), \end{aligned} \tag{6.38}$$

Before formulating magnetic constitutive equation, let us transform vectors of magnetization and magnetic field into the corresponding second-rank skew symmetric tensors. This is done by means of the product with material convective Ricci third-rank permutation tensor $\mathcal{E}$ in the following way:

$$\mathbf{H} \equiv \mathcal{E}\boldsymbol{\mathcal{H}} = -\mathbf{H}^T, \qquad \mathbf{M}_a \equiv \mathcal{E}\boldsymbol{\mathcal{M}}_a = -\mathbf{M}_a^T,\ a \in \{r, R\}. \tag{6.39}$$

In the above replacements

$$\boldsymbol{\mathcal{M}}_r := \boldsymbol{\mathcal{M}} - \boldsymbol{\mathcal{M}}_R \tag{6.40}$$

is the *reversible magnetization vector* while the skew symmetric second-rank tensors $\mathbf{H}, \mathbf{M}_r$ and $\mathbf{M}_R$ should replace corresponding vectors.

As for the magnetic constitutive equation we first note that we have accepted Heaviside–Lorentz form of Maxwell equations (cf. table on page 59. of [Mau88]). Thus from the last of these equations, i.e., (6.21), we see that the relationship between magnetic induction field vector $\boldsymbol{\mathcal{B}}$ and the *internal magnetic field vector* $\boldsymbol{\mathcal{H}}$ (opposing the local magnetic field vector) holds as follows:

$$\boldsymbol{\mathcal{H}} := \boldsymbol{\mathcal{B}} - J^{-1}\mathbf{C}\boldsymbol{\mathcal{M}}. \tag{6.41}$$

Due to Assumption 6.2.1 the magnetic constitutive equation must have the form

$$\mathbf{H} = c_7\mathbf{M}_r + c_8(\mathbf{E}_P\mathbf{M}_r + \mathbf{M}_r\mathbf{E}_P) + c_9(\mathbf{E}_P^2\mathbf{M}_r + \mathbf{M}_r\mathbf{E}_P^2), \tag{6.42}$$

which is linear in reversible magnetization. The above constitutive expression for $\mathbf{H}$ has been derived from (6.33) under Assumption 6.1.1 by means of tensorial representations for the proper orthogonal group [Spe71].[8]

Equation (6.38) is the generalized Hooke's law accounting for plastic strain-induced mechanical anisotropy. It is worth noting that the constitutive equation for internal magnetic field predicts *magnetic anisotropy induced by the same cause.*

According to Assumption 6.2.1 the free energy function generating (6.38) and (6.42) must be quadratic in elastic strain and reversible magnetization, i.e.,

$$
\begin{aligned}
g = {} & \frac{1}{2}c_1\, i_1^2 + \frac{1}{2}c_2\, i_2^2 + \frac{1}{2}c_3\, i_3^2 + c_4\, i_4 + c_5\, i_5 \\
& + c_6\, i_6 + \frac{1}{2}c_7\, i_7 + \frac{1}{2}c_8\, i_8 + \frac{1}{2}c_9\, i_9,
\end{aligned}
\tag{6.43}
$$

where the following proper and mixed invariants appearing in the above scalar function must be introduced (cf. [Spe71]):

$$
\begin{aligned}
i_1 &= \mathrm{tr}\widetilde{\mathbf{E}}_e,\ i_2 = \mathrm{tr}\{\mathbf{E}_P\widetilde{\mathbf{E}}_e\},\ i_3 = \mathrm{tr}\{\mathbf{E}_P^2\widetilde{\mathbf{E}}_e\}, \\
i_4 &= \mathrm{tr}\{\widetilde{\mathbf{E}}_e^2\},\ i_5 = \mathrm{tr}\{\mathbf{E}_P\widetilde{\mathbf{E}}_e^2\},\ i_6 = \mathrm{tr}\{\mathbf{E}_P^2\widetilde{\mathbf{E}}_e^2\}, \\
i_7 &= \mathrm{tr}\{\mathbf{M}_r^2\},\ i_8 = \mathrm{tr}\{\mathbf{E}_P\mathbf{M}_r^2\},\ i_9 = \mathrm{tr}\{\mathbf{E}_P^2\mathbf{M}_r^2\}.
\end{aligned}
\tag{6.44}
$$

In the sequel, inverse forms of (6.38) and (6.42) will be useful. They can be written in the following way:

$$
\begin{aligned}
\widetilde{\mathbf{E}}_e = {} & (\gamma_1\mathbf{1} + \gamma_2\mathbf{E}_P + \gamma_3\mathbf{E}_P^2)\ tr\mathbf{T}_\kappa + 2\gamma_4\ \mathbf{T}_\kappa \\
& + \gamma_5\ (\mathbf{E}_P\mathbf{T}_\kappa + \mathbf{T}_\kappa\mathbf{E}_P) + \gamma_6\ (\mathbf{E}_P^2\mathbf{T}_\kappa + \mathbf{T}_\kappa\mathbf{E}_P^2),
\end{aligned}
\tag{6.45}
$$

and

$$
\mathbf{M}_r = \gamma_7\mathbf{H} + \gamma_8(\mathbf{E}_P\mathbf{H} + \mathbf{H}\mathbf{E}_P) + \gamma_9(\mathbf{E}_P^2\mathbf{H} + \mathbf{H}\mathbf{E}_P^2). \tag{6.46}
$$

The relationships between sets $\{c_1, \ldots, c_9\}$ and $\{\gamma_1, \ldots, \gamma_9\}$ can be found as follows. Let us multiply (6.38) as well as (6.45) by the tensors $\mathbf{1}$, $\mathbf{E}_P$ and $\mathbf{E}_P^2$ finding traces of both sides. If we introduce notations

$$
\begin{aligned}
s_1 &= tr\mathbf{T}_\kappa,\ s_2 = tr\{\mathbf{E}_P\mathbf{T}_\kappa\},\ s_3 = tr\{\mathbf{E}_P^2\mathbf{T}_\kappa\}, \\
s_4 &= tr\{\mathbf{T}^2\},\ s_5 = tr\{\mathbf{E}_P\mathbf{T}_\kappa^2\},\ s_6 = tr\{\mathbf{E}_P^2\mathbf{T}_\kappa^2\}, \\
s_7 &= tr\{\mathbf{H}^2\},\ s_8 = tr\{\mathbf{E}_P\mathbf{H}^2\},\ s_9 = tr\{\mathbf{E}_P^2\mathbf{H}^2\},
\end{aligned}
\tag{6.47}
$$

---

[8] The skew symmetric tensors are favored instead of the corresponding vectors for convenience and more compact representation. Of course, an equivalent formulation using products of vectors $\boldsymbol{\mathcal{M}}_r$ and $\boldsymbol{\mathcal{M}}_R$ with symmetric second-rank tensor $\mathbf{E}_P$ is also possible.

then such a procedure will allow for finding relationships between $\{c_1, \ldots, c_6\}$ and $\{\gamma_1, \ldots, \gamma_6\}$. Of course, the same procedure applied to (6.42) as well as to (6.46) would connect sets $\{c_7, \ldots, c_9\}$ and $\{\gamma_7, \ldots, \gamma_9\}$.

Similarly, the evolution equations for plastic strain rate and residual magnetization rate are explicitly stated by the following formulae:

$$\begin{aligned} D_t\mathbf{E}_P = <f>\Big(& d_1\mathbf{1} + d_2\widetilde{\mathbf{E}}_e + d_3(\widetilde{\mathbf{E}}_e\mathbf{E}_P + \mathbf{E}_P\widetilde{\mathbf{E}}_e) \\ & + d_4(\widetilde{\mathbf{E}}_e\mathbf{E}_P^2 + \mathbf{E}_P^2\widetilde{\mathbf{E}}_e) + d_5\mathbf{E}_P + d_6\mathbf{E}_P^2 \\ & + d_7(\mathbf{M}_r\mathbf{E}_P - \mathbf{E}_P\mathbf{M}_r) + d_8(\mathbf{M}_r\mathbf{E}_P^2 - \mathbf{E}_P^2\mathbf{M}_r) \\ & + d_9(\mathbf{M}_R\mathbf{E}_P - \mathbf{E}_P\mathbf{M}_R) + d_{10}(\mathbf{M}_R\mathbf{E}_P^2 - \mathbf{E}_P^2\mathbf{M}_R) \\ & + d_{11}(\mathbf{E}_P\mathbf{M}_r\mathbf{E}_P^2 - \mathbf{E}_P^2\mathbf{M}_r\mathbf{E}_P) \\ & + d_{12}(\mathbf{E}_P\mathbf{M}_R\mathbf{E}_P^2 - \mathbf{E}_P^2\mathbf{M}_R\mathbf{E}_P)\Big), \end{aligned} \tag{6.48}$$

$$\begin{aligned} D_t\mathbf{M}_R = {} & e_1\mathbf{M}_r + e_2(\mathbf{M}_r\mathbf{E}_P + \mathbf{E}_P\mathbf{M}_r) + e_3\,(\mathbf{M}_r\mathbf{E}_P^2 \\ & + \mathbf{E}_P^2\mathbf{M}_r) + e_4\mathbf{M}_R + e_5(\mathbf{M}_R\mathbf{E}_P + \mathbf{E}_P\mathbf{M}_R) \\ & + e_6(\mathbf{M}_R\mathbf{E}_P^2 + \mathbf{E}_P^2\mathbf{M}_R) + e_7(\mathbf{E}_P\widetilde{\mathbf{E}}_e - \widetilde{\mathbf{E}}_e\mathbf{E}_P) \\ & + e_8(\mathbf{E}_P^2\widetilde{\mathbf{E}}_e - \widetilde{\mathbf{E}}_e\mathbf{E}_P^2) + e_9(\mathbf{E}_P^2\widetilde{\mathbf{E}}_e\mathbf{E}_P - \mathbf{E}_P\widetilde{\mathbf{E}}_e\mathbf{E}_P^2). \end{aligned} \tag{6.49}$$

It should be noted here that all the scalar coefficients in above relationships (6.38)–(6.42) and (6.48)–(6.49) are functions of the principal invariants of the plastic strain tensor $\mathbf{E}_P$. Of course, if plastic strain itself is small, then the corresponding complete linearization of constitutive and evolution equations is straightforward which might be of interest especially if dynamic effects are considered, i.e., wave equations of the linearized problem written (cf. [MM-89]). Evolution equations then would reduce to Onsager–Casimir reciprocity relations.

6.3 Generalized normality applied to small magnetoelastic-viscoplastic strains of isotropic insulators

Let us see what consequences could result from an introduction of a generalized loading function Ω with the following orthogonality properties (cf. [MSC87, Mic01])

$$D_t\mathbf{E}_P = D_t\Lambda\frac{\partial\Omega}{\partial\mathbf{T}_\kappa} \quad \text{and} \quad D_t\mathbf{M}_R = D_t\Lambda\frac{\partial\Omega}{\partial\mathbf{H}}. \tag{6.50}$$

where the material time rate of a scalar function Λ vanishes if the yield functions f as well as $f^{\#}$ are either negative or zero (cf. (6.26)).[9] Suppose for simplicity that Assumption 6.2.2 holds whereas Assumption 6.2.1 is replaced by means of:

Assumption 6.3.1 *Elastic and plastic strain, reversible and irreversible magnetization, as well as plastic strain rate and irreversible magnetization rate are all small of the same order.*

Then we may assume the loading function in the following polynomial form:

$$\Omega = \frac{1}{2}\omega_1\ s_1^2 + \frac{1}{2}\omega_2\ s_4 + \frac{1}{2}\omega_3\ s_7, \tag{6.51}$$

leading by means of (6.50) into the following two evolution equations:

$$D_t\mathbf{E}_P = D_t\Lambda\,(\omega_1\mathbf{1}\ \mathrm{tr}\mathbf{T}_\kappa + \omega_2\ \mathbf{T}_\kappa), \tag{6.52}$$

$$D_t\mathbf{M}_R = D_t\Lambda\ \omega_3\mathbf{H}, \tag{6.53}$$

whose simplicity follows from the above very special loading scalar function Ω. In addition, the free energy function (6.43) might be reduced to

$$g = \frac{1}{2}c_1\ i_1^2 + c_4\ i_4 + \frac{1}{2}c_7\ i_7 + g^*(\mathbf{E}_P, \mathbf{M}_R) \tag{6.54}$$

where g^* would depend on proper and mixed invariants of $\mathbf{E}_P$ and $\mathbf{M}_R$. Such a function allows for the following two mostly simplified constitutive equations:

$$\mathbf{T}_\kappa = c_1\mathbf{1}\ \mathrm{tr}\widetilde{\mathbf{E}}_e + 2c_4\ \widetilde{\mathbf{E}}_e, \tag{6.55}$$

$$\mathbf{H} = c_7\mathbf{M}_r, \tag{6.56}$$

whose material constants are easily recognized to be *Lame constants*

$$c_1 \equiv \lambda \ \text{ and } \ c_2 \equiv \mu, \tag{6.57}$$

as well as the constant of *magnetic susceptibility* (cf. [Mau88]) being identified with the inverse of c_7, i.e.,

$$\chi \equiv \frac{1}{c_7}. \tag{6.58}$$

It should be noted that, if the tensor of magnetostriction constants $\mathbb{L}$, defined in (6.36), is introduced into (6.55), then magnetostriction process can be shown explicitly.[10]

[9] The constitutive equations (6.50) are, in fact, an extension of the MAM mechanical model, given in Section 5.6, to include magnetomechanical reversible and irreversible interactions. To emphasize that the evolution equations are of quasi-rate-independent type we used $D_t\Lambda$ instead of Λ. Of course, this is only another notation and not a physical difference.

[10] This section being too simplified is aimed to present a simple example of tensor representation of magnetomechanical interaction at an irreversible process. In the recent paper [Mic06] such an approach is developed to account for susceptibilities dependent on stress by means of Boltzmann distribution of magnetic domains as well as grains inside a representative volume element.

6.4 Magneto-viscoplastic evolution equations by endochronic thermodynamics

1. According to the discussion in Section 4.3 and the paper [Mic02a] we will first briefly discuss *purely mechanical inelastic irreversible behaviour of steels.* The specific free energy of the considered body is taken to be of the form

$$g = g_E\left(\mathbf{E}_E, T\right) + g_P\left(\lambda, T\right), \tag{6.59}$$

where λ is the *isotropic hardening parameter*. Its time rate is given by

$$D_t\lambda := \mathbf{T}_\kappa : D_t\mathbf{E}_P, \tag{6.60}$$

having the meaning of plastic power (cf. (3.20)). Since the free energy is assumed in the form (6.59) we have the plastic part of dissipation[11]

$$\aleph^P = \left(1 - \rho\partial_\lambda g\right) D_t\lambda.$$

According to (4.4) and (4.5), the total thermoplastic dissipation appearing in the second law of thermodynamics is denoted by $\aleph$, namely, $\aleph \equiv T\big(\rho D_t s + \operatorname{div}(\mathbf{q}/T)\big) \geq 0$, where ρ is the mass density, T is the absolute temperature, $\mathbf{q}$ is the heat flux vector and s is the specific entropy.

The plastic dissipation served Vakulenko to introduce his *thermodynamic time* (cf. [Vak70]) by the following hereditary function:

$$\zeta(t) := \int_0^t \psi\big(\aleph^P(t')\big)\, dt' \tag{6.61}$$

(cf. Section 4.3, page 74). The function $\zeta(t)$ is piecewise continuous and non-decreasing in the way that $D_t\zeta(t) = 0$ within elastic ranges and $D_t\zeta(t) > 0$ when plastic deformation takes place. Splitting the whole time history into a sequence of infinitesimal segments Vakulenko represented the plastic strain tensor as a functional of stress and stress rate history.

Moreover, in the paper [Mic02a] the accumulated plastic strain $\varepsilon^p_{eq}(\zeta) \equiv \int_0^\zeta \|D_t\mathbf{E}_P(\xi)\|\, d\xi$, as the important inelastic history parameter, was included in the memory kernel, extending in such a way formerly mentioned Vakulenko's arguments. Another important generalization of his model in [Mic02a] was extension of the function ψ to have the nonlinear power form:

$$\psi(\aleph^P) = \left(\aleph^P\right)^a. \tag{6.62}$$

The exponent a is of great importance since it shows the speed of ageing. For example, $a < 1$ may be called *decelerated ageing* whereas $a > 1$ would define

[11] A more detailed derivation of total, plastic and thermal dissipation is given by (7.27) in the subsequent chapter devoted to pure mechanical inelastic behaviour of polycrystals.

accelerated ageing. By such a classification Vakulenko's value $a = 1$ might be termed *steady ageing.*

Now, according to Vakulenko's postulate we have

$$\mathbf{E}_P(\zeta) = \int_0^\zeta \mathbf{\Psi}\left[\zeta - \xi, \mathbf{T}_\kappa(\xi), D_\xi \mathbf{T}_\kappa(\xi), \pi(\xi)\right] d\xi. \tag{6.63}$$

Of course, this integral equation is adapted to our case of finite plastic strains and absence of plastic rotation. Differentiation of (6.63) with respect to the thermodynamic time gives

$$\begin{aligned}\partial_\zeta \mathbf{E}_P &= \mathbf{\Psi}\left[0, \mathbf{T}_\kappa(\zeta), D_\zeta \mathbf{T}_\kappa(\zeta), \pi(\zeta)\right] \\ &+ \int_0^\zeta \partial_\zeta \mathbf{\Psi}\left[\zeta - \xi, \mathbf{T}_\kappa(\xi), D_\xi \mathbf{T}_\kappa(\xi), \pi(\xi)\right] d\xi.\end{aligned} \tag{6.64}$$

Further analysis of the above integral equation is given in the next chapter (Section 7.2.2).

2. Let us apply now the above-explained concept to evolution of irreversible magnetization. Again we have nonsteady ageing speed defined by the exponent a by means of

$$D_t\zeta = (\aleph^{PM})^a \equiv \left(\boldsymbol{\mathcal{H}} D_t \boldsymbol{\mathcal{M}}_R + \mathbf{T}_\kappa : D_t \mathbf{E}_P\right)^a, \tag{6.65}$$

but now irreversible power induced by magnetization must be taken into account. It is included in the magnetoplastic dissipation $\aleph^{PM}$. Suppose now that magnetomechanical interaction is only through equivalent plastic strain history. Then the magnetic evolution equation in its integral form may be taken as

$$\boldsymbol{\mathcal{M}}_R(\zeta) := \int_0^\zeta \mathbf{\Psi}(\varepsilon^p_{eq}(z), \zeta - z, \boldsymbol{\mathcal{H}}(z))\, dz, \tag{6.66}$$

where the corresponding endochronic memory is characterized by the thermodynamic time (6.65). Choosing a special form of the integral kernel as follows[12]:

$$\mathbf{\Psi} = \boldsymbol{\mathcal{H}}(z)\, \omega(\varepsilon^p_{eq}) \exp\left\{-\beta\,(\zeta - z)\right\} \tag{6.67}$$

we would arrive at the following simple explicit evolution equation for residual magnetization vector:

$$D_\zeta \boldsymbol{\mathcal{M}}_R = \omega \boldsymbol{\mathcal{H}} - \beta \boldsymbol{\mathcal{M}}_R. \tag{6.68}$$

However, in this equation derivative is taken not with respect to real but to thermodynamic time. In order to transform it to real time, let us first introduce *irreversible magnetic power* by means of

[12] Such an exponential kernel is typical for endochronic theories (cf. [Val71]).

$$D_t\lambda_{\text{mag}} := \boldsymbol{\mathcal{H}} D_t \boldsymbol{\mathcal{M}}_R, \tag{6.69}$$

following the same notation in (6.60). Now, when we substitute (6.69) into (6.65) and multiply this by magnetic field vector and time derivative of thermodynamic time we get a nonlinear algebraic equation:

$$\left(\frac{D_t\lambda_{\text{mag}}}{\omega|\boldsymbol{\mathcal{H}}|^2 - \boldsymbol{\mathcal{H}}\boldsymbol{\mathcal{M}}_R}\right)^{\frac{a}{1-a}} - D_t\lambda_{\text{mag}} := D_t\lambda. \tag{6.70}$$

This equation explicitly characterizes magnetoplastic interaction. Its validity should be checked by experiments where simultaneously stress, plastic strain, magnetic field and residual magnetization are measured. Two interesting special cases may be drawn from this equation:

- If plastic power is approximately equal to zero, then we would have a *thermoelastic irreversible magnetization*. Since $D_t\lambda \approx 0$, (6.70) gives a simplified time rate of the thermodynamic time:

$$D_t\zeta = (D_t\lambda_{\text{mag}})^a = \left(\omega|\boldsymbol{\mathcal{H}}|^2 - \boldsymbol{\mathcal{H}}\boldsymbol{\mathcal{M}}_R\right)^{\frac{a}{1-a}}. \tag{6.71}$$

- Another special case of interest would be choice $a = 1$ which might be called *Vakulenko's coupled magneto-viscoplasticity*. For such a choice time rate of ζ reads

$$D_t\zeta = D_t\lambda_{\text{mag}} + D_t\lambda = \frac{D_t\lambda}{1 - \omega|\boldsymbol{\mathcal{H}}|^2 - \boldsymbol{\mathcal{H}}\boldsymbol{\mathcal{M}}_R}. \tag{6.72}$$

In both cases the evolution equation for residual magnetization in real time domain has the following form:

$$D_t\boldsymbol{\mathcal{M}}_R = (\omega\boldsymbol{\mathcal{H}} - \beta\boldsymbol{\mathcal{M}}_R)\, D_t\zeta. \tag{6.73}$$

Suppose for simplicity that coefficients $\gamma_8 = 0$ and $\gamma_9 = 0$ are equal to zero whereas $\gamma_7 =$ constant, i.e., that (6.56) holds. Then $\boldsymbol{\mathcal{M}}_r \approx \chi\boldsymbol{\mathcal{H}}$ and using (6.41) we arrive at the integral evolution equation connecting magnetic induction and magnetic field vectors:

$$\begin{aligned}\mathbf{B}(\zeta) = \boldsymbol{\mathcal{H}}(\zeta) + J(\zeta)^{-1}\mathbf{C}(\zeta)\Big(\chi\boldsymbol{\mathcal{H}}(\zeta) \\ + \int_0^{\zeta} \boldsymbol{\Psi}(\varepsilon^p_{eq}(z), \zeta - z, \boldsymbol{\mathcal{H}}(z))\, dz\Big),\end{aligned} \tag{6.74}$$

where the total deformation tensor has the explicit form

$$\mathbf{C} = \mathbf{1} + 2\mathbf{E}_P + 2\widetilde{\mathbf{E}}_e + 2\boldsymbol{\Pi}^{pT}\mathbf{E}_P^{\text{mag}}\boldsymbol{\Pi}^p.$$

Obviously, the oversimplified equation (6.53) might hold only in the case of negligible β. We believe that the above integral equation could be used for some nondestructive experimental check of order of magnitude of magnetomechanical interactions at low-cycle fatigue or for some other experiments designed to establish characteristic points of inelastic behaviour of steels or some other ferromagnetic materials. However, as we will see in the next section the linear approximation $\mathbf{M}_r = \chi \mathbf{H}$ holds only for values of magnetization much lower than magnetization saturation value.

6.5 Low-cycle fatigue of ferromagnetics

The constitutive and evolution equations described in previous sections might be used to describe piezomagnetic behaviour induced by low-cycle fatigue of ferromagnetics.

Such a process has been investigated in the paper [EGD97]. A cylindrical specimen of AISI 1018 was uniaxially treated by push-pull tests on an MTS-810 servo-hydraulic testing machine such that total strain was periodic and triangularly shaped strain $\|\mathbf{E}\| \in \{0,\ 0.009\}$ with cycle duration of 2 s. Magnetic induction due to piezomagnetic effect was also almost periodic with very slight changes of periodicity with increase of relative number of cycles N/N_f and cumulation of phase delay with respect to strain with growth of accumulated plastic strain. Maxima and minima of total Lagrangian strain $\mathbf{E}$ are displaced with respect to minima and maxima of the magnetic induction vector.[13] Let us calculate the *accumulated plastic strain* by means of[14]

$$\varepsilon^p_{eq}(t) := \int_0^t \|D_t \mathbf{E}_P(\tau)\| \; d\tau. \tag{6.75}$$

Now, if uniaxial components of tensors $\mathbf{E}, \mathbf{B}, \boldsymbol{\mathcal{M}}_r, \boldsymbol{\mathcal{M}}_R$ are denoted by means of $E_{11}, B_1, M_{r1}, M_{R1}$, then the following memory-type equation emerging from (6.53) and (6.66)

$$B_1(t) := \int_0^t J(\epsilon^p_{eq}, t-\tau)\, D_\tau \lambda(\tau)\, H_1(\tau)\; d\tau, \tag{6.76}$$

could cover the delay between the measured functions $E_{11}(t)$ and $B_1(t)$. Time differentiation of the above relationship gives rise to the expression

$$\begin{aligned} D_t B_1(t) &:= J(\epsilon^p_{eq}, 0)\, D_t \lambda(t) H_1(t) \\ &+ \int_0^t \frac{\partial}{\partial t} J(\epsilon^p_{eq}, t-\tau)\, D_\tau \lambda(\tau)\, H_1(\tau)\; d\tau. \end{aligned} \tag{6.77}$$

[13] Here for convenience again magnetic induction is represented by the vector $\mathbf{B}$.

[14] This definition is different from (3.16) since here representation in material convective coordinates of (χ_t) configuration is used.

In the above integro-differential equation the second term on the right-hand side is responsible for the above-mentioned change of time delay and the deflection of pure periodicity of $B_1(t)$. Therefore, it is much smaller than the first part. On the other hand, if the constitutive equation $B_1 = \mu H_1$ (where μ is *magnetic permeability*) is used, then we have

$$\begin{aligned} D_t B_1 &= \frac{\mu}{\chi}(D_t M_1 - D_t M_{R1}), \\ D_t E_{11} &= D_t E_{e11} + D_t E_{P11} + D_t E_{P11}^{\mathrm{mag}}. \end{aligned} \tag{6.78}$$

Since in the paper [EGD97] such a splitting has not been made, a more specific comment on simultaneous zeros of $D_t \mathbf{E}_P$ and $D_t \boldsymbol{\mathcal{M}}_R$ (following from (6.52) and (6.53)) is not possible.

In order to obtain a more explicit endochronic kernel in (6.77) let us remind ourselves that the Langevin function is suggested in many references as the best approximant for anhysteretic curve[15] (e.g., [MF-90]). We recall that the steel AISI 316H analyzed in Section 5.7 has about 12.4% of Ni belonging in such a way to soft ferromagnetics. Then using the data from [Chi87, pages 284, 379, 382] for such cubic crystals[16] it is possible to depict the following figure by translating the anhysteretic curve by *coercive* magnetic field H_c either to the right or to the left depending on the sign of the time rate of magnetic field. Dropping indices for simplicity this may be represented by the following kernel:

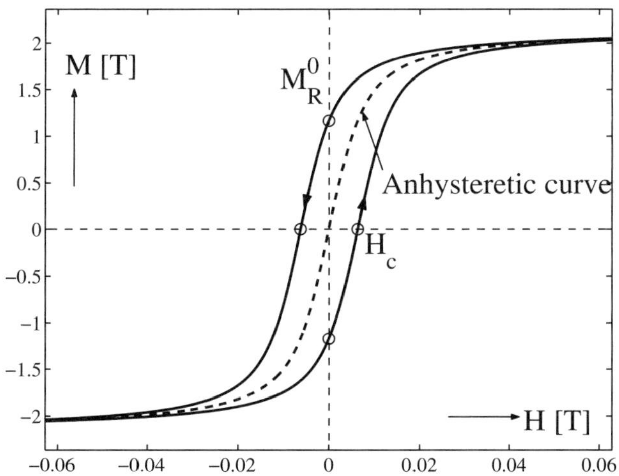

Fig. 6.1. Soft ferromagnetic steel behaviour approximated by Langevin function according to [Chi87]

[15] Let us recall that this function has the form $\mathfrak{L}(H) = \coth(H) - 1/H$.

[16] In the case of a soft ferromagnetic steel this author suggests the following data: $(M)|_{H=0} = 0.832 M_{\mathrm{sat}}$, $(dM/dH)|_{\max} = 0.6 M_{\mathrm{sat}}/H_c$ where $H_c = 0.0063\,\mathrm{T}$ is the *coercive* magnetic field and $M_{\mathrm{sat}} = 2.15\,\mathrm{T}$ is the *saturated magnetization value*.

$$\begin{aligned}\Psi(\zeta, z) = M_0 \Big[& \delta\big(H(z) - H(\zeta) + H_c(\lambda)\text{sgn}(D_t H)\big) \\ & - \delta\big(H(z) - H(\zeta)\big)\Big] \mathfrak{L}\big(H(z)\big) D_z H,\end{aligned} \tag{6.79}$$

where $\text{sgn}(x) = 1$ for $x > 0$ and $\text{sgn}(x) = -1$ for $x < 0$. If this kernel is inserted into the integral equation (6.66), it gives rise to the following explicit expression for residual magnetization:

$$\begin{aligned}M_R &= M - M_r \\ &= M_0\, \mathfrak{L}\big(H - H_c\, \text{sgn}(D_t H)\big) - M_0\, \mathfrak{L}(H).\end{aligned} \tag{6.80}$$

It is worth noting that herein the magnetomechanical interaction is taken into account by *dependence of coercive field on plastic power* λ whose time rate is given by (6.60).

Turning again to the paper [EGD97] it may be concluded that the dependence $H_c(\lambda)$ and (6.80) permit taking into account complex disturbances of shape from simple periodicity of $E_{11}(t)$ towards more complicated shape of $B_1(t)$ as well as their relative delays of minima and maxima. It may thus be concluded that such an equation could be fruitful for description of magneto-viscoplastic phenomena at low-cycle fatigue.

6.6 Some concluding comments

This chapter has dealt with viscoplasticity of ferromagnetic materials. Evolution equations were derived either from inelastic materials of differential type or from loading function generalized normality. In both cases tensor representation was applied to such a set of evolution equations. Restrictions to the set of field equations were established by means of the extended irreversible thermodynamics (version which follows exposition in [CLJ01]). Small magnetoelastic strains of isotropic insulators were considered in detail in two special cases of finite as well as small plastic strain. As one example, low-cycle fatigue of ferromagnetics was considered with special account of time delay between stress and magnetic field histories. To describe such an experimental evidence an integro-differential equation was proposed whose equivalent plastic strain dependent kernel covers the observed delay.

Concluding this chapter it is inevitable to compare the presented results to existing achievements in the field. The major contributions to viscoplasticity of ferromagnetic materials have been given by Maugin and his collaborators in [MSC87, Mau88]. The principal assumptions accepted in this chapter are closer to the scope of the first of these references where

- small strain case together with absence of exchange forces and gyromagnetic effects has been assumed;
- the emphasis on hysteresis effects has been given and

- evolution equations have been derived by normality of plastic strain rate and residual magnetization rate onto a loading surface.

On the other hand, we presented here the following results:

- in the case of finite plastic strains magnetic anisotropy induced by plastic strain is predicted by (6.49) where development of residual magnetization by mechanical terms is also evident;
- the influence of magnetization on plastic strain rate is obtained even in the case of isotropic ferromagnetic materials;
- the extended thermodynamics procedure allows for more general history effects with inhomogeneities of magnetization taken into account;
- the obtained relationships with couplings allow for magnetic measurements of inelastic phenomena but the measurements will show their order of magnitude and practical measurability of these phenomena;
- in general, the developed theory is of nonassociate type for plastic strain rate and residual magnetization rate are not perpendicular to the yield surface;
- although a generalized normality is much simpler with smaller number of material constants, a careful examination of the experiments on piezomagnetism and magnetostriction processes would give the final judgement on which theory should be applied;
- endochronic thermodynamical approach is less general than the approach by extended thermodynamics but it is much more suitable for explicit description and calibration of inelastic magnetomechanical experiments like low-cycle fatigue stress–strain–induction histories.

Acknowledgment. The author is grateful to Professors Gerard Maugin and Vukota Babović for valuable suggestions concerning some aspects of electromagnetism dealt with in this chapter.

7

Self-Consistent Method and Quasi-Rate-Dependent Polycrystals

The principal objective in this chapter is to present a rational thermodynamic approach to inelasticity of polycrystalline materials in a simplified version which should serve primarily to describe multiaxial experimental results on austenitic steels like AISI 316H having face-centered cubic lattice (compare [Mic97]). In this regard it is essential to reduce the number of material constants to be found from the available experiments. In other words, the general desire is always to make evolution equations with minimal number of material constants even if these equations originate from very general functionals like in [Mic93]. The evolution equations usually comprise plastic stretching (often reffered to by experimentalists as plastic strain rate tensor) and plastic spin. Some authors claim that this spin has a triggering role for localization behaviour while some others like [Daf84] require independence of these two evolution equations which greatly complicates identification problem. Although this issue is treated in the second chapter, it will be tackled here as well.

7.1 Free meso-rotations and constrained micro-rotations

Since a very detailed analysis of deformation geometry has been given in the first chapter, it is necessary to add in this section some new consideration specific for polycrystals.

Now we consider a polycrystalline body in a real configuration (χ_t) with dislocations and an inhomogeneous temperature field $T(X,t)$ (where t stands for time and X for the considered particle of the body) subject to surface tractions. Everything stated about total deformation gradient and component distortions, connected to Fig. 1.4, page 11, holds. To shorten the discussion we just recall the formula

$$\mathbf{F}_P(.,t) = \mathbf{\Phi}(.,t)^{-1} \;\; \mathbf{G}(.,t) \;\; \mathbf{\Phi}(.,t_0) \tag{7.1}$$

M.V. Mićunović, *Thermomechanics of Viscoplasticity: Fundamentals and Applications*, Advances in Mechanics and Mathematics 20, DOI: 10.1007/978-0-387-89490-4_7,
© Springer Science + Business Media, LLC 2009

which follows from (cf. (1.65), page 23). Here $\mathbf{G}$ is found by comparison of material fibres in (χ_0) and (χ_t) while $\mathbf{\Phi}$ is determined by crystallographic vectors in (ν_t) and (χ_t). It is worthy of note that $\operatorname{curl} \mathbf{\Phi}(.,t)^{-1} \neq 0$ and this incompatibility is commonly attributed to an asymmetric second-order tensor *of dislocation density*.

Let us imagine that a typical (ν_t) element (called in the sequel *representative volume element* (RVE)) is composed of N single crystal grains, such that each Λ-th grain has N_s slip systems $\mathbf{A}_{\alpha\Lambda} = \mathbf{s}_{\alpha\Lambda} \otimes \mathbf{n}_{\alpha\Lambda}$. For instance, for fcc crystals $N_s = 12$. Here $\mathbf{s}_{\alpha\Lambda}$ is the unit slip vector and $\mathbf{n}_{\alpha\Lambda}$ is the unit vector normal to the slip plane. For convenience and easier description of climb and cross slip, let us introduce a third unit vector $\mathbf{z}_{\alpha\Lambda}$ normal to the considered slip plane (cf. [Asa83]) with dyads $\mathbf{A}_{\alpha\Lambda 1} = \mathbf{n}_{\alpha\Lambda} \otimes \mathbf{z}_{\alpha\Lambda}$ and $\mathbf{A}_{\alpha\Lambda 2} = \mathbf{z}_{\alpha\Lambda} \otimes \mathbf{s}_{\alpha\Lambda}$. Comparing an RVE in (ν_t) and (ν_0) we may write a formula similar to (7.1) for the *plastic micro-distortion tensor*:

$$\mathbf{\Pi}_\Lambda := \mathbf{\Pi}_{\Lambda E} \mathbf{\Pi}_{\Lambda P}, \tag{7.2}$$

whose components are the *residual elastic micro-distortion tensor* $\mathbf{\Pi}_{\Lambda E}$ and *plastic micro-distortion tensor* $\mathbf{\Pi}_{\Lambda P}$. By means of the polar decomposition we get $\mathbf{\Pi}_{\Lambda E} = \mathbf{R}_\Lambda \mathbf{U}_{\Lambda E}$. Here micro-rotation satisfies the relations $\mathbf{R}_\Lambda^T \mathbf{R}_\Lambda = \mathbf{1}$ and its rate equals

$$D_t \mathbf{R}_\Lambda = \mathbf{\Omega}_\Lambda \mathbf{R}_\Lambda. \tag{7.3}$$

Therefore, slip systems dyadics evolve according to

$$\mathbf{A}_{\alpha\Lambda}(t) = \mathbf{R}_\Lambda(t)\, \mathbf{A}_{\alpha\Lambda}(t_0)\, \mathbf{R}_\Lambda^T(t)$$

since micro-rotations must be constrained inside each RVE. By making use of these dyadics as well as micro-rotations we may write

$$\mathbf{U}_{\Lambda E} = \operatorname{diag}(1 + \lambda_{k\Lambda}),\ k \in \{1, 2, 3\} \tag{7.4}$$

as well as

$$\mathbf{\Pi}_{\Lambda P} := \mathbf{1} + \sum_\alpha \gamma_{\alpha\Lambda} \mathbf{A}_{\alpha\Lambda}, \tag{7.5}$$

where $\gamma_{\alpha\Lambda}\, (\alpha \in \{1, N_s\},\ \Lambda \in \{1, N_g\})$ are plastic micro-shears inside the Λ-th grain.

If an RVE has the volume $\Delta V = \sum_\Lambda \Delta V_\Lambda$ and the micro-plastic deformation tensors for individual grains are

$$\begin{aligned} \mathbf{C}(\mathbf{\Pi}_\Lambda) &= \mathbf{\Pi}_{\Lambda P}^T \mathbf{U}_{\Lambda E}^2 \mathbf{\Pi}_{\Lambda P} \\ &\equiv \big(\mathbf{1} + \sum_\alpha \gamma_{\alpha\Lambda} \mathbf{A}_{\alpha\Lambda}^T\big)\, \mathbf{U}_{\Lambda E}^2 \,\big(\mathbf{1} + \sum_\alpha \gamma_{\alpha\Lambda} \mathbf{A}_{\alpha\Lambda}\big), \end{aligned} \tag{7.6}$$

then their volume average termed *plastic meso-deformation tensor* $\mathbf{C}_P := \mathbf{F}_P^T \mathbf{F}_P$ has the following form:

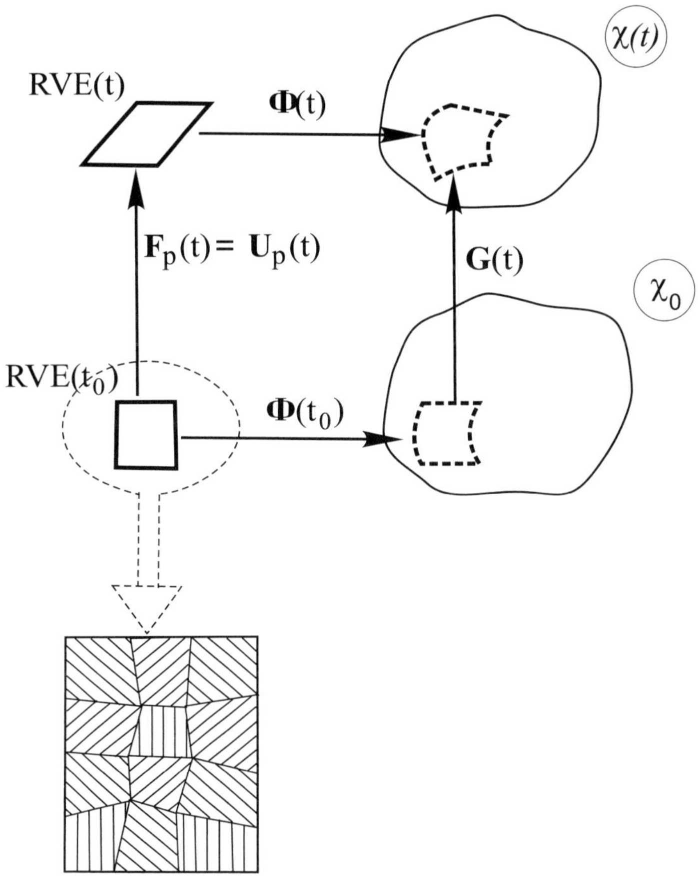

Fig. 7.1. Principal configurations of a polycrystalline body with illustration of free meso and constrained micro-rotation

$$\mathbf{C}_P = \langle \mathbf{C}_{\Pi\Lambda} \rangle = \langle \mathbf{\Pi}_\Lambda^T \mathbf{\Pi}_\Lambda \rangle \equiv \frac{1}{\Delta V} \sum_\Lambda \mathbf{\Pi}_\Lambda^T \mathbf{\Pi}_\Lambda \Delta V_\Lambda. \tag{7.7}$$

Moreover, in the corresponding polar meso-decomposition

$$\mathbf{F}_P = \mathbf{R}_P \mathbf{U}_P$$

the plastic meso-rotation tensor $\mathbf{R}_P$ is free to be taken arbitrary ([Zrw74]) and might be fixed either to be a unit tensor or to have Mandel's isoclinicity property (details are given in the second chapter as well as in [Mic93]). For a definition of isoclinicity we should have to find average crystal directions in RVE(t) and RVE(t_0) and to make them coincident (see Remark 1.7 on page 20). Accepting the first choice in the sequel we acquire the relationship

$$\mathbf{R}_P = \mathbf{1} \qquad \Rightarrow \qquad \mathbf{F}_P = \mathbf{U}_P = \mathbf{C}_P^{1/2}, \tag{7.8}$$

of all the grains inside an RVE is taken into account when determining the effective strain.

On the other hand, the same author in a previous paper ([Lev76]) by applying such a self-consistent method has found the effective stiffness fourth-rank tensor. It may be written as follows (index "M" stands for matrix while the notation $\langle\bullet\rangle_\omega$ means averaging by orientation over the manifold of inclusions only):

$$\mathbb{D}^{\text{eff}}_\Lambda = \mathbb{D}_M + [\mathbb{D}_\Lambda]\left(\mathbb{I} - \langle\mathbb{A}\,\mathbb{P}\rangle_\omega[\mathbb{D}_\Lambda]\right)^{-1}\langle\mathbb{A}\rangle_\omega. \tag{7.16}$$

It is worthy of note that if contact among grains is assumed to be ideal without holes, then index "ω" may be dropped since the matrix is composed by the same grains of different orientations. Here $\mathbb{D}_M = \langle\mathbb{D}\rangle_\omega$ is the average RVE-stiffness 4-tensor, $\mathbb{I} = (\mathbf{1}\,\square\,\mathbf{1} + \mathbf{1}\diamond\mathbf{1})/2 \equiv (\mathbf{1}\,\square\,\mathbf{1} + \mathbf{1}\,\overset{T}{\square}\,\mathbf{1})/2$ (cf. also (3.22)) or, equivalently, $(\mathbb{I})_{abcd} = (\delta_{ac}\delta_{bd} + \delta_{ad}\delta_{bc})/2$ is the unit fourth-rank tensor whereas

$$\begin{aligned}
\mathbb{S}_\Lambda\mathbb{D}_M^{-1} &\equiv \mathbb{P}_\Lambda = -\int_{\Delta V_\Lambda}\mathbb{K}\left(x - x'\right)\, dV',\\
(\mathbb{K})_{abcd} &= \left(\partial_a\partial_d G_{bc}\right)_{(ab)},\\
\mathbb{A}_\Lambda &= \left(\mathbb{I} + \mathbb{P}_\Lambda[\mathbb{D}_\Lambda]\right)^{-1} \qquad \text{and}\\
[\mathbb{D}_\Lambda] &\equiv \mathbb{D}_\Lambda - \mathbb{D}_M.
\end{aligned} \tag{7.17}$$

In these relations $\mathbb{S}_\Lambda$ is Eshelby's tensor at the Λ-grain and $\mathbf{G}$ is the Green's function for the considered anisotropic crystal. The above expressions may be used for an analytical determination of the effective elastic constants.

For our purposes instead of an infinite medium we employ the explained reasoning to the considered RVE.

Suppose, moreover, that the considered polycrystalline RVE differs by a small amount from a single crystal having the same structure as individual grains composing the RVE. Let a grain orientation be obtained by three consecutive rotations about three Cartesian axes X_1, X_2, X_3 equal to $\varphi_1 \in [0, \alpha_1]$, $\varphi_2 \in [0, \alpha_2]$ and $\varphi_3 \in [0, \alpha_3]$ (notation by [Cla97]). These axes coincide with principal crystallographic axes. Then choosing $\varphi_1, \varphi_2, \varphi_3$ randomly with α_1, α_2, α_3 kept sufficiently small we acquire a polycrystal with *slight disorder* differing slightly from the single crystal having $\alpha_1 = 0$, $\alpha_2 = 0$, $\alpha_3 = 0$ aligned along the X_3 axis (cf. Fig. 7.2). For the case when grains are cubic crystals a small value of α_3 corresponds to approximate orthotropy, while $\alpha_3 = 2\pi$ would give rise to approximate transverse isotropy.

Introduce now the notation $\mathfrak{D}_\Lambda \equiv \mathbb{D}_M^{-1}[\mathbb{D}_\Lambda]$, and let us assume that for a slightly disordered RVE such a relative jump of grain stiffness is sufficiently small such that $\left\|\mathfrak{D}^3_\Lambda\right\| \ll \left\|\mathfrak{D}^2_\Lambda\right\|$ holds. Then developing fourth-rank tensors appearing in (7.16) into a power series we get a quadratic approximation for the effective stiffness of grains in the RVE as follows:

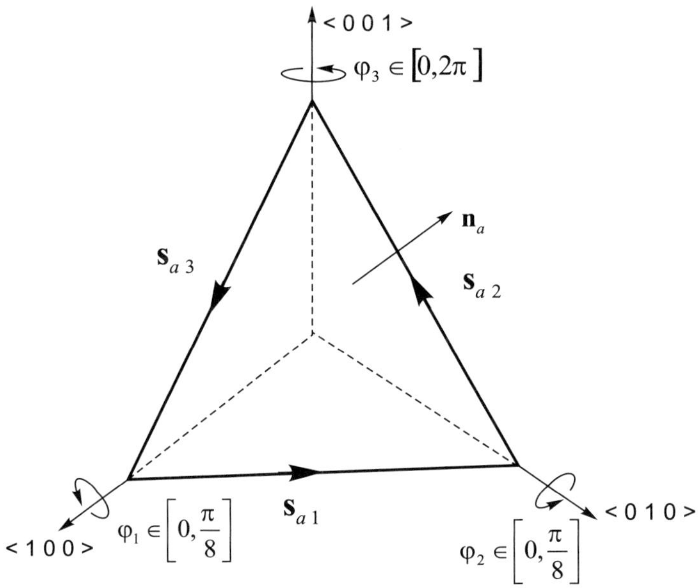

Fig. 7.2. Illustration of statistical generation of slight disorder of all the grains inside an RVE

$$\mathbb{D}_{\Lambda}^{\text{eff}} = \mathbb{D}_M \left\{ \mathbb{I} + \mathfrak{D}_{\Lambda} \left(\mathbb{I} + \langle \mathbb{S} \rangle_{\omega} \mathfrak{D}_{\Lambda} - \langle \mathbb{S}\, \mathfrak{D} \rangle_{\omega} \right) \right\}. \tag{7.18}$$

For making computations with many grains inside an RVE the advantage of this formula is that we need only one inverse of the fourth-rank tensor $\mathbb{D}_M$ while in (7.16) such an inverting for tensors $(\mathbb{I} + \mathbb{P}_{\Lambda}[\mathbb{D}_{\Lambda}])^{-1}$ and $(\mathbb{I} - \langle \mathbb{A}\,\mathbb{P} \rangle_{\omega}[\mathbb{D}_{\Lambda}])^{-1}$ must be repeated for each grain.[1] Anyway, analytical explicit results for anisotropic grains are not available and a numerical determination of Eshelby tensor must be applied. This issue is elaborated in more detail in the next section.

It is worth noting that the simplest linear approximation for the effective grain stiffness tensors is worthless since it does not include Eshelby tensor at all. Namely, the term inside the interior bracket on the right-hand side in (7.18) is then simply replaced by the unit fourth-rank tensor.

Finally, it is important to note here that a majority of researchers analyzing inelasticity of polycrystals apply the self-consistent method to total strain with oversimplified constitutive models (for instance, compare [MCA87]). In our viewpoint (cf. also [Mur88]) the Eshelbian approach to eigen-strains combined with Kröner's incompatibility is more naturally connected to elastic strains permitting more sophisticated constitutive models and anisotropy of grains.

[1] This saves very much computation time which is important especially when this is done by means of a PC.

7.2.2 Evolution equations. Accelerated ageing by endochronic thermodynamics

1. In Chapter 5 devoted to experimental issues as well as to calibration of evolution equations we have seen that tensor function representation combined with use of the universal viscoplastic constant $\mathcal{M}$ permits very good agreement with experimental data even when performed with a very small number of material constants (cf. Figs. 5.5–5.7). For instance, in the case of the reduced MAM model, if we accept that plastic meso-stretching is a third-order power function of stress, then it follows from (5.25) (neglecting a_4 and a_5)

$$\mathbf{D}_P = \Lambda\, \mathrm{dev}\left\{ \left(a_1 + a_2 \mathrm{tr}\mathbf{S} + a_3 \mathrm{tr}\{\mathbf{S}_d^2\}\right) \frac{\mathbf{S}}{Y_0} - \frac{3}{2} a_2 \frac{\mathbf{S}_d^2}{Y_0^2} \right\}, \tag{7.19}$$

where the coefficient Λ is specified by (5.24), page 91, i.e.,

$$\Lambda = \eta(\sigma_{eq} - Y) \left(\frac{\sigma_{eq}}{Y_0} - 1\right)^{\lambda} D_t \sigma_{eq} \exp(-\mathcal{M}) \tag{5.24'}$$

under the assumption that the loading function type normality derived in [Ric71] and discussed in detail in Section 5.6 holds, i.e., that $\mathbf{D}_P = \Lambda\, \partial_{\mathbf{S}} \Omega$ and taking into account that the loading function Ω depends on some proper and mixed stress and plastic strain invariants.

Here the initial yield stress Y_0 is dependent on the initial equivalent stress rate. It is assumed that plastic straining initiates at a time t^* when the equivalent stress reaches the critical stress rate dependent initial yield stress value $Y\left(D_t \sigma_{eq}\right)$, such that $\sigma_{eq}\left(t^*\right) = Y\left(D_t \sigma_{eq}\left(t^*\right)\right)$. As already mentioned in Section 5.6 for this *quasi-rate-independent* viscoplastic material the rate dependence appears in stress rate dependent value of the initial yield stress which has triggering role for inelasticity onset.

Special attention is due the normality assumed in the derivation of (7.19). Indeed, under the conditions that if time rate of residual elastic micro-strains are negligible, i.e., that the first term on the RHS of (7.13) disappears, the volumes of all the grains inside the considered RVE are the same, and the micro-spins are very small (i.e., $\mathbf{\Omega}_\Lambda = \mathbf{0}$), then

$$\mathbf{D}_P \approx \frac{1}{N} \sum_{\Lambda} \sum_{\alpha} \left(\mathbf{A}_{\alpha\Lambda}^T + \mathbf{A}_{\alpha\Lambda}\right) D_t \gamma_{\alpha\Lambda}. \tag{7.20}$$

If, moreover, there exists a scalar function of resolved shear stresses $\tau_{\alpha\Lambda} = \mathrm{tr}\{\mathbf{S}\mathbf{A}_{\alpha\Lambda}\}$, i.e., $\Omega = \Omega\left(\tau_{\alpha\Lambda}\right)$ such that

$$D_t \gamma_{\alpha\Lambda} = \Lambda \frac{\partial \Omega}{\partial \tau_{\alpha\Lambda}} \qquad \text{it follows that} \qquad \Lambda \frac{\partial \Omega}{\partial \tau_{\alpha\Lambda}} \frac{\partial \tau_{\alpha\Lambda}}{\partial \mathbf{S}} = \mathbf{D}_P. \tag{7.21}$$

Therefore, the mentioned normality holds. Such a function is called the *loading function.* Even if the mentioned simplifying assumptions are valid, this normality could fail. To show this for a single crystal let us introduce two types of additional resolved shear stresses $\tau_{\alpha\Lambda 1} = \mathrm{tr}\{\mathbf{S}\mathbf{A}_{\alpha\Lambda 1}\}$, $\tau_{\alpha\Lambda 2} = \mathrm{tr}\{\mathbf{S}\mathbf{A}_{\alpha\Lambda 2}\}$ (defined by $\mathbf{n}_{\alpha\Lambda} \otimes \mathbf{z}_{\alpha\Lambda}$ and $\mathbf{z}_{\alpha\Lambda} \otimes \mathbf{s}_{\alpha\Lambda}$, respectively) and suppose that the loading function has the form (summation over slip systems is assumed and index "Λ" omitted)

$$\Omega = \frac{1}{2} h_{\alpha\beta}\tau_\alpha\tau_\beta + A\,\tau_\alpha\tau_{\alpha 1} + B\,\tau_\alpha\tau_{\alpha 2}. \tag{7.22}$$

The application of (7.22) and left part of (7.21) yields the result

$$\begin{aligned} \mathbf{D}_P - \sum_\alpha (\mathbf{A}_{\alpha\Lambda}^T + \mathbf{A}_{\alpha\Lambda}) D_t\gamma_{\alpha\Lambda} \\ = \sum_\alpha [A(\mathbf{A}_{\alpha 1}^T + \mathbf{A}_{\alpha 1}) + B(\mathbf{A}_{\alpha 2}^T + \mathbf{A}_{\alpha 2})]\tau_\alpha \end{aligned} \tag{7.23}$$

with

$$D_t\gamma_\alpha = \Lambda\ (\ h_{\alpha\beta}\tau_\beta + A\,\tau_{\alpha 1} + B\,\tau_{\alpha 2})\,, \tag{7.24}$$

obtained by [Asa83] for cross-slip of screw dislocations. The above functions $h_{\alpha\beta}$ serve to introduce latent hardening, while A, B are responsible for cross-slip effect. Evidently, (7.23) is different from (7.20). Thus, the above brief analysis shows that loading function is allowed to depend only on shear stresses resolved in slip planes if simultaneous validity of (7.20) and (7.21) is required.

Anyway, as we can see from Fig. 5.6, the very simple equation (7.19) with just four material constants, namely, c_1, c_2, c_3, λ and one *universal viscoplastic constant* $\mathcal{M}$ ([Mic97]) showed considerably high agreement with multiaxial experiments from very low ($D_t\varepsilon^p_{eq} \sim 0.001\,\mathrm{s}^{-1}$) to very high strain rates ($D_t\varepsilon^p_{eq} \sim 1000\,\mathrm{s}^{-1}$).

It is worth noting that the assumed normality gives good results due to the fact that the tensor representation is applied to the loading function by means of some proper and mixed stress and plastic strain invariants. On the contrary, an oversimplifying assumption that Ω depends only on $J_2 \sim \sigma^2_{eq}$ (so-called J_2 theory) has problems not only with identification but with nonproportional stress histories as well (cf. [Mic88]).

2. It would be of interest to consider some kind of nonlocality as well. This has been done in [Mic02a] for the case when cross-slip is neglected but nonlocality taken into account. Special attention in the mentioned paper has been devoted to the notion of Kröner's *perfect disorder*. This issue is discussed in some detail in the next chapter devoted to inelastic micromorphic polycrystals.

3. Let the specific free energy of the considered body be of the form ([Mic02a])

$$\begin{aligned} g = g_E(\mathbf{E}_E, T) + g_P(\lambda, T), \\ D_t\lambda := \mathrm{tr}\{\mathbf{T}\mathbf{D}\} - \mathrm{tr}\{\mathbf{S}D_t\mathbf{E}_E\} \end{aligned} \tag{7.25}$$

$$= \mathbf{S} : \mathrm{sym}\left\{\mathbf{C}_E\, D_t\mathbf{U}_P\, \mathbf{U}_P^{-1}\right\} \equiv \mathrm{tr}\{\mathbf{S}_U D_t \mathbf{U}_P\}, \tag{7.26}$$

where λ is the isotropic hardening parameter having the meaning of plastic power.[2] By means of the dissipation function appearing in the second law of thermodynamics, namely,

$$\aleph \equiv T\big(\rho\, D_t s + \mathrm{div}\,(\mathbf{q}/T)\,\big) \geq 0,$$

it is possible to introduce inelastic part of dissipation as follows:

$$\aleph - \aleph^T \equiv \aleph^P = \rho T\, D_t s + \mathrm{div}\,\mathbf{q},$$

where ρ is the mass density, T the absolute temperature, $\mathbf{q}$ the heat flux vector, and s the specific entropy. Since the free energy is assumed in the form (7.25) we have

$$\aleph^P = (1 - \rho\,\partial_\lambda f)\; D_t\lambda. \tag{7.27}$$

By making use of this inelastic dissipation Vakulenko introduced a concept of *thermodynamic time* (in the paper [Vak70]) by the following hereditary function:

$$\zeta(t) = \int_0^t \left(\aleph^P(t')\right)^\alpha\, dt' \equiv \int_0^t \psi\left(\aleph^P(t')\right)\, dt', \tag{7.28}$$

where in his approach the exponent α equals unity. The function $\zeta(t)$ is piecewise continuous and nondecreasing in the way that $D_t\zeta(t) = 0$ within elastic ranges and $D_t\zeta(t) > 0$ whenever plastic deformation takes place.

It has been shown in the paper [Mic02a] that if we want to cover creep-plasticity interaction, then his definition must be extended by taking the positive exponent α to be different from unity. This is discussed below.

Splitting the whole time history into a sequence of infinitesimal segments Vakulenko claimed that a superposition and causality exist in such a way that the plastic strain tensor (for instance, $\boldsymbol{\varepsilon}_P = \mathbf{V}_P - \mathbf{1}$ (like in (3.9), page 49)) is a functional of stress and stress rate history as follows:

$$\varepsilon_P = \int_0^\zeta \mathbf{\Phi}\big[\zeta - z,\, \mathbf{S}\,(z)\,,\, D_t\mathbf{S}\,(z)\,,\, \varepsilon^p_{eq}\,(z)\big]\, dz \tag{7.29}$$

holds. In the above formula the accumulated plastic strain (being the integral of (5.5) with the substitution $t \to \zeta$) is added to Vakulenko's list of arguments. He considered small strains. Of course, the above integral equation is adapted to our case of finite plastic strains and absence of plastic meso-rotation. Differentiation of (7.29) with respect to ζ gives

[2] It was introduced with reference to material convective coordinates of (χ_t) configuration in the previous section by (6.60). Herein the same power is defined in structural coordinate frame.

$$\begin{aligned}\partial_\zeta \varepsilon_p = {} & \mathbf{\Phi}[0, \mathbf{S}(\zeta), D_t\mathbf{S}(\zeta), \varepsilon^p_{eq}(\zeta)] \\ & + \int_0^\zeta \partial_\zeta \mathbf{\Phi}[\zeta - z, \mathbf{S}(z), D_t\mathbf{S}(z), \varepsilon^p_{eq}(z)]dz.\end{aligned} \tag{7.30}$$

When the tensorial kernel in (7.29) is chosen in such a way that

$$\begin{aligned}&\mathbf{\Phi}\left[\zeta - z, \mathbf{S}\left(z\right), D_t\mathbf{S}\left(z\right), \varepsilon^p_{eq}\left(z\right)\right] \\ &\qquad = J(\zeta - z)\, \partial_\zeta \sigma_{eq}\left(\zeta\right) \mathbf{\Phi}_1\left[\mathbf{S}\left(z\right), \varepsilon^p_{eq}\left(z\right)\right] \\ &\qquad + \mathbf{\Phi}_2\left[\mathbf{S}\left(z\right), \varepsilon^p_{eq}\left(z\right)\right]\end{aligned}$$

holds and $\partial_\zeta J(\zeta - z) = 0$, then the integral on the right-hand side of (7.30) vanishes. If, moreover, the function in (7.28) is of the power type, i.e., $\psi(x) = x^\alpha$, then a multiplication of (7.30) by the rate $D_t\zeta$ transforms this equation into

$$D_t\mathbf{V}_P = \mathbf{\Phi}_1\, J(0) D_t\sigma_{eq} + \mathbf{\Phi}_2 D_t\zeta. \tag{7.31}$$

Multiplying it by $\mathbf{S}_U$ allows further

$$D_t\lambda - i_2(1 - \rho\partial_\lambda f)^\alpha (D_t\lambda)^\alpha = i_1 J(0) D_t\sigma_{eq} \tag{7.32}$$

where $i_\alpha = \mathrm{tr}\{\mathbf{S}_U \mathbf{\Phi}_\alpha\}, \alpha = 1, 2$.

The exponent α is of great importance since it shows the speed of ageing. Vakulenko accepted only the value $\alpha = 1$. In such a case we get from (7.32) the following relationship:

$$\alpha = 1 \quad \Rightarrow \quad D_t\zeta = \frac{1 - \rho\partial_\lambda f}{1 - (1 - \rho\partial_\lambda f)\, i_2} i_1 J(0) D_t\sigma_{eq}. \tag{7.33}$$

Therefore, the correction introduced by means of the tensor $\mathbf{\Phi}_2$ seems unnecessary apart from the stress rate dependent kernel $J(\zeta - z)$ (cf. (5.17)). Moreover, this relation excludes the possibility of creep since for constant stress, thermodynamic time is also constant. Thus it follows from (7.31) that inelastic strain rate must be zero as well.

However, taking some value $\alpha \neq 1$ different from Vakulenko's value, the difference becomes significant. First of all, from the equation we may have $D_t\zeta \neq 0$ even if $D_t\sigma_{eq} = 0$. Thus, the constitutive theory allows for creep as well. This has been shown in the paper [Mic02a] where the values $\alpha = 1/2$ and $\alpha = 2$ permiting explicit solutions were especially analyzed.

Let us suppose that the plastic dissipation is greater than unity, i.e., $\aleph^P > 1$. Then we may talk about *decelerated ageing* for $\alpha < 1$ whereas $\alpha > 1$ means *accelerated ageing*. Accepting such language, Vakulenko's value $\alpha = 1$ might be named *steady ageing*.

Finally, let us underline that (7.31) has the same structure as (7.19) with (5.24), which is in accordance with Vakulenko's endochronic approach to irreversible thermodynamics presented in Section 4.3.

7.3 Numerical procedure of integration of the field equations

7.3.1 Evolution equations at micro-level

A transition from a micro-evolution equation to the corresponding meso-evolution equation would be very valuable showing us more physical motivation for diverse constitutive models. In this chapter two microscopic models are going to be applied:

1. Model of the so-called J_2 *theory* (applied in the paper [NTJ90] and denoted here as the *NTJ model*) described usually by the equation

$$D_t \gamma_{\alpha \Lambda} = \dot{\gamma}_0 \left(\tau_{\alpha \Lambda} / \tau_0 \right) \left| \tau_{\alpha \Lambda} / \tau_0 \right|^{m-1}, \tag{7.34}$$

where the resolved shear stresses $\tau_{\alpha\Lambda} = \mathrm{tr}\{\, \mathbf{S}\mathbf{A}_{\alpha\Lambda}\}$ are built by means of slip dyadics and the second Piola–Kirchhoff stress tensor defined in (2.10). The material constants $\dot{\gamma}_0$, τ_0, m are, respectively, reference shear rate and reference shear stress as well as the exponent m called *strain rate sensitivity index* having a special importance in this model. It is usually assumed that very large values of this exponent are responsible for rate-independent plasticity. In the aforementioned paper the authors accepted the values $m \in \{0.05, 0.125\}$ for low and medium strain rates. Such a choice is dictated by the restriction that $\dot{\gamma}_0 = \text{const}$ in their model. However, the problem with such a choice is that it gives the slope of equivalent stress-equivalent plastic strain curves contrary to the experimental evidence. Namely, smaller strain rate sensitivity index gives as a consequence a sharp increase in the equivalent stress before yielding and its almost constant value at advanced strains. From the overwhelming majority of experiments it is known that such a behaviour is characteristic for higher strain rates. Thus, the choice that $\dot{\gamma}_0 = \text{const}$ restricts severely the capacity of the model.

2. Another micro-model emerging from experimental evidence, briefly reviewed in Section 5.4, takes into account the stress rate dependence of initial yield stress. This dependence shows that at higher stress rates initial yield stress is considerably larger than at lower stress rates. It is found to have the form ([Mic97])

$$Y = Y_0 \Big(1 + b \ln (D_t \sigma_{eq}) \Big)^{\lambda}, \tag{7.35}$$

where Y_0 is "static" yield stress (included in (7.19)) and experimentally detected flow curves of equivalent stress versus equivalent plastic strain are almost parallel. Moreover, equivalent stress rate and equivalent plastic strain rate are linearly connected by a universal material constant $\mathcal{M}$ being the same for tension, shear and biaxial stress–strain histories covering strain rates in the range $[0.001, 1000]\ \mathrm{s}^{-1}$ (for details see Section 5.3 and the paper [Mic97]). All these histories are covered if we keep the microscopic evolution equation in its simplest form (7.34) for each grain but assume that $\dot{\gamma}_0 \equiv D_t\gamma_0$ is not a constant any more and that its value is specified by the following equation:

$$D_t\gamma_0 = \exp(-\mathcal{M})\ D_t\sigma_{eq}\,. \tag{7.36}$$

Such a relationship is motivated by (5.15). This second quasi-rate-independent model is referred to in this chapter as the *MAM model*. It gives diverse values for $D_t\gamma_0$ at low, medium and high strain rates proportional to the equivalent stress rate $D_t\sigma_{eq} \equiv \dot{\sigma}_{eq}$. This is based on the very well known fact that initial yield stress increases with initial equivalent stress rate as clearly follows from (7.35). It is worth noting that such a choice leads to quite opposite character of the strain rate sensitivity index. In fact, by choosing $m \in \{0.2, 0.18, 0.05\}$ for equivalent plastic strain rates in the range $D_t\varepsilon^p_{eq} \in \{0.005, 1, 200\}$ s^{-1} we get a very good agreement with experiments. This is illustrated for the MAM model by Figs. 7.3–7.5 for the case of perfectly aligned fcc single crystal at low, medium and high strain rates (in the local reference frame $\langle\bar{1}10\rangle$ and $\langle 001\rangle$).[3] In these figures it is seen that the tension curve is between an easy slip direction shear stress curve (having much larger strains for the same equivalent stress) and a curve for shear stress in the perpendicular plane. Fig. 7.6 is added to show that the NTJ model at low strain rate shows the behaviour which is met at high strain rate experiments due to their choice of $m = 0.2$.

It should be noted that for the considered perfect crystal all the rotation angles, i.e., φ_1, φ_2, φ_3 in Fig. 7.2 vanish while the coordinate frame x^k, $k \in \{1, 2, 3\}$ is rotated around the axis $\langle\ 0\ 0\ 1\ \rangle$ by $\pi/4$.

3. Nevertheless, accepting either of these two micro-evolution equations requires a simplification of the problem of passage from a grain to the RVE where the considered grain belongs. The simplest approach for both models would be to accept Taylor's assumption (cf. [Tay38]) requiring that all the grains inside an RVE have same plastic strains, i.e., that

$$\mathbf{C}_{\Pi\Lambda} = \mathbf{C}_P,\ \Lambda \in \{\ 1, N\}. \tag{7.37}$$

However, although meso-rotation is specified by (7.8) a subsequent analysis requires taking into account diverse micro-rotations and residual elastic micro-strains (cf. (7.2)–(7.7)). This is caused by the fact that micro-rotations are not free but constrained.

7.3.2 Computational procedure and results

In this subsection a numerical procedure serving to find effective material constants as well as crystal grain orientations is given under the assumption

[3] These notations are standard for crystallography. Here fcc means face-centered cubic crystal whereas triplet numbers give components of two orthogonal vectors defining shear directions in the local crystallographic frame.

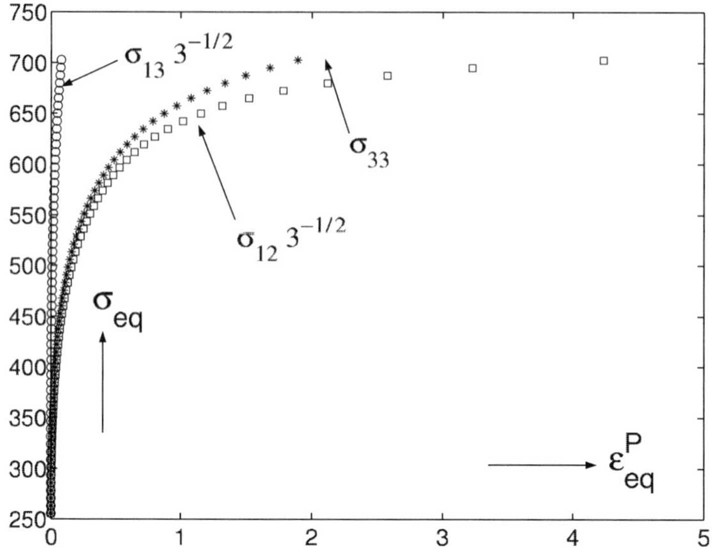

Fig. 7.3. Simulation of micro-constitutive equation in a single crystal for MAM model $m = 0.2$ caused by a shear stress at easy slip direction by low stress rate

that evolution law for micro-strains is known. Some special tension and shear meso-stress histories are chosen to test transition from micro to meso-level (from grains to their RVE). Since all the calculations have been performed on a PC the computing time had to be reduced. For this reason the computations were split into three parts.

1. The most essential assumption in the procedure is the so-called *"relaxed" Taylor's assumption* (7.37) stating simply that plastic micro and meso-strains are the same whereas micro-rotations of grains are not equal mutually. Thus, the essential difference between the two models for micro-evolution equations appearing in the second box is twofold as explained above.

In the following table consisting of four boxes this first part of the procedure (already applied in [KM-01]) is shown schematically. Grains are randomly oriented but within limits shown in Fig. 7.2. In this first part of calculations the influence of *eigen-strains* (cf. [Mur88]) to residual stresses and change of corresponding effective material constants ([Lev76]) is neglected.

The output of these calculations gives us for 12 slip systems and 125 grains: rotations, stresses, plastic and elastic strains of individual grains at low, medium and high strain rates (for MAM model) and the same for low strain rate for NTJ model.

2. In the second part the formula (7.18) is applied to 125 ellipsoidal inclusions made of cubic crystal under the simplifying assumption that for each grain principal crystallographic planes coincide with ellipsoid semi-axes. The slightly oblate ellipsoids semi axes ratios have been taken to be $\{1.1,\ 1,\ 0.9\}$

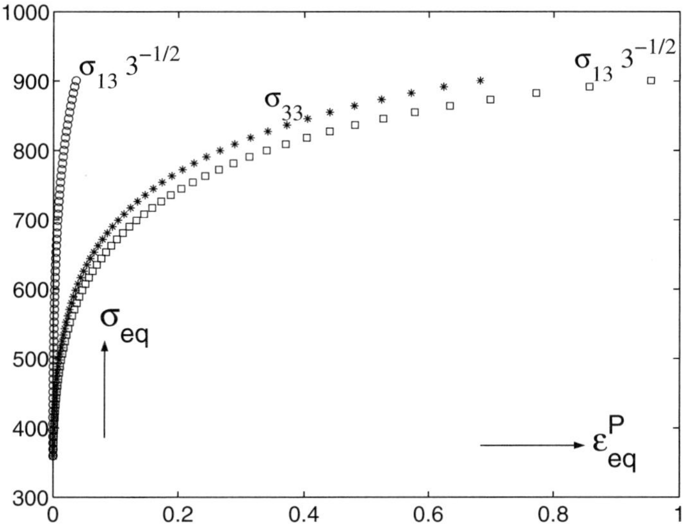

Fig. 7.4. Simulation of micro-constitutive equation in a single crystal for MAM model $m = 0.18$ caused by a shear stress at easy slip direction by medium stress rate

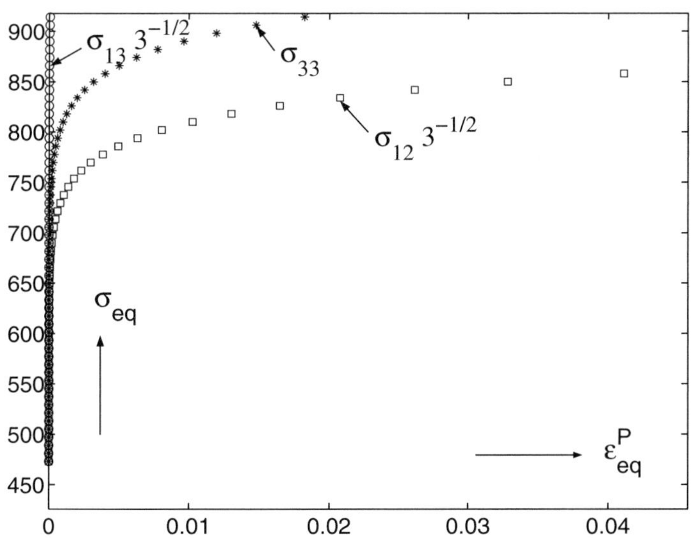

Fig. 7.5. Simulation of micro-constitutive equation in a single crystal for MAM model $m = 0.05$ caused by a shear stress at easy slip direction by high stress rate

keeping in such a way the same volume as original spheres. Of course, some other values are also possible. Since, according to the author's knowledge,

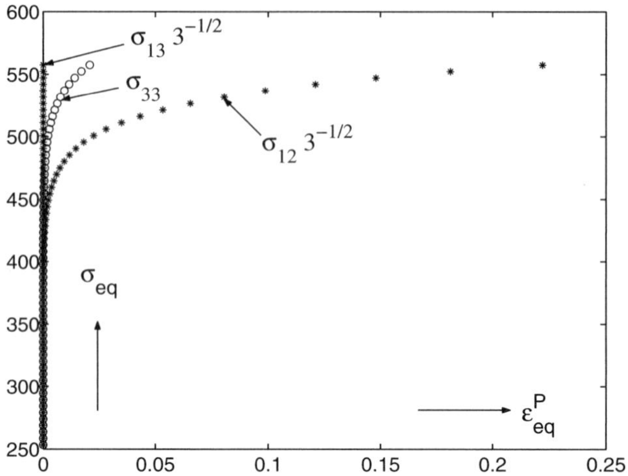

Fig. 7.6. Simulation of micro-constitutive equation in a single crystal for NTJ model $m = 0.05$ caused by a shear stress at easy slip direction by low stress rate

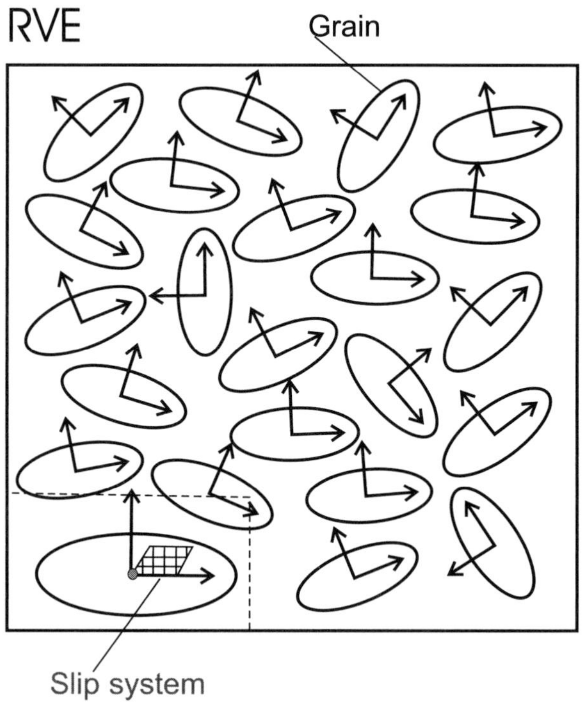

Fig. 7.7. A random distribution of Eshelby's ellipsoidal cubic grains inside an RVE

there do not exist explicit expressions for the Eshelby 4-tensor of cubic crystals its components in this section have been numerically calculated applying the approach of [KS-71]. The mentioned coaxiality of an ellipsoid and crystal directions of the corresponding grain, into which the ellipsoid is immersed, can be abandoned but then computational time would be considerably larger. Since the averaged terms $\langle \mathbb{S} \rangle_\omega$ and $\langle \mathbb{S}\, \mathfrak{D} \rangle_\omega$ appearing in (7.18) depend on orientations of individual grains inside the considered RVE history of grain rotations must be known in advance. In other words, before passing to step 3 we must calculate effective stiffness of the considered grain for all its orientations inside the fixed RVE taking into account that disorder is slight.

It is worthy of note that calculation of the effective stiffness $\mathbb{D}_{\Lambda}^{\text{eff}}$ depending on relative disorientation between the grain and the matrix (i.e., set of all the other grains in the RVE) permits establishing of an approximate symmetry of the considered slightly disordered polycrystal.

3. The last part of the computational procedure consists in finding residual stresses of individual grains by making use of the local constitutive equation (7.14) as well as the homogenized equation (7.18) for effective elastic constants of each grain. The residual stresses at all the grains have been determined by means of *Eshelby's equivalent inclusion method* ([Mur88]) where the eigen-strain is found from the equality of two products: (i) product of inclusion stiffness and elastic strain of the inclusion and (ii) product of effective stiffness and the sum of elastic and eigen-strain of the considered grain. The global residual stress of the RVE is then found as the volume average value of residual stresses of the grains.

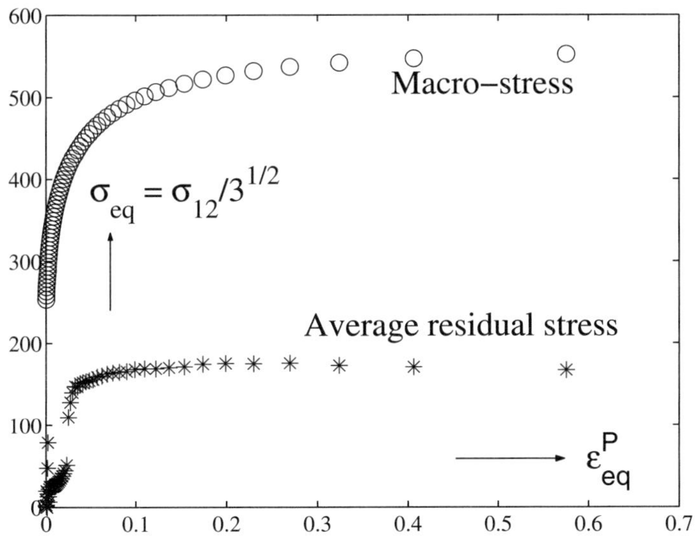

Fig. 7.8. Average external and residual stress versus equivalent plastic strain for MAM model of slightly disordered cubic polycrystal caused by a slow shear stress at easy direction (at the direction $\langle\bar{1}10\rangle$ in the plane $\{110\}$ if undisturbed)

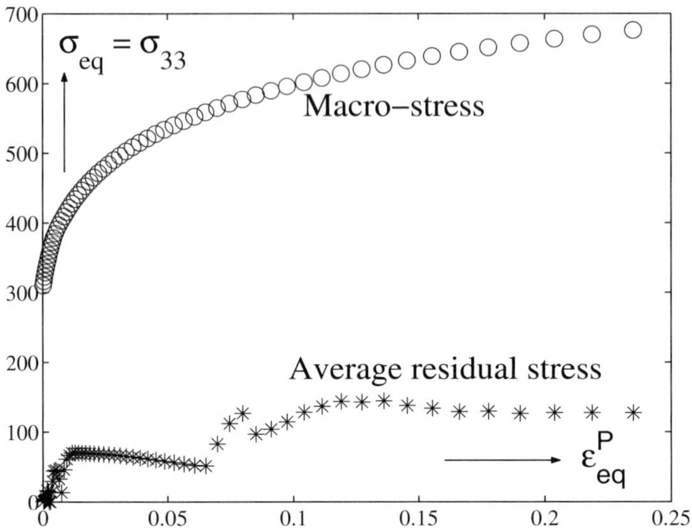

Fig. 7.9. Average external and residual stress versus equivalent plastic strain for MAM model of slightly disordered cubic polycrystal caused by a slow uniaxial tension stress perpendicular to easy direction ($\langle 001 \rangle$ direction if undisturbed)

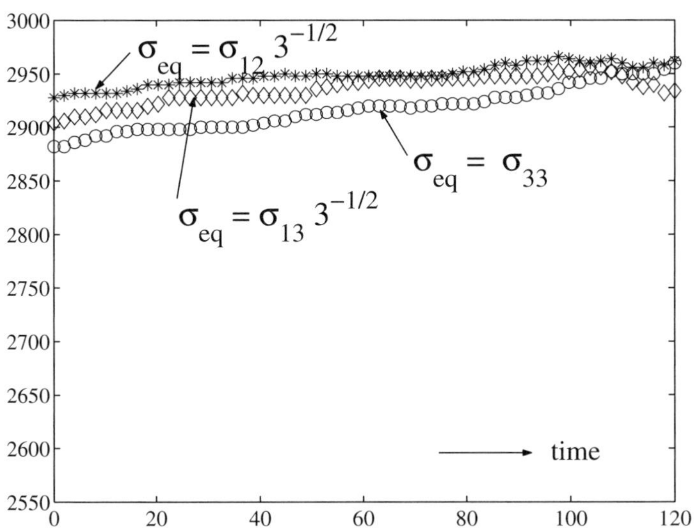

Fig. 7.10. Evolution of the number of active slip systems for MAM model of slightly disordered cubic polycrystal at low stress rate and $m = 0.2$

After all three steps are completed a lot of figures may be shown. For the sake of illustration residual stresses by MAM model with slow history are here described by Figs. 7.8 and 7.9 with shear and uniaxial tension along the

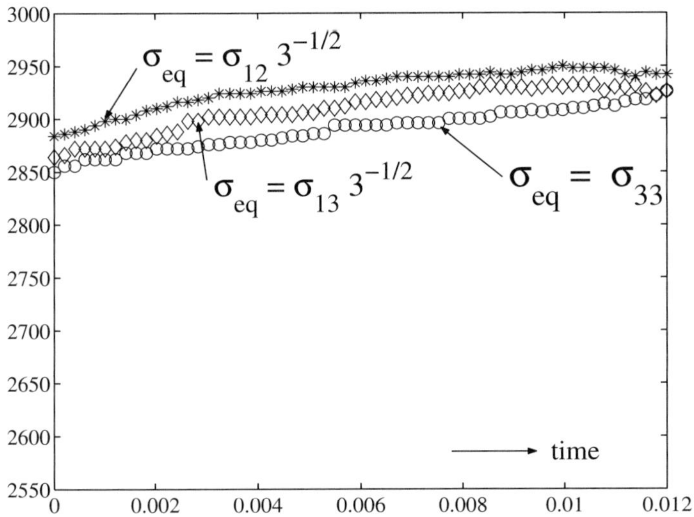

Fig. 7.11. Number of active slip systems for MAM model of slightly disordered cubic polycrystal at medium stress rate and $m = 0.18$

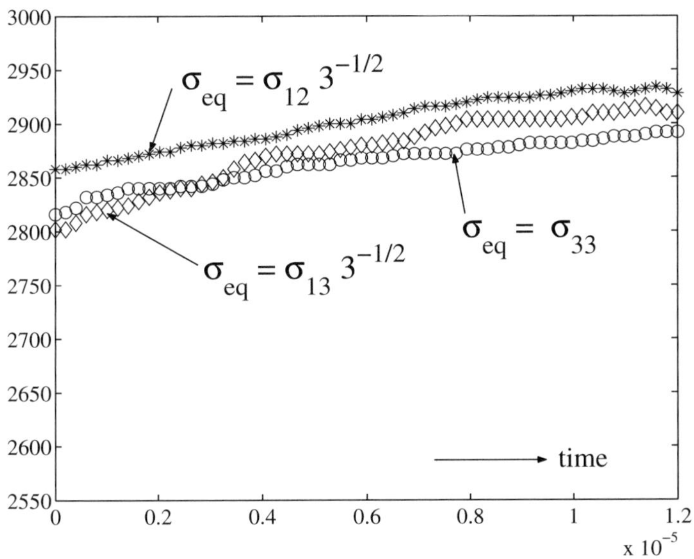

Fig. 7.12. Number of active slip systems for MAM model at medium stress rate and $m = 0.05$

symmetry axis. Moreover, Figs. 7.10–7.13 show evolution of the number of slip systems for slow, medium and high strain rates by MAM model as well as the number of slip systems by NTJ model and slow strain rate. As expected:

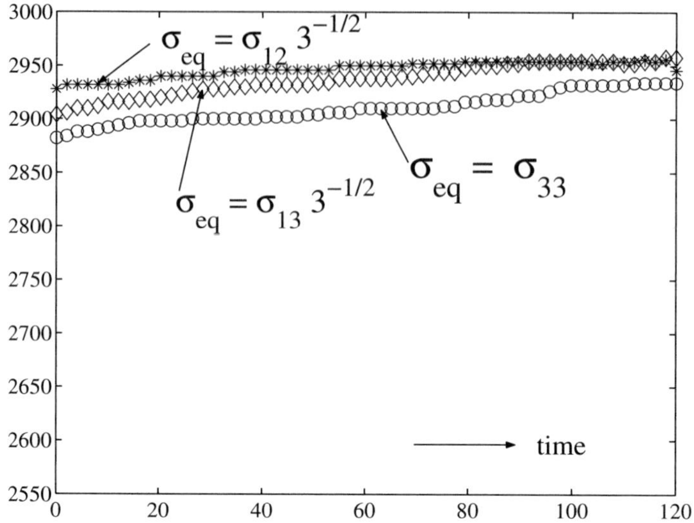

Fig. 7.13. Number of active slip systems for NTJ model at low stress rate and $m = 0.05$

- number of slip systems at shear is larger than at uniaxial tension and
- at higher strain rates the number of slip systems is smaller than at lower strain rates.

For the sake of brevity the obtained pole diagrams are not shown here. The interested reader could find them in the presentation [MicWWW].

Table 7.1. Numerical procedure of integration inside RVE by SC method

Step 1.

Initial RVE configuration:

Number of grains and slip systems are $N_g = 125, N_s = 12.$

Random initial grain orientations:

$$\mathbf{R}^0_\Lambda,\ \Lambda \in \{1, N_g\}\ \Rightarrow\ \mathbf{A}^0_{\alpha\Lambda} = \mathbf{s}_{\alpha\Lambda} \otimes \mathbf{n}_{\alpha\Lambda},\ \ \alpha \in \{1, N_s\}.$$

Step 2.

The loading process is defined by means of the stress history which is given by linear function h and direction $\mathbf{T}^0$

$$T^0_{11} - tension, \ldots, T^0_{12} - shear$$

$$\mathbf{T}^0 = \begin{Bmatrix} 1\,0\,0 \\ 0\,0\,0 \\ 0\,0\,0 \end{Bmatrix}, \ \ldots, \ \mathbf{T}^0 = \begin{Bmatrix} 0\,1\,0 \\ 1\,0\,0 \\ 0\,0\,0 \end{Bmatrix}$$

$$\mathbf{T}(t) = \mathbf{T}^0 h(t) Y_0, \quad t_{\max} \in \{10^2, 10^{-1.5}, 10^{-5}\}s, \quad h_{\min} = 2, \ h_{\max} = 3$$

$$\mathbf{R}_\Lambda = \mathbf{R}^0_\Lambda \quad \Rightarrow \quad \mathbf{T}_{local} = \mathbf{R}_\Lambda \mathbf{T} \mathbf{R}^T_\Lambda, \qquad \tau_{\alpha\Lambda} = \mathbf{T}_{local} : \mathbf{A}_{\alpha\Lambda} \qquad (a)$$

First the evolution equation (7.34) is chosen.

$$\Delta\gamma_{\alpha\Lambda} = D_t\gamma_0 \left(\tau_{\alpha\Lambda}/\tau_0\right) \left|\tau_{\alpha\Lambda}/\tau_0\right|^{m-1} \Delta t \qquad (b)$$

Two micro-models are distinguished by $D_t\gamma_0$ and values of m:

$$\textit{MAM model:} \qquad D_t\gamma_0 = \exp(-M)\ D_t\sigma_{eq} \qquad (c1)$$

$$\textit{NTJ model:} \qquad D_t\gamma_0 = \text{const} \qquad (c2)$$

Step 3.

The increment of plastic distortion is found in two steps

Iteration $= 0 \quad \Rightarrow \quad \Delta\mathbf{A}_{\alpha\,\Lambda} = 0$

$$\mathbf{\Pi}_{\Lambda P}\,(t + \Delta t) = \mathbf{\Pi}_{\Lambda P}\,(t) + \sum_\alpha \Delta\gamma_{\alpha\Lambda}\mathbf{A}_{\alpha\Lambda} + \boxed{\sum_\alpha \gamma_{\alpha\Lambda}\Delta\mathbf{A}_{\alpha\Lambda}} \qquad (d)$$

Step 4.

Iteration > 0

$$\mathbf{C}(\boldsymbol{\Pi}_\Lambda) \approx \boldsymbol{\Pi}_\Lambda^T \boldsymbol{\Pi}_\Lambda \Rightarrow \mathbf{C}_P = \langle \mathbf{C}(\boldsymbol{\Pi}_\Lambda) \rangle \text{ (volume averaging)} \qquad (e)$$

$$\mathbf{F}_P = \mathbf{U}_P = \mathbf{C}_P^{1/2} \Leftarrow \mathbf{R}_P = \mathbf{1} \text{ (plastic meso-rotation eliminated)}$$

$$\boldsymbol{\Pi}_{\Lambda E} = \mathbf{U}_P \boldsymbol{\Pi}_{\Lambda P}^{-1} \text{ (relaxed Taylor's assumption)}$$

$$\mathbf{U}_{\Lambda E} = (\boldsymbol{\Pi}_{\Lambda E}^T \boldsymbol{\Pi}_{\Lambda E})^{1/2}$$

$$\mathbf{R}_\Lambda = \boldsymbol{\Pi}_{\Lambda E} \mathbf{U}_{\Lambda E}^{-1} \text{ (grain micro-rotation)} \qquad (f)$$

$$\boldsymbol{\Omega}_\Lambda = [\mathbf{R}_\Lambda(t + \Delta t) - \mathbf{R}_\Lambda(t)]/\Delta t \text{ (micro-spin)} \qquad (g)$$

$$\boxed{\Delta \mathbf{A}_{\alpha\Lambda} = \boldsymbol{\Omega}_\Lambda \mathbf{A}_{\alpha\Lambda} + \mathbf{A}_{\alpha\Lambda}^T \boldsymbol{\Omega}_\Lambda^T} \qquad (h)$$

Replace (h) in (d) until it becomes balanced, i.e., difference between LHS and RHS of (d) becomes smaller than a prescribed tolerance:

$$\text{if } \|LHS - RHS\| > TOL, \qquad (h) \Rightarrow (d), \qquad \text{else } \mathbf{R}_\Lambda^0 = \mathbf{R}_\Lambda \Rightarrow (a).$$

7.4 A brief summary of the chapter

In this chapter further geometric and kinematic aspects of intragranular as well as intergranular plastic deformation of polycrystals are discussed. Main points of this chapter are:

- On the basis of the quasi-rate-independent meso-evolution equation an incremental inelastic micro-evolution equation is postulated. The rate dependence takes place by means of stress rate dependent value of the initial yield stress. Both mesoscopic as well as microscopic evolutions obey Vakulenko's concept of thermodynamic time
- An essential extension of concept of M. Zorawski applied here is that deformation geometry is based on constrained micro and free meso-rotations in intermediate reference configuration. This has as a consequence that evolution equation for plastic spin of RVE is an outcome of evolution equation for plastic stretching. In the papers [FFP91] and [Mic01] the authors connected to the natural state elements magnetization vectors in such a way

that they are isoclinic in (ν_t) and inhomogeneous in (χ_t) the inhomogeneity being responsible for magnetostrictive strains. A similar approach was made in [Mic74b] for pure thermoelastic strains. In this chapter another approach is accepted: free RVE plastic rotations and constrained grain plastic rotations have given clear understanding of intra- and intergrain behaviour.

- Elastic strain is covered by a homogenization method inside a representative volume element (RVE). Under the assumption that residual micro-strains are negligible, the *effective field* version of the *self-consistent method* has been applied by making use of results of [Lev82] leading to effective stiffness tensor for individual grains. This theory is afterwards specialized to slightly disordered fcc polycrystals modeled by randomly distributed oblate ellipsoidal inclusions. Anisotropic Eshelby tensor is calculated on the basis of [KS-71]. The results for effective stiffness tensors of individual grains are then used for finding residual stresses. For some characteristic given stress histories (at low, medium and high strain rates) the number of active slip systems for RVE with 125 grains is found and compared to so-called J_2 approach.
- Eshelby 4-tensor is determined numerically taking account of cubic anisotropy and the theory is applied to slightly disordered fcc polycrystals.
- For some characteristic given stress histories (leading to low, medium and high strain rates), the number of active slip systems for RVE with 125 grains are found and compared to so-called J_2 approach. In order to illustrate the proposed micro-meso-approach, the deformation geometry and micro-evolution equations have been used for computer simulation of inelastic behaviour of an RVE composed of $N = 5^3$ fcc grains with randomly oriented but slightly disordered crystal orientations and 12 potential slip systems. The RVE was loaded by slow, medium and fast stress history with three typical stress states: uniaxial tension, uniaxial easy shear and more difficult shear (with one principal stress vanishing and the other two being opposite). In all cases smooth increase of active slip systems has been noted during stress growth. The NTJ constitutive J_2 model was unable to distinguish among fast and slow processes being in such a way rate independent by its very nature.
- The hereinabove analysis has opened a way for treating symmetry issues of slightly disordered inelastic polycrystals. Namely, before applying correct tensor generators ([Spe71]) we must know which type of symmetry the considered polycrystal possesses. Such materials appear, for instance, after rolling and, consequently, forming limit diagrams should benefit from the results along this line of reasoning.

8

Inelastic Micromorphic Polycrystals

The principal task here is to find the simplest yet realistic way of describing polycrystal behaviour of metals taking into account inhomogeneous strains and stresses throughout a typical RVE. In such a problem grains of diverse orientations meet at their boundaries where most dislocations are concentrated. Intergranular and intragranular plastic sliding must be accompanied by thermoelastic straining in order to preserve continuity of the body. Even without external forces, residual stresses do exist and due to discontinuous change of orientation of neighboring grain lattices it is natural to expect appearance of couple stresses. Of special interest would be to connect material constants for stress and couple stress achieving their minimal number to be calibrated from specially designed experiments. Another already mentioned issue of great importance is Eshelby's problem: how to insert a grain larger than its available "hole" into the material of the considered body. The usual answer to this question is obtained by the so-called self-consistent methods. Again the question arises as to which part of strain do we apply such an approach. The third issue which must be analyzed is proper geometry of the considered thermo-inelastic strain history for such a polycrystalline body.

The presentation first gives a short amendment to up to this point given geometrical analysis of finite thermo-inelastic strains of polycrystalline bodies. Here issue of micro and meso-rotations is especially considered. Then conditions for homogeneous total and/or elastic and plastic strains are formulated leading to balance laws. The same analysis has been applied to materials homogenized in such a way that deformation gradient, elastic and plastic distortion are linear functions of relative position inside the RVE. Constitutive equations for stress and its moment are formulated by the homogenization method. Finally, a brief account to evolution equations following mainly [Mic02a, Mic02b] is given.

M.V. Mićunović, *Thermomechanics of Viscoplasticity: Fundamentals and Applications*, Advances in Mechanics and Mathematics 20, DOI: 10.1007/978-0-387-89490-4_8,
© Springer Science + Business Media, LLC 2009

8.1 Further details on polycrystal micro-strains

Since the first chapter of this monograph contains a detailed explanation of geometrical background and in the previous chapter some new details characteristic for polycrystals have been added, here only new details necessary for the subsequent theory are presented. Nevertheless, in order to accomplish a clarity of exposition some facts are briefly repeated.

For further considerations let us introduce vectors $d\mathbf{x} \in (\chi_t), d\boldsymbol{\xi} \in (\nu_t)$ and $d\mathbf{x}_0 \in (\kappa)$ connecting two infinitesimally adjacent particles:

$$d\mathbf{x} = \mathbf{F}\, d\mathbf{x}_0,\ d\mathbf{x} = \boldsymbol{\Phi}\, d\boldsymbol{\xi},\ d\boldsymbol{\xi} = \boldsymbol{\Pi}\, d\mathbf{x}_0. \tag{8.1}$$

8.1.1 Micro and meso-rotations of RVE

Although the definition of plastic distortion (1.64) is more correct, the original Kröner's decomposition (1.16) is preferred here due to its simplicity. The mapping tensors in it are represented by double tensor fields (cf. [Sto72, Mic74b])

$$\mathbf{F} = F^{k}_{\cdot K}\mathbf{g}_k \otimes \mathbf{g}_0^K, \quad \boldsymbol{\Phi} = \Phi^{k}_{\cdot \lambda}\mathbf{g}_k \otimes \mathbf{h}^\lambda, \quad \boldsymbol{\Pi} = \Pi^{\lambda}_{\cdot K}\mathbf{h}_\lambda \otimes \mathbf{g}_0^K, \tag{8.2}$$

where a tacit dependence on *spatial* $\mathbf{g}_k$, *structural* $\mathbf{h}_\kappa$ (1.59) and *material* $\mathbf{g}_{0K}$ base vectors has been taken into account.

Like in the previous chapter being motivated by [Mic02a] let us imagine that a typical *representative volume element* in its (ν_t) state is composed of N single crystal grains, such that each Λ-th grain has N_s slip systems $\mathbf{A}_{\alpha\Lambda} \equiv \mathbf{s}_{\alpha\Lambda} \otimes \mathbf{n}_{\alpha\Lambda},\ \alpha \in \{1, N_s\}$. For instance, for fcc crystals $N_s = 12$. Here $\mathbf{s}_{\alpha\Lambda}$ is the unit slip vector and $\mathbf{n}_{\alpha\Lambda}$ is the unit vector normal to the slip plane. For convenience, let us introduce a third unit vector $\mathbf{z}_{\alpha\Lambda}$ normal to the considered slip plane (cf. [Asa83]) with dyads $\mathbf{A}^1_{\alpha\Lambda} \equiv \mathbf{n}_{\alpha\Lambda} \otimes \mathbf{z}_{\alpha\Lambda}$ and $\mathbf{A}^2_{\alpha\Lambda} \equiv \mathbf{z}_{\alpha\Lambda} \otimes \mathbf{s}_{\alpha\Lambda}$ useful when either cross-slip or climb of dislocations has to be taken into account.

Let structural vectors attached to lattices of grains in the configuration (ν_t) be denoted by

$$\mathbf{h}^{\alpha}_{\cdot\Lambda}(t) = \mathbf{R}_{\Lambda\Pi}(t)\, \mathbf{h}^\alpha(t), \tag{8.3}$$

where $\alpha \in \{1, 3\}$ and $\Lambda \in \{1, N\}$. Suppose that an RVE has the volume $\Delta V = \sum_\Lambda \Delta V_\Lambda$. Then introducing grain *concentration* factors $c_\Lambda \equiv \Delta V_\Lambda / \Delta V$ for whole RVE the mean structural vector equals

$$\mathbf{h}^\alpha(t) := \sum_\Lambda c_\Lambda \mathbf{h}^{\alpha}_{\cdot\Lambda}(t). \tag{8.4}$$

Clearly, the tensors $\mathbf{R}_{\Lambda\Pi},\ \ (\Lambda \in \{1, N\})$ describe the *relative plastic microrotations* of all grains with respect to average orientation of RVE. For single crystal each of these vectors reduces to unit tensor. Similar formulae hold true for RVE in the initial natural state configuration (ν_t).

By comparing an RVE in (ν_t) and (ν_0) we may write a formula analogous to (1.16) for the *plastic micro-distortion tensor* (the same as (7.2)):

$$\mathbf{\Pi}_{\Lambda} := \mathbf{\Pi}_{\Lambda E}\, \mathbf{\Pi}_{\Lambda P}, \tag{7.2$'$}$$

whose components are the *residual elastic micro-distortion tensor* $\mathbf{\Pi}_{\Lambda E}$ and *plastic micro-distortion tensor* $\mathbf{\Pi}_{\Lambda P}$. Having in mind that unit slip vectors satisfy the relationships

$$\mathbf{s}_{\alpha\Lambda}(t) = S^{\alpha}_{\Lambda\beta}\, \mathbf{h}^{\beta}_{\cdot\,\Lambda}(t), \qquad \mathbf{n}_{\alpha\Lambda}(t) = N^{\alpha}_{\Lambda\beta}\mathbf{h}^{\beta}_{\cdot\,\Lambda}(t), \tag{8.5}$$

where $S^{\alpha}_{\Lambda\beta} = \text{const}$ and $N^{\alpha}_{\Lambda\beta} = \text{const}$, we may write a representation for the plastic micro-distortion as follows:

$$\mathbf{\Pi}_{\Lambda P} = \mathbf{1} + \mathbf{R}_{\Lambda\Pi}(t)\Big(\sum_{\alpha} \gamma_{\alpha\Lambda}\, S^{\alpha}_{\Lambda\beta}N^{\alpha}_{\Lambda\gamma}\mathbf{h}^{\beta}(t)\otimes\mathbf{h}^{\gamma}(t)\Big)\mathbf{R}^{T}_{\Lambda\Pi}(t). \tag{8.6}$$

On the other hand, assuming that residual elastic micro-rotation is negligibly small (cf. also [Mic02a]),[1] i.e., that by means of the polar decomposition $\mathbf{\Pi}_{\Lambda E} = \mathbf{U}^{res}_{\Lambda E} = \mathbf{V}^{res}_{\Lambda E}$ we may finally write

$$\mathbf{\Pi}_{\Lambda}(t) = \mathbf{V}^{res}_{\Lambda E}(t)\mathbf{R}_{\Lambda\Pi}(t)\Big(\,\mathbf{1} + \sum_{\alpha} \gamma_{\alpha\Lambda}\, S^{\alpha}_{\Lambda\beta}N^{\alpha}_{\Lambda\gamma}\mathbf{h}^{\beta}(t)\otimes\mathbf{h}^{\gamma}(t)\Big)\mathbf{R}^{T}_{\Lambda\Pi}(t). \tag{8.7}$$

It is natural to connect plastic meso (for RVE) and micro-distortions (for individual grains) by means of spatial averaging

$$\mathbf{\Pi}(t) = \langle\mathbf{\Pi}_{\Lambda}(t)\rangle \equiv \sum_{\Lambda} c_{\Lambda}\mathbf{\Pi}_{\Lambda}(t). \tag{8.8}$$

Let us apply the polar decomposition $\mathbf{\Pi} = \mathbf{R}_P\mathbf{U}_P$ to the plastic meso-distortion introducing the plastic meso-rotation tensor $\mathbf{R}_P$ which is arbitrary (according to [Zrw74]) and might be fixed either to be a unit tensor or to have Mandel's isoclinicity property (cf. Remark 1.3 on page 13). For a definition of isoclinicity we should have to find average crystal directions in RVE(t) and RVE(t_0) and to make them equal. The first choice, i.e., $\mathbf{R}_P = \mathbf{1}$, seems more appropriate for polycrystals.

Remark 8.1 (Microplastic rotations). In this way all the necessary ingredients for a discussion on micro and meso-rotations are prepared. The two mentioned approaches are very useful to fix plastic meso-rotations:
(a1) The first way of eliminating plastic meso-rotation by means of

$$\mathbf{\Pi} = \mathbf{U}_P = \mathbf{C}_P^{1/2} \tag{8.9}$$

[1] This does not mean that all the elastic residual rotations are exhausted since elastic meso-residual rotation has not been discussed until now.

greatly simplifies plastic meso-spin issue (cf. Remark 1.3, page 13, and the papers [Mic93, Mic02a]).

(a2) The second way could be called Mandel's average meso-isoclinicity by the identification

$$\mathbf{h}^{\alpha}(t) = \mathbf{h}^{\alpha}(t_0), \tag{8.10}$$

which approximately aligns plastically deformed RVE elements in (ν_t) and (ν_0). For instance, average structural vectors could be aligned to base $\mathbf{g}_0^K$ if such an identification is accepted.
However, for both approaches relative plastic micro-rotations *must not be eliminated* unless further tearing of RVE elements is performed to the size of single crystal grains which is, by assumption, much smaller than the typical dimension of RVE.

Joining the RVE elements of (ν_t) into a continuous body requires for each grain elastic micro-distortion $\mathbf{\Phi}_{\Lambda}$. Again polar decomposition allows: $\mathbf{\Phi}_{\Lambda} = \mathbf{R}_{\Lambda E}\mathbf{U}_{\Lambda E}$. Here the right elastic micro-stretch $\mathbf{U}_{\Lambda E}$ does not include the left residual micro-stretch $\mathbf{V}_{\Lambda E}^{res}$.

Recalling Kröner's decomposition (1.16) we could make a new grouping of terms in the following way[2]

$$\mathbf{F}_{\Lambda} = \mathbf{R}_{\Lambda E}\, \mathbf{U}_{\Lambda E}\, \mathbf{V}_{\Lambda E}^{res}\, \mathbf{\Pi}_{\Lambda P} \equiv \mathbf{\Phi}_{\Lambda E}\, \mathbf{\Pi}_{\Lambda P}. \tag{8.11}$$

Here $\mathbf{\Phi}_{\Lambda E}$ encompasses rotation of an RVE element in (χ_t) configuration and residual as well as external forces induced elastic stretches whereas $\mathbf{\Pi}_{\Lambda P}$ includes pure plastic distortion and relative plastic micro-rotations.

8.1.2 Some additional notes on inelastic micro-strains

If the plastic micro-deformation tensors for individual grains are (cf. also (8.6))

$$\begin{aligned}\mathbf{C}_{\Lambda\Pi} = \mathbf{\Pi}_{\Lambda P}^{T}\mathbf{\Pi}_{\Lambda P} \equiv \mathbf{1} + \mathbf{R}_{\Lambda\Pi}\sum_{\alpha}\gamma_{\alpha\Lambda}\left(\mathbf{A}_{\alpha\Lambda} + \mathbf{A}_{\alpha\Lambda}^{T}\right)\mathbf{R}_{\Lambda\Pi}^{T} \\ + \mathbf{R}_{\Lambda\Pi}\sum_{\alpha}\gamma_{\alpha\Lambda}\mathbf{A}_{\alpha\Lambda}^{T}\sum_{\beta}\gamma_{\beta\Lambda}\mathbf{A}_{\beta\Lambda}\mathbf{R}_{\Lambda\Pi}^{T},\end{aligned} \tag{8.12}$$

then their volume average termed *plastic meso-deformation tensor* $\mathbf{C}_P$ (7.7) has the following form:

$$\mathbf{C}_P = \langle \mathbf{C}_{\Lambda\Pi} \rangle = \left\langle \mathbf{\Pi}_{\Lambda}^{T}\mathbf{\Pi}_{\Lambda} \right\rangle \equiv \sum_{\Lambda} c_{\Lambda}\mathbf{\Pi}_{\Lambda}^{T}\mathbf{\Pi}_{\Lambda} \neq \left\langle \mathbf{\Pi}^{T} \right\rangle \langle \mathbf{\Pi} \rangle \tag{8.13}$$

and is different from product of averages of plastic distortions. This fact must be taken into account in all of the subsequent derivations.

[2] The following decomposition differs from the approach in [Mic02a] where plastic micro-distortion also contains residual elastic micro-strains.

Suppose that we accept meso-isoclinicity assumption. Then taking $\mathbf{h}^\alpha(t) = \mathbf{h}^\alpha(t_0) = \text{const}$ and by making use of the relation $D_t \mathbf{R}_{\Lambda\Pi} \mathbf{R}_{\Lambda\Pi}^T = \mathbf{\Omega}_{\Lambda\Pi}$ for *relative plastic micro-spin* the relation (8.6) gives

$$\begin{aligned} D_t \mathbf{\Pi}_{\Lambda P} &= \mathbf{R}_{\Lambda\Pi} \sum_\alpha \mathbf{A}_{\alpha\Lambda}\; D_t \gamma_{\alpha\Lambda} \mathbf{R}_{\Lambda\Pi}^T \\ &\quad + \mathbf{\Omega}_{\Lambda\Pi} \left(\mathbf{\Pi}_{\Lambda P} - \mathbf{1}\right) + \left(\mathbf{\Pi}_{\Lambda P} - \mathbf{1}\right) \mathbf{\Omega}_{\Lambda\Pi}^T . \end{aligned} \tag{8.14}$$

Then, the velocity gradient at Λ grain may be expressed by means of

$$\begin{aligned} \mathbf{L}_\Lambda &= D_t \mathbf{\Phi}_{\Lambda E}\, \mathbf{\Phi}_{\Lambda E}^{-1} + \mathbf{\Phi}_{\Lambda E} D_t \mathbf{\Pi}_{\Lambda P} \mathbf{\Pi}_{\Lambda P}^{-1} \mathbf{\Phi}_{\Lambda E}^{-1} \\ &\equiv \mathbf{L}_{\Lambda E} + \mathbf{\Phi}_{\Lambda E} \mathbf{L}_{\Lambda P} \mathbf{\Phi}_{\Lambda E}^{-1}, \end{aligned} \tag{8.15}$$

such that the product $\mathbf{\Phi}_{\Lambda E}^T \mathbf{L}_\Lambda \mathbf{\Phi}_{\Lambda E}$ may be split into symmetric and antisymmetric parts as follows:

$$\mathbf{\Phi}_{\Lambda E}^T\, \mathbf{L}_\Lambda\, \mathbf{\Phi}_{\Lambda E} \;=\; D_t \mathbf{E}_{\Lambda E} \;+ \mathcal{E}_\nu\, \boldsymbol{\omega}_{\Lambda E} + \left(2\mathbf{E}_{\Lambda E} + \mathbf{1}\right) \mathbf{L}_{\Lambda P}, \tag{8.16}$$

where $\mathcal{E}_\nu = \mathcal{E}^{\alpha\beta\gamma} \mathbf{h}_\alpha \otimes \mathbf{h}_\beta \otimes \mathbf{h}_\gamma$ is the Ricci tensor related to the structural coordinates of (ν_t) configuration (cf. also (1.95)), $\boldsymbol{\omega}_{\Lambda E}$ is the elastic micro-spin and $2\mathbf{E}_{\Lambda E} = \mathbf{\Phi}_{\Lambda E}^T \mathbf{\Phi}_{\Lambda E} - \mathbf{1}$ is the elastic micro-strain tensor.

On the other hand, if we accept eliminating plastic meso-rotation, then plastic meso-stretching and plastic meso-spin are not independent. They are determined by the time rate of plastic stretch tensor as explicitly given in (7.12), page 124.

8.2 Balance laws. Nonproportionality and microsymmetry

8.2.1 Notion of low-order polycrystals

Let us introduce a *mass distribution function* by means of

$$\varphi(\mathbf{x}') = \{\varphi_\Lambda(\mathbf{x}') \,|\, \mathbf{x}' \in \Delta V,\ \Lambda \in \{1, N\}\}, \tag{8.17}$$

with the property

$$\varphi_\Lambda(\mathbf{x}') = \begin{cases} 1, & \mathbf{x}' \in \Delta V_\Lambda, \\ 0, & \text{otherwise.} \end{cases}$$

Then positions of grain centers and the RVE center are determined by

$$\int_{\Delta V} \varphi(\mathbf{x}')\mathbf{x}' dm' = \mathbf{x}\Delta m, \quad \int_{\Delta V_\Lambda} \varphi_\Lambda(\mathbf{x}'_0)\mathbf{x}_0^* dm' = \mathbf{x}_{0\Lambda}^* \Delta m_\Lambda, \tag{8.18}$$

where the notation $\mathbf{x}' \equiv \mathbf{x} + \mathbf{x}^*$ will be used in the sequel.

Assumption 8.2.1 *Suppose that individual grains have the same density and are subject to homogeneous micro-strains.*

Such materials are called in the paper by [Mic02b] *low-order micromorphic polycrystals.*[3] Moreover, let the center of an RVE element be occupied in (χ_0) and (χ_t) by the same material point. According to this assumption the deformation gradient is split into its average value and fluctuations as follows:

$$\mathbf{F}' = \langle \mathbf{F}' \rangle + \mathbf{F}^* \equiv \mathbf{F} + \varphi_{\Lambda}(\mathbf{x}')\mathbf{F}^*_{\Lambda}, \qquad (\text{where } \mathbf{x}' \in \Delta V_{\Lambda}). \tag{8.19}$$

Then momentum, moment of momentum and kinetic energy of such an RVE element read

$$\mathbf{v}\,\Delta m = \int_{\Delta V} \varphi(\mathbf{x}')\, D_t\mathbf{x}'\, dV' \equiv D_t\mathbf{x}\,\Delta m, \tag{8.20}$$

$$\begin{aligned} \Delta \mathbf{l}_O &= \int_{\Delta V} \varphi(\mathbf{x}')\,\mathbf{x}' \times D_t\mathbf{x}'\, dV' \\ &= \Big(\mathbf{x} \times \mathbf{v} + \mathcal{E} : \sum_{\Lambda} c_{\Lambda} \mathbf{F}_{\Lambda} \mathbf{J}_{0\Lambda} D_t \mathbf{F}^T_{\Lambda}\Big)\Delta m, \end{aligned} \tag{8.21}$$

$$\begin{aligned} 2\Delta T &= \int_{\Delta V} \varphi(\mathbf{x}')\, D_t\mathbf{x}' D_t\mathbf{x}'\, dV' \\ &= \Big(\mathbf{v}\mathbf{v} + \sum_{\Lambda} c_{\Lambda} tr \left\{ D_t\mathbf{F}_{\Lambda}\mathbf{J}_{0\Lambda} D_t\mathbf{F}^T_{\Lambda} \right\} \Big)\Delta m. \end{aligned} \tag{8.22}$$

Here the *material grain microinertia tensor* (cf. also [Sto72]) for a Λ-th grain equals

$$\mathbf{J}_{0\Lambda}\Delta m_{\Lambda} := \int_{\Delta V_{\Lambda}} \varphi_{\Lambda}(\mathbf{x}'_0)\mathbf{x}^*_0 \otimes \mathbf{x}^*_0 dm'. \tag{8.23}$$

Remark 8.2 (Nonproportionality and microsymmetry).

It is interesting to note here that due to symmetry properties of $\mathbf{J}_{0\Lambda}$ and antisymmetry of Ricci tensor $\mathcal{E} = \mathcal{E}_{ijk}\,\mathbf{g}^i \otimes \mathbf{g}^j \otimes \mathbf{g}^k$ the second term in (8.21) disappears for proportional strain paths whenever $\mathbf{F}_{\Lambda}$ and $D_t\mathbf{F}_{\Lambda}$ have the same directions, i.e., when $D_t\mathbf{F}_{\Lambda} = \alpha\mathbf{F}_{\Lambda}$, $\alpha \in \mathcal{R}$. It should be recalled that in general nonproportionality in strain histories affects very much experimental results for characterization of inelastic behaviour of metals (cf., e.g., [Mic97]).

Consider now stresses and their moments in an RVE element. Although Assumption 8.2.1 does not require homogeneity of grain distortions but only of the micro-deformation gradient of a grain inside the considered RVE we will assume that for $\mathbf{x}' \in \Delta V_{\Lambda}$ their homogeneity

$$\begin{aligned} \boldsymbol{\Phi}' &= \langle \boldsymbol{\Phi}' \rangle + \boldsymbol{\Phi}^* \equiv \boldsymbol{\Phi} + \varphi_{\Lambda}(\mathbf{x}')\boldsymbol{\Phi}^*_{\Lambda}, \\ \boldsymbol{\Pi}' &= \langle \boldsymbol{\Pi}' \rangle + \boldsymbol{\Pi}^* \equiv \boldsymbol{\Pi} + \varphi_{\Lambda}(\mathbf{x}')\boldsymbol{\Pi}^*_{\Lambda} \end{aligned} \tag{8.24}$$

[3] In other words, we are dealing with a single phase material.

also holds. At the place $\mathbf{x}' \in \Delta V_\Lambda$ let second Piola Kirchhoff (ν_t) related and Cauchy stress be related by means of $\mathbf{\Phi}'\mathbf{S}'\mathbf{\Phi}'^T = \mathbf{T}' \det\mathbf{\Phi}'$. Then Hooke's law at such a point has the form

$$\mathbf{S}' = \mathbb{D}' : \mathbf{E}'_E = \frac{1}{2}\varphi_\Lambda(\mathbf{x}')\,\mathbb{D}_\Lambda : \left(\mathbf{\Phi}^T_{\Lambda E}\mathbf{\Phi}^T_{\Lambda E} - \mathbf{1}_\nu\right) \equiv \varphi_{\Lambda}(\mathbf{x}')\,\mathbf{S}_\Lambda \tag{8.25}$$

and the meso-stress of the RVE is obtained by the averaging[4]

$$\mathbf{S} = \langle \mathbf{S}' \rangle = \sum_\Lambda \mathbb{D}_\Lambda : \mathbf{E}_{\Lambda E}. \tag{8.26}$$

In our case of low-order micromorphic polycrystals the fourth-rank tensor of elasticity has the form

$$\mathbb{D}_\Lambda = D^0_{\alpha\beta\gamma\delta}\mathbf{h}^\alpha_{\cdot\,\Lambda} \otimes \mathbf{h}^\beta_{\cdot\,\Lambda} \otimes \mathbf{h}^\gamma_{\cdot\,\Lambda} \otimes \mathbf{h}^\delta_{\cdot\,\Lambda} \tag{8.27}$$

with

$$D^0_{\alpha\beta\gamma\delta} = \text{const},$$

which means that grains are identical but their crystallographic orientations change throughout the RVE.[5] Let us introduce the notation

$$\left(\mathbb{R}_{\Lambda\Pi}\right)_{\alpha\gamma\beta\delta} := \left(\mathbf{R}_{\Lambda\Pi} \,\square\, \mathbf{R}_{\Lambda\Pi}\right)_{\alpha\gamma\beta\delta} \equiv \left(\mathbf{R}_{\Lambda\Pi}\right)_{\alpha\beta}\left(\mathbf{R}_{\Lambda\Pi}\right)_{\gamma\delta}.$$

With respect to average RVE base in (ν_t) configuration we have the equivalent of (8.26) as follows:

$$\mathbf{S} = \sum_\Lambda \left(\mathbb{R}_{\Lambda\Pi} : \mathbb{D}^0 : \mathbb{R}^T_{\Lambda\Pi}\right) : \mathbf{E}_{\Lambda E}. \tag{8.28}$$

Remark 8.3 (Hill–Mandel average).

In this and subsequent sections the so-called *Hill–Mandel principle of macrohomogeneity* ([Kro71, Mau92]) would be of great practical use allowing replacement of average of a product by the corresponding product of averages. To our regret there is not much justification for its application. Kröner calls it the *ergodicity property* in [Kro71]. In statistical theories it is widely applied.

The vector of stress moments for an RVE is given by

$$\begin{aligned}\Delta\mathbf{m}_O &= \int_{\partial\Delta V} \varphi(\mathbf{x}')\mathbf{x}' \times \mathbf{T}'\mathbf{n}'ds' = \int_{\Delta V} \varphi(\mathbf{x}')\mathrm{div}\left(\mathbf{x}' \times \mathbf{T}'\right)dV' \\ &= \Delta V\left(\mathbf{x} \times \mathrm{div}\mathbf{T} - \boldsymbol{\tau}\right) + \sum_\Lambda \Delta S_\Lambda\left(\mathbf{x}^* \times [\![\mathbf{T}^*_\Lambda]\!]\;\mathbf{n}_\Lambda\right)|_{\partial\Delta V_\Lambda}. \end{aligned} \tag{8.29}$$

[4] Even in the case of meso-homogeneous elastic distortion throughout the RVE due to different orientations of grain structures, elastic meso-constants differ from their elastic micro-counterparts.

[5] In other words base $\mathbf{h}^\gamma_{\cdot\,\Lambda}$ is variable since we do not have a single crystal but polycrystal.

Here again we have splitting $\mathbf{T}' = \langle \mathbf{T}' \rangle + \mathbf{T}^* = \mathbf{T} + \varphi_\Lambda(\mathbf{x}')\, \mathbf{T}^*_\Lambda$ for $\mathbf{x}' \in \Delta V_\Lambda$ and $[\![\mathbf{T}^*_\Lambda]\!]$ denotes jump of the fluctuation of Cauchy stress on boundaries of Λ-th grain (cf. also [Mau92]). The average Cauchy stress is here divided into its antisymmetric and symmetric parts by means of

$$\boldsymbol{\tau} = \frac{1}{2}\mathcal{E} : \mathbf{T}, \quad \mathbf{T}_a = \mathcal{E}\,\boldsymbol{\tau} \equiv \mathrm{skw}\mathbf{T}, \quad \mathbf{T}_s = \mathrm{sym}\mathbf{T} \equiv \boldsymbol{\sigma}. \tag{8.30}$$

In the same way the mechanical working done by micro-stresses throughout an RVE equals

$$\begin{aligned} \Delta P = \int_{\partial \Delta V} \varphi(\mathbf{x}')\, D_t\mathbf{x}'\, \mathbf{T}'\, \mathbf{n}' ds' &= \sum_\Lambda \Delta S_\Lambda \left(\mathbf{x}^* \times [\![\mathbf{T}^*_\Lambda]\!]\ \mathbf{n}_\Lambda\right)|_{\partial \Delta V_\Lambda} \\ &+ \Delta V \Big(\mathbf{T} : \mathrm{grad}\, D_t\mathbf{x} + D_t\mathbf{x}\, \mathrm{div}\mathbf{T} + \sum_\Lambda c_\Lambda \mathbf{T}^*_\Lambda : \mathrm{grad}\, D_t\mathbf{x}^*_\Lambda\Big). \end{aligned} \tag{8.31}$$

Now we are ready to write the balance equations for typical RVE. First, balance of momentum leads to the traditional nonpolar equation by means of

$$D_t \int_{\Delta V} \varphi(\mathbf{x}')\, D_t\mathbf{x}'\, dV' = \int_{\partial \Delta V} \varphi(\mathbf{x}')\mathbf{T}'\mathbf{n}' ds' \;\Rightarrow\; \rho D_t^2 \mathbf{x} = \mathrm{div}\mathbf{T}, \tag{8.32}$$

whereas balance of moment of momentum $D_t\, \Delta \mathbf{l}_O = \Delta \mathbf{m}_O$ gives the antisymmetric part of Cauchy stress as follows:

$$\begin{aligned} 2\boldsymbol{\tau} &= -\rho\,\mathcal{E} : \sum_\Lambda c_\Lambda \mathbf{F}_\Lambda \mathbf{J}_\Lambda D_t^2 \mathbf{F}^T_\Lambda \\ &+ \frac{1}{\Delta V} \sum_\Lambda \Delta S_\Lambda \left(\mathbf{x}^* \times [\![\mathbf{T}^*_\Lambda]\!]\ \mathbf{n}_\Lambda\right)|_{\partial \Delta V_\Lambda}\,. \end{aligned} \tag{8.33}$$

Looking at the above formula we may draw the conclusion that for low order micromorphic polycrystals stress is symmetric only if the two following conditions are satisfied:

Condition 8.2.1 (Intergranular continuity) *Stress vector is continuous on grain boundaries.*

Condition 8.2.2 (Proportional paths) *Deformation gradient of each grain follows a proportional path.*

Concerning the first sum on the RHS of (8.33) we note that the second condition 8.2.2 is approximately satisfied also when either *higher order inertial terms* in (8.21) or grain micro-accelerations (i.e., second-order material time derivatives of micro-deformation gradients) are negligibly small.

The last balance law is the first law of thermodynamics, which for the RVE reads

$$D_t\, \Delta T + D_t\, \Delta U = \Delta P + \Delta Q, \tag{8.34}$$

where $D_t\,\Delta U$ is time rate of internal energy, while ΔP is mechanical working.[6] The last term, namely ΔQ, is nonmechanical working including thermal and other effects mentioned in previous chapters. If the average velocity gradient tensor $\mathbf{L} \equiv \operatorname{grad} D_t \mathbf{x}$ is split into its symmetric and antisymmetric parts:

$$\begin{aligned} \boldsymbol{\omega} = \frac{1}{2}\mathcal{E}\colon \mathbf{L}, \quad \mathbf{W} &= \frac{1}{2}\left(\mathbf{L} - \mathbf{L}^T\right) \equiv \operatorname{skw}\mathbf{L} = \mathcal{E}\,\boldsymbol{\omega}, \\ \mathbf{D} &= \frac{1}{2}\left(\mathbf{L} + \mathbf{L}^T\right) \equiv \operatorname{sym}\mathbf{L}, \end{aligned} \tag{8.35}$$

then the energy conservation law (8.34) in its local form may be written as follows:

$$\begin{aligned} \rho\, D_t u - \boldsymbol{\sigma} : \mathbf{D} =\ & \boldsymbol{\tau}\,\omega + \sum_{\Lambda} c_\Lambda \left(\boldsymbol{\sigma}^*_\Lambda\colon \mathbf{D}^*_\Lambda + 2\boldsymbol{\tau}^*_\Lambda\,\boldsymbol{\omega}^*_\Lambda\right) \\ & + 2\frac{1}{\Delta V}\sum_{\Lambda} \Delta S_\Lambda \left(D_t\mathbf{x}^* \left[\!\left[\mathbf{T}^*_\Lambda\right]\!\right]\ \mathbf{n}_\Lambda\right)|_{\partial\Delta V_\Lambda} \\ & - \rho\sum_{\Lambda} c_\Lambda tr\left\{\operatorname{sym}\left\{D_t D_t \mathbf{F}_\Lambda \mathbf{J}_\Lambda D_t \mathbf{F}^T_\Lambda\right\}\right\}. \end{aligned} \tag{8.36}$$

8.2.2 A comment on homogenization

Let us now describe approximately grain behaviour by some smooth functions. The easiest way to do that is to suppose that deformation gradient and distortions are linear functions of position throughout the considered RVE, i.e.,

$$\bar{\mathbf{F}}' = \bar{\mathbf{F}} + \mathcal{F}\,\mathbf{x}^*_0, \quad \bar{\boldsymbol{\Phi}}' = \bar{\boldsymbol{\Phi}} + \mathbf{x}^*\,\mathcal{F}_E, \quad \bar{\boldsymbol{\Pi}}' = \bar{\boldsymbol{\Pi}} + \mathcal{F}_P\,\mathbf{x}^*_0. \tag{8.37}$$

Now, the approximating "constants" $\bar{\mathbf{F}}$ and $\mathcal{F}$ are found by minimization of the functional (cf. also [For02], [Mic02b]):

$$I = \frac{1}{\Delta V}\int_{\Delta V} \left\|\bar{\mathbf{F}}' - \mathbf{F}'\right\|^2 dV', \tag{8.38}$$

which means that I has the role of so-called *chi-square function*:

$$\min_{(\bar{\mathbf{F}},\mathcal{F})} I \Rightarrow \left(\bar{\mathbf{F}},\mathcal{F}\right). \tag{8.39}$$

Taking the inner product as a means to form the norm used in (8.38) we get

$$\bar{\mathbf{F}} = \mathbf{F} \qquad \text{as well as} \qquad \mathcal{F} = \left\langle \mathbf{F}^* \otimes \mathbf{x}^*_0\right\rangle \left\langle \mathbf{J}_0\right\rangle^{-1}. \tag{8.40}$$

---

[6] The mechanical working is related to work W by $\Delta P = \mathfrak{D}_t W/dt$ where the sign $\mathfrak{D}_t$ stands to show that infinitesimal increment of work is not an exact differential (cf. also Table 4.1 on page 62). It is calculated by integration over the RVE. To show this sign Δ is chosen. Similar holds true for nonmechanical working.

By the same minimization procedure we arrive at approximating "constants" for the distortions[7]:

$$\bar{\mathbf{\Pi}} = \mathbf{\Pi} \quad \text{as well as} \quad \mathcal{F}_P = \langle \mathbf{\Pi}^* \otimes \mathbf{x}_0^* \rangle \langle \mathbf{J}_0 \rangle^{-1}, \tag{8.41}$$

$$\bar{\mathbf{\Phi}} = \mathbf{\Phi} \quad \text{as well as} \quad \mathcal{F}_E = \langle \mathbf{J} \rangle^{-1} \langle \mathbf{x}^* \otimes \mathbf{\Phi}^* \rangle, \tag{8.42}$$

where the *spatial grain microinertia tensor* has the following form (cf. also (8.23)):

$$\mathbf{J}_\Lambda \Delta m_\Lambda := \int_{\Delta V_\Lambda} \varphi_\Lambda(\mathbf{x}')\mathbf{x}^* \otimes \mathbf{x}^* dm'. \tag{8.43}$$

This tensor has been used in the above balance laws.

Remark 8.4 (Dislocation densities).
Antisymmetric parts of $\mathcal{F}_E$ and $\mathcal{F}_P$ are used to build the asymmetric second-rank true spatial and material dislocation density tensors (cf. [Sto62, Teo70, Mic74b]) in the following way:

$$\mathbf{A}_E = \mathcal{E} : \mathcal{F}_E, \quad \mathbf{A}_P = \mathcal{F}_P : \mathcal{E}_0, \tag{8.44}$$

where Ricci permutation tensor in material coordinates of (κ) configuration $\mathcal{E}_0 = \mathcal{E}_{KLM}\mathbf{g}_0^K \otimes \mathbf{g}_0^L \otimes \mathbf{g}_0^M$ is used in the second relationship. The two representations of dislocation density are connected by (cf. also (1.41) and (1.50))

$$\mathbf{A}_E \det\mathbf{\Phi} = \mathbf{A}_P \mathbf{F}^T.$$

It must be noted here that early papers of Kondo, Bilby, Kröner, Stojanović and their collaborators on non-Euclidean geometry of natural state space are so fertile that the subject would require a separate analysis. Of special interest is here non-Euclidean geometry of oriented materials like Cosserat media. A lot of other important remarks about non-Euclidean aspects of geometry of deformation could be made but such notes have to be omitted here for the sake of brevity.[8]

For further analysis of constitutive equations we will need balance laws for such a smoothed RVE. On the other hand, the moment of momentum[9] becomes

---

[7] It is worth noting that although we apply the same type of approximation functions for $\mathbf{F}'$ and $\mathbf{\Phi}', \mathbf{\Pi}'$ it would be possible to have $\mathcal{F} = 0$, but $\mathcal{F}_P \neq 0, \mathcal{F}_E \neq 0$ for instance. In fact, as will be shown soon, this is typical lowest approximation for dislocated bodies where it is not allowed to assume that either $\mathcal{F}_P$ or $\mathcal{F}_E$ vanish.

[8] A more comprehensive review is given in [Mic74b].

[9] A shorter notation $D_t\mathcal{A} \equiv \mathcal{A}_t$ for an arbitrary tensor $\mathcal{A}$ is applied only in the sequel.

$$\frac{\Delta \mathbf{l}_O}{\Delta m} = \frac{1}{\Delta m}\int_{\Delta V} \varphi(\mathbf{x}')\,\mathbf{x}' \times \mathbf{x}'_t\, dV' = \mathbf{x} \times \mathbf{v} \tag{8.45}$$
$$+ \mathcal{E} : \left(\mathbf{F}\mathbf{J}_0\mathbf{F}_t^T + \mathbf{F}\mathcal{J}_0 : \mathcal{F}_t^T + \mathcal{F} : \mathcal{J}_0\mathbf{F}_t^T + \mathcal{F} : \mathbb{J}_0 : \mathcal{F}_t^T\right),$$

where third- and fourth-rank microinertia moments of an RVE are explicitly given by (cf. also [Kro71])

$$\mathcal{J}_0 := \int_{\Delta V} \varphi(\mathbf{x}'_0)\; \mathbf{x}_0^* \otimes \mathbf{x}_0^* \otimes \mathbf{x}_0^*\; dm', \quad \mathbb{J}_0 := \int_{\Delta V} \varphi(\mathbf{x}'_0)\; \mathbf{x}_0^* \otimes \mathbf{x}_0^* \otimes \mathbf{x}_0^* \otimes \mathbf{x}_0^*\; dm'.$$

The balance equation for the RVE momentum has the same form as (8.32) while the corresponding balance equation of moment of momentum with the above linear approximation becomes

$$2\boldsymbol{\tau}\;\Delta V - \sum_{\Lambda}\int_{\partial \Delta V_\Lambda} (\mathbf{x}^* \times [\![\mathbf{T}_\Lambda^*]\!]\;\mathbf{n}_\Lambda)\; dS'_\Lambda$$
$$= -\rho\mathcal{E}\colon \left(\mathbf{F}\mathbf{J}_0\mathbf{F}_{tt}^T + \mathcal{F} : \mathbb{J}_0 : \mathcal{F}_{tt}^T\right) \tag{8.46}$$
$$- \rho\mathcal{E} : D_t\left(\mathbf{F}\mathcal{J}_0 : \mathcal{F}_t^T + \mathcal{F} : \mathcal{J}_0\,\mathbf{F}_t^T\right).$$

The expression for kinetic energy of an RVE is obtained in the same way:

$$2\frac{\Delta T}{\Delta m} = \frac{1}{\Delta m}\int_{\Delta V} \varphi(\mathbf{x}')\mathbf{x}'_t\mathbf{x}'_t dV' = \mathbf{v}\,\mathbf{v} + \mathrm{tr}\left\{\mathbf{F}_t\mathbf{J}_0\mathbf{F}_t^T\right\}$$
$$+ \mathrm{tr}\left\{\mathbf{F}_t\mathcal{J}_0 : \mathcal{F}_t^T + \mathcal{F}_t : \mathcal{J}_0 D\mathbf{F}_t^T + \mathcal{F}_t : \mathbb{J}_0 : \mathcal{F}_t^T\right\}. \tag{8.47}$$

In order to obtain an expression for mechanical working we suppose that velocity gradient $\mathbf{L}'$ is also a linear function throughout the RVE element:

$$\mathbf{D}' = \mathbf{D} + \mathbf{x}^*\mathrm{grad}\mathbf{D}, \qquad \boldsymbol{\omega}' = \boldsymbol{\omega} + \mathbf{x}^*\,\mathrm{grad}\,\boldsymbol{\omega}, \tag{8.48}$$

such that (with notation $\mathcal{A} \circ \mathcal{B} = \mathcal{A}^{abc}\mathcal{B}_{abc}$)

$$\frac{\Delta P}{\Delta V} - \mathbf{x}_t\,\mathrm{div}\mathbf{T} - \boldsymbol{\sigma} : \mathbf{D} = 2\boldsymbol{\tau}\,\boldsymbol{\omega} + \mathcal{M} \circ \mathrm{grad}\mathbf{D} + 2\mathbf{M} : \mathrm{grad}\boldsymbol{\omega}$$
$$+ \frac{1}{\Delta V}\sum_{\boldsymbol{\Lambda}} \int_{\partial \boldsymbol{\Delta} \mathbf{V}_{\boldsymbol{\Lambda}}} (\mathbf{x_t}\;[\![\mathbf{T}_{\boldsymbol{\Lambda}}^*]\!]\;\mathbf{n}_{\boldsymbol{\Lambda}})\; dS'_\Lambda, \tag{8.49}$$

where moments of the symmetric and antisymmetric stresses are, respectively,

$$\mathcal{M}\;\Delta V = \int_{\Delta V} \boldsymbol{\sigma}^* \otimes \mathbf{x}^*\; dV', \quad \mathbf{M}\;\Delta V = \int_{\Delta V}^{*} \boldsymbol{\tau}^* \otimes \mathbf{x}^*\; dV' \tag{8.50}$$

and vanish for homogeneous stress inside the RVE element. Finally, by means of (8.47) and (8.49), energy conservation equation (8.34) for an RVE obtains the following form:

$$
\begin{aligned}
&\rho\, u_t - \boldsymbol{\sigma} : \mathbf{D} - \operatorname{div} \mathbf{q} \\
&\quad = 2\boldsymbol{\tau}\, \boldsymbol{\omega} + \frac{1}{\Delta V} \sum_{\Lambda} \int_{\partial \Delta V_\Lambda} \left(\mathbf{x}_t \, [\![\mathbf{T}^*]\!] \, \mathbf{n} \right) dS'_\Lambda \\
&\quad + \rho D_t \operatorname{tr} \left\{ \mathbf{F}_t (\mathcal{J}_0 : \mathcal{F}_t^T) + (\mathcal{F}_t : \mathcal{J}_0) \mathbf{F}_t^T + \mathcal{F}_t : \mathbb{J}_0 : \mathcal{F}_t^T \right\} \\
&\quad + \mathcal{M} \circ \operatorname{grad} \mathbf{D} + 2\mathbf{M} : \operatorname{grad} \boldsymbol{\omega} + \rho D_t \operatorname{tr} \left\{ \mathbf{F}_t \mathbf{J}_0 \mathbf{F}_t^T \right\} .
\end{aligned}
\tag{8.51}
$$

Herein by $\mathbf{q}$ the RVE average heat flux vector is denoted, i.e., $\mathbf{q} = \langle \mathbf{q}' \rangle$.

8.3 Evolution and constitutive equations

8.3.1 Constitutive equations for stress and its moment

For the analysis in this section it is useful to express the energy equation (8.51) in terms of intermediate local reference (ν_t) configuration. Moreover, let us suppose that Condition 8.2.1 on intergranular continuity is fulfilled, i.e., that stress vector is continuous on grain boundaries.

Recalling the definition of the micro-stress $\mathbf{S}'$ we denote its symmetric and antisymmetric parts by $\boldsymbol{\sigma}'_S$ and $\boldsymbol{\tau}'_S$. On the other hand, elastic stretching is nothing but $D_t \mathbf{E}'_E$ while antisymmetric part of elastic "velocity gradient," denoted by $\mathbf{L}'_E := \boldsymbol{\Phi}'_t \left(\boldsymbol{\Phi}' \right)^{-1}$, is given by the vector $\boldsymbol{\omega}'_E$. Then the energy conservation equation may be transformed into

$$
\begin{aligned}
\rho_0\, u_t - \operatorname{div}_\nu \mathbf{q}_\nu - \boldsymbol{\sigma}_S : D_t \mathbf{E}_E &= \mathcal{M}_S \circ \operatorname{grad}_\nu D_t \mathbf{E}_E \\
&+ 2\boldsymbol{\tau}_S\, \boldsymbol{\omega}_E + 2\mathbf{M}_S : \operatorname{grad}_\nu \boldsymbol{\omega}_E + \rho_0 D_t \operatorname{tr} \left\{ \mathbf{F}_t \mathbf{J}_0 \mathbf{F}_t^T \right\} \\
&+ \rho_0 D_t \operatorname{tr} \left\{ \mathbf{F}_t (\mathcal{J}_0 : \mathcal{F}_t^T) + (\mathcal{F}_t : \mathcal{J}_0) \mathbf{F}_t^T + \mathcal{F}_t : \mathbb{J}_0 : \mathcal{F}_t^T \right\} \\
&+ R \left(\boldsymbol{\Pi}_t, \operatorname{grad}_\nu \boldsymbol{\Pi}_t \right),
\end{aligned}
\tag{8.52}
$$

where the last term on the RHS is linear in material time rates of plastic distortion and plastic distortion gradient tensors whereas

$$
\mathbf{q}_\nu = \left\langle \left(\boldsymbol{\Phi}' \right)^{-1} \mathbf{q}' \det \boldsymbol{\Phi}' \right\rangle
$$

is the average heat flux vector related to (ν_t) configuration, $\operatorname{grad}_\nu \mathcal{A} := \boldsymbol{\Phi} \operatorname{grad} \mathcal{A} \equiv \nabla_\nu \otimes \mathcal{A}$, $(\forall \mathcal{A})$ (cf. (1.71), page 24) $\rho_0 = \rho \det \boldsymbol{\Phi}$ and

$$
\mathcal{M}_S\, \Delta V_0 = \int_{\Delta V} \boldsymbol{\sigma}_S^* \otimes \boldsymbol{\xi}^*\, dV'_0, \quad \mathbf{M}_S\, \Delta V_0 = \int_{\Delta V}^{*} \boldsymbol{\tau}_S^* \otimes \boldsymbol{\xi}^*\, dV'_0. \tag{8.53}
$$

We want to exploit the above energy equation in order to get some constitutive restrictions for stresses and their moments. First the internal energy function must be analyzed in order to see on which arguments it depends. Let us start with Hooke's law for each grain and compose them into an RVE. For an arbitrary point inside RVE we obtain by means of (8.25) the following expression:

$$2u' = \mathbf{E}'_E : \mathbb{D}' : \mathbf{E}'_E = (\mathbf{E}_E + \mathbf{E}^*_E) : (\mathbb{D} + \mathbb{D}^*) : (\mathbf{E}_E + \mathbf{E}^*_E)\,, \tag{8.54}$$

where again $\mathbb{D}' = \mathbb{R}'_\Pi : \mathbb{D}_0 : \mathbb{R}'^T_\Pi = \varphi_\Lambda(\mathbf{x}')\,\mathbb{D}_\Lambda$ depends on grain orientations. Averaging the above relationship leads to

$$\begin{aligned} 2u = 2u_{HM} &+ \left(\mathbf{J}_\nu \mathrm{grad}_\nu \mathbf{E}_E : \mathbb{D}_0^T\right) \circ \mathrm{grad}_\nu \mathbf{E}_E \\ &+ \mathbf{E}_E : \langle \mathbb{D}^* \otimes \boldsymbol{\xi}^* \rangle \circ (\mathrm{grad}_\nu \mathbf{E}_E)^T \\ &+ \mathrm{grad}_\nu \mathbf{E}_E \circ \langle \boldsymbol{\xi}^* \otimes \mathbb{D}^* \rangle : \mathbf{E}_E \\ &+ \mathrm{grad}_\nu \mathbf{E}_E \circ \langle \boldsymbol{\xi}^* \otimes \mathbb{D}^* \otimes \boldsymbol{\xi}^* \rangle \circ (\mathrm{grad}_\nu \mathbf{E}_E)^T \end{aligned} \tag{8.55}$$

with notations

$$\mathbf{J}_\nu \equiv \langle \boldsymbol{\xi}^* \otimes \boldsymbol{\xi}^* \rangle \quad \text{and} \quad (\mathrm{grad}_\nu \mathbf{E}_E)^T_{\lambda\alpha\beta} = (\mathrm{grad}_\nu \mathbf{E}_E)_{\alpha\beta\lambda}\,.$$

Here by means of

$$2u_{HM} = \mathbf{E}_E : \mathbb{D}_0 : \mathbf{E}_E \tag{8.56}$$

we denote the *Hill–Mandel approximation of the internal energy function.* Taking into account Remark 8.3, page 151 it is worthy of note that the last term on the RHS of (8.55) contains also third-order products.

Suppose that inertial higher order terms in (8.52) are negligible. Thus, now our averaged internal energy function (8.55) contains only elastic strain and its gradient. Its differentiation and replacement into (8.52) leads to constitutive restrictions for symmetric part of stress tensor and its moment as follows:

$$\begin{aligned} \frac{1}{\rho_0}\boldsymbol{\sigma}_S &= \partial_{\mathbf{E}_E} u \\ &= \mathbb{D}_0 : \mathbf{E}_E + \langle \mathbb{D}^* \otimes \boldsymbol{\xi}^* \rangle \circ (\mathrm{grad}_\nu \mathbf{E}_E)^T\,, \end{aligned} \tag{8.57}$$

$$\begin{aligned} \frac{1}{\rho_0}\mathcal{M}_S &= \partial_{\mathrm{grad}_\nu \mathbf{E}_E} u \\ &= \mathbf{J}_\nu \mathrm{grad}_\nu \mathbf{E}_E : \mathbb{D}_0^T + \langle \boldsymbol{\xi}^* \otimes \mathbb{D}^* \rangle : \mathbf{E}_E \\ &\quad + \langle \boldsymbol{\xi}^* \otimes \mathbb{D}^* \otimes \boldsymbol{\xi}^* \rangle \circ (\mathrm{grad}_\nu \mathbf{E}_E)^T\,. \end{aligned} \tag{8.58}$$

As already remarked, antisymmetric stress and its gradient are negligible if higher order inertial terms can be disregarded. Two special cases of (8.57) and (8.58) are of special interest to us.

Remark 8.5 (Isotropic grains).

Suppose that each grain is elastically isotropic, which means that $\mathbb{D}' = \mathbb{D}_0\,(\lambda, \mu)$, or, equivalently, $\mathbb{D}^* = 0$. Then

$$
\begin{aligned}
2u &= \lambda\left(\mathrm{tr}\mathbf{E}_E\right)^2 + 2\mu\,\mathrm{tr}\left\{\mathbf{E}_E^2\right\} \\
&\quad + \lambda\mathbf{J}_\nu : \left[\mathrm{grad}_\nu\left(tr\mathbf{E}_E\right)\otimes\mathrm{grad}_\nu\left(tr\mathbf{E}_E\right)\right] \\
&\quad + 2\mu\left(\mathbf{J}_\nu\mathrm{grad}_\nu\mathbf{E}_E\right)\circ\mathrm{grad}_\nu\mathbf{E}_E .
\end{aligned}
\tag{8.59}
$$

The main advantage of this expression is that it needs only the Lame constants λ and ν. In other words it does not include any new material constant. The second-order microinertia tensor $\mathbf{J}_\nu$ may be calculated explicitly and in the special case when tessellation of an RVE into grains is by means of cubes we would have its diagonal terms equal to 1/12 whereas its off-diagonal terms would be 1/16, i.e.,

$$
\mathbf{J}_\nu = \frac{1}{48}\left\{\begin{matrix} 4\,3\,3 \\ 3\,4\,3 \\ 3\,3\,4 \end{matrix}\right\}.
$$

For instance, in the paper of Lubarda and Markenscoff (cf. [LM-00, equation (42)]) they have one additional constant and an unknown scaling length. As another example, Fleck and Hutchinson in [FH-97] have five additional material constants and apply so-called J_2 theory attempting to discuss their influence on overall inelastic behaviour of an RVE.

Remark 8.6 (Small grain rotations).

The second case of interest is *slight disorder* in Kröner's terminology (cf. [Kro80]). In other words, unlike Kröner's perfect disorder here relative grain rotations $\mathbf{R}' = \mathbf{1} + \mathbf{R}^*$ are so small that higher order products of $\mathbf{R}^*$ in $\mathbb{D}'$ may be neglected. Then

$$
D'^{\alpha\beta\gamma\delta} = D_0^{\alpha\beta\gamma\delta} + R^{*\alpha}_{\cdot\kappa}\,D_0^{\kappa\beta\gamma\delta} + R^{*\beta}_{\cdot\kappa}\,D_0^{\alpha\kappa\gamma\delta} + R^{*\gamma}_{\cdot\kappa}\,D_0^{\alpha\beta\kappa\delta} + R^{*\delta}_{\cdot\kappa}\,D_0^{\alpha\beta\gamma\kappa}.
$$

In this case an approximate symmetry of the whole RVE may be established around average structural vectors. It would be of special interest to make an inquiry about so-called *defective invariants* with respect to such average directions.[10]

8.3.2 Nonlocal hardening and microscopic evolution equation

Let us recall that in this chapter the intention has been to consider the simplest inelastic polycrystals composed of individual grains having the same density and are subject to homogeneous micro-strains. They have been approximated by linear distributions of total, elastic and plastic strains inside each RVE. Following this approach at this point it seems plausible to pose the following question:

[10] Recent papers by Rychlewski, Ostrowska-Maciejewska and Kowalczyk seem to be very helpful in treating this issue.

Which is the simplest yet reasonable approach to gradient viscoplasticity of polycrystals?

1. Let us recall the structure of microstructural evolution equations in the paper [KG-04] treating rate independent plasticity of polycrystals. In this paper such equations read (for α-slip system in Γ-grain)

$$D_t\gamma_\alpha = \frac{\lambda_G}{Y_\alpha}\left(\frac{\tau_\alpha}{Y_\alpha}\right)^{2m-1}, \tag{8.60}$$

with the yield condition for Γ-grain

$$f_\Gamma(\mathbf{S},\varphi_\Gamma) = \frac{1}{2m}\left(\sum_\alpha \left(\frac{\tau_\alpha}{Y_\alpha}\right)^{2m} - m_G\right). \tag{8.61}$$

Herein φ_Γ are Euler angles of crystallographic directions of the grain, $m_G \gg 1$ is a material constant (cf. also model in Section 7.3.1), τ_α is the resolved shear stress, Y_α is *critical resolved shear stress* for α-th slip system. The coefficient λ_G is found from the so-called consistency condition when simultaneously we have $f_\Gamma = 0$ and $D_t f_\Gamma = 0$. When this condition holds, then all the slip systems are active but those $D_t\gamma_\alpha$ with $\tau_\alpha < Y_\alpha$ are negligible due to large value of the exponent m_G. For notational simplicity index Γ is suppressed in τ_α and Y_α. This approach gives smooth yield surfaces without corners lying between Tresca and Huber-Mises yield surfaces. Hardening has a self and latent component according to the hardening law ([Asa83]):

$$D_tY_\alpha = \sum_\beta H^0_{\alpha\beta}|D_t\gamma_\beta|, \tag{8.62}$$

where $H_{\alpha\beta} > H_{\alpha\alpha}$ ($\alpha \neq \beta$). The model predicts interaction of slip systems but within a single grain without grain-to-grain interaction.

2. Mutual influence of slip systems in different grains may be covered by the *latent nonlocal hardening* introduced in [MBK93]. Instead of the differential equation (8.62) the authors proposed the following integro-differential equation:

$$D_t\tau_\alpha(\mathbf{x}) = \int_V H_{\alpha\beta}(\mathbf{x},\mathbf{x}')D_t\gamma_\beta(\mathbf{x}')dV', \tag{8.63}$$

connecting in such a way rate of resolved shear stress at a point with all the points of the considered body.

3. However, such an equation is too general and not practical for our low-order inelastic polycrystals. Instead of (8.63) in the paper [Mic02a] the critical resolved shear stress is governed by a similar integro-differential equation:

$$\begin{aligned} D_tY_{\alpha\Gamma}(\mathbf{x}) = \int_V \Big(&H^0_{\alpha\beta}(\mathbf{x}')\delta(\mathbf{x}-\mathbf{x}') \\ &+ \eta_{\alpha\beta}\exp(-|\mathbf{x}-\mathbf{x}'|^p/d^p)\Big) D_t\gamma_{\beta\Gamma}(\mathbf{x}')dV', \end{aligned} \tag{8.64}$$

with $p = 1$. In the special case when the *characteristic distance* d (analogous to *correlation length* in [Kro80]) vanishes, i.e., when $d \mapsto 0$, the pure locality described by (8.62) is restored. Then only the delta-function in (8.64) is used to describe interaction of slip systems. From the statistical point of view such a situation is called in [Kro80] *perfect disorder*.

Otherwise nonlocality is not negligible and increases with growth of d. If we want the nonlocality restricted within an RVE, the choice of d and p is to be such that outside the RVE the second term in (8.64) becomes negligible. In this case larger values for p might be chosen and d is approximately typical RVE diameter.[11] Obviously not only exponential function is acceptable. The other functions are also quite legitimate. For instance, if tessellation of an RVE is such that it is composed of cubes, then a piecewise linear function which vanishes outside RVE could be also convenient for numerical purpose.

Suppose now that the arguments of Section 7.3.1 hold and that a plastic shear rate at Γ-grain and α slip system is expressible by

$$D_t\gamma_{\alpha\Lambda} = D_t\sigma_{eq}\exp(-\mathcal{M})\ (\tau_{\alpha\Lambda}/Y_{\alpha\Lambda})^m \operatorname{sgn}(\tau_{\alpha\Lambda}) \tag{8.65}$$

which follows from (7.34) and (7.36) for quasi-rate-independent polycrystals. Substituting (8.65) into (8.64) this becomes very complex:

$$\begin{aligned} &\exp(\mathcal{M})\, D_t Y_{\alpha\Gamma}(\mathbf{x}) \\ &\quad = \int_V \Big(H^0_{\alpha\beta}(\mathbf{x}')\delta(\mathbf{x}-\mathbf{x}') + \eta_{\alpha\beta}\exp(-|\mathbf{x}-\mathbf{x}'|^p/d^p)\Big) \\ &\quad \times D_t\sigma_{eq}(\mathbf{x}') \left(\frac{\tau_{\alpha\Lambda}(\mathbf{x}')}{Y_{\alpha\Lambda}(\mathbf{x}')}\right)^m sgn\big(\tau_{\alpha\Lambda}(\mathbf{x}')\big)dV', \end{aligned} \tag{8.66}$$

unless only a very slight nonlocality is kept. Now, the simplest version of evolution equation is obtained by means of the formula

$$\mathbf{D}_P \approx \frac{1}{N}\sum_{\Lambda}\sum_{\alpha} \left(\mathbf{A}^T_{\alpha\Lambda} + \mathbf{A}_{\alpha\Lambda}\right) D_t\gamma_{\alpha\Lambda} \tag{7.20'}$$

derived in the previous section. Of course, a lot of improvements and discussions about possible normality starting from a micro-potential behaviour $D_t\gamma_{\alpha\Gamma} = \Lambda\, \partial\Omega/\partial\tau_{\alpha\Gamma}$ are possible.[12]

[11] Let $\Delta V_\Gamma^{1/3} \equiv l_\Gamma$ be the typical RVE *diameter*. If we choose $p = 2$, then for $d = 3\, l_\Gamma$ the interaction on the RVE boundary is 1.25×10^{-4} less than in its center. If the exponent p has the value 10, then the same interaction on the RVE boundary would be for $d = 1.25\, l_\Gamma$ and the shape of the exponential function in (8.64) is almost rectangular.

[12] The potential behaviour following (5.19) and (7.21) seems here almost improbable if the highly nonlinear interaction (8.66) holds. Thus, it should be concluded that in the case of elastic heterogeneity and plastic nonlocality an associativity of flow rule is even less correct than in the case of local viscoplastic materials discussed in Section 3.1.2 on page 53.

8.4 Notes on micromorphic polycrystals

This chapter adds new finer microscopic details to the issue of polycrystals. The following general conclusions might be drawn from the above analysis:

- The simplest case of higher gradient theory with couple stresses is discussed. Homogeneous grain strains are composed into the resulting behaviour of representative volume element and the corresponding homogenization of total, plastic and elastic strains has been done.
- Besides the constitutive equation (derived in the previous section) relating the RVE effective stress with the RVE effective elastic strain, being derived by means of the effective field method, some new constitutive equations for symmetric part of stress and its moment have been derived. It is important to note that for the special case of isotropic grains inside RVE new material constants apart from Lame constants are not necessary.
- Nonproportionality of strain path is connected to microsymmetry properties of the microinertia tensor. Such a conclusion seems interesting in light of the existence of couple stresses.
- The transition from micro inelastic behaviour to the corresponding mesobehaviour is performed. Balance laws as well as constitutive equations with minimal number of material constants have been derived with special account of confrontation of homogeneous grain strains and a linear smoothing homogenization results.
- Although successfully compared to experimental results, evolution equations based on Vakulenko's concept of thermodynamic time and tensor representation theory have to be connected to inelastic slip micromechanisms following the same approach as for constitutive equations for stress and moment stress.
- It is challenging to apply the self-consistent approach to strain gradient viscoplasticity of polycrystals with couple stresses. Then a comparison with results of Chapter 7 could be fruitful.

Conclusions Related to Parts I and II

As a continuous attempt in each section, geometry has been a central link serving to build an appropriate treatment of the problem treated. Another support must be an irreversible thermodynamics. In the opinion of the author, viscoplasticity has two legs—geometry and thermodynamics. Without either of them it would be stumbling—disabled for walking forward properly. The cruel judge who looks at such a walk is the experiment. It decides whether the theory is just blind or approximately correct. For this reason the section devoted to experiments has a very important role in this book.

An attempt has been made here to formulate the fundamental issues of viscoplasticity in a comprehensive yet concise way. The following general conclusions might be drawn from the analysis given in preceding sections:

- Throughout the text three approaches for treating meso-rotations of an RVE in the intermediate local reference configuration are given. The first is Mandel's approach by adopting isoclinicity of the crystal structure with some fixed directions (Mandel's "directors"). The second is the approach of the author of this book. It consists of eliminating plastic meso-rotation of the whole RVE. The third consists of elimination of elastic rotations. The first two approaches are much better for a proper formulation of material symmetries. Both, when properly formulated, clearly lead to the conclusion that evolution equation for plastic meso-spin is just a consequence of the evolution equation for plastic meso-stretching. Such a conclusion is very important since it eliminates troublesome introduction of new material constants for the evolution equation for plastic meso-spin.

- The principle of determinism (or, in other words, causality) based on damage-plastic deformation history is formulated. As a consequence, viscoplastic materials of integral type as well as of differential type are obtained from description based on functionals. A general nonassociate flow rule for materials of viscoplastic differential type is formulated on the basis of tensor function representation.
  The causality principle should not have only total deformation history since such a history is blind for structural rearrangements and defects, which are essential players in the process of inelasticity. This is in agreement with the fact that by its very definition elastic deformation is not hereditary.
- A careful analysis has shown that normality in all its versions is just a handy assumption simplifying evolution equations. It is shown that such a normality cannot be a consequence from postulate of extremal plastic power on real trajectories in stress-plastic strain space. An associativity of flow rule could be approximately applied only for very small plastic strains. It is made clear by introducing the nonassociativity tensor. For metal forming problems the tensor representation approach is inevitable if we want to have a proper description of plastic strain induced anisotropy.
- Considerable attention has been paid to irreversible thermodynamics: classical, rational, two versions of extended and endochronic as well as applications of statistical thermodynamics to inelasticity. It seems that rational thermodynamics is too formal applying beautiful methods of analysis but without many results in inelasticity. However, its method of analysis is really enlightening. Extended thermodynamics has also many equations but small practicality. In the author's opinion the endochronic thermodynamics with properly chosen thermodynamic time is the real answer to modeling needs—especially in coupled field problems like inelasticity of ferromagnetics, neutron irradiation, etc. Statistical thermodynamics is a mighty tool in treating choice of internal variables.
- From the section devoted to experiments we may conclude that introduction of the universal viscoplastic constant into a standard viscoplastic evolution equation for plastic stretching is very important permitting its scaling from static until almost impact values. It turns out that inelastic materials behave according to quasi rate independence—incremental evolution equation with stress rate dependent initial yield stress. It is remarkable that tensor representation applied to such materials gives rise to a model with small number of material constants which shows a very good agreement with uni-tension, shear and bi-tension tests from very low until very high strain rates.
  It must be mentioned that the so-called *universal flow curve* even when corrected by kinematic hardening is not capable of covering multiaxial stress–strain histories. A modification of such theories towards an inclusion of the thermodynamic time would give them a strong breadth. Without the concept of thermodynamic time the ability to describe low, medium

and high strain rates within a single evolution equation is very poor. On the other hand, it is not clear how to find the direction tensor for cruciform specimen when such theories are applied. Thus they cannot describe properly directionality or real inelastic stress–strain–strain rate histories.

- Viscoplasticity of ferromagnetics has been treated by tensor representation applying either nonassociativity with extended thermodynamics or generalized normality which includes orthogonality of residual magnetization rate on some loading surface. Small magneto-elasto-viscoplastic strains were considered in detail in order to analyze magnetomechanical interaction at low-cycle fatigue. Furthermore, endochronic thermodynamics with Vakulenko's thermodynamic time helped us to account for (experimentally observed) time delay between stress and magnetic field histories. Although at this stage it is a rather modest contribution to the subject[13] it is believed that such results could be useful in inelastic testing with magnetic fields either induced or applied. Some other results are listed at the end of the sixth section.
  The deformation geometry is extended here to include magnetostriction connecting to the natural state intermediate reference configuration isoclinic magnetization vectors. The approach is the same as in continuum theory of dislocations. It is worthy of note that the approach permits a promising treatment of viscoplastic polycrystalline ferromagnetics.
- The essential difference between micro and meso-spin has as the origin constrained micro-rotations of grains inside a representative volume element. Since an RVE, having volume of an infinitesimal volume element, cannot be disintegrated any more, micro-spin does not follow from plastic micro-stretching. In treating viscoplastic polycrystals such an idea has been proved to be very successful. Namely, elastic micro-strains have been assumed to be covered by the self-consistent method (effective field approach by Levin) whereas for plastic stretching a quasi-rate-independent incremental meso-evolution equation is postulated. The rate dependence occurs by means of stress rate dependent value of the initial yield stress. The meso-scopic inelastic evolution obeys Vakulenko's concept of thermodynamic time. The meso-evolution equation for plastic spin of the RVE is an outcome of evolution equation for plastic stretching. The same does not hold for plastic micro-spin.
  The theory is afterwards applied to some characteristic stress histories of slightly disordered fcc polycrystals. Numerical comparison with J_2 theory has proved that such a theory is not able to distinguish among fast and slow processes being in such a way rate independent by its very nature. However, it could be improved if thermodynamic time is introduced but the problem of directionality cannot be cured for such a theory.

[13] As has already been noted, the major contributions to diverse issues of continuum mechanics of ferromagnetic materials have been given by Maugin and his collaborators.

This paper has opened a way for treating symmetry issues of slightly disordered inelastic polycrystals. Namely, before applying correct tensor generators [Spe71, Ryc83] we must know which type of symmetry the considered polycrystal possesses. Such materials appear, for instance, after rolling and, consequently, forming limit diagrams should benefit from the results along this line of reasoning.

- The last section of this part of the book adds new finer microscopic details to the issue of polycrystals. A simplest case of gradient viscoplasticity theory of polycrystals with couple stresses is discussed. The simplicity derives from homogeneous grain strains inside an RVE. New constitutive equations for symmetric part of stress and its moment have been derived. Balance laws as well as constitutive equations with minimal number of material constants have been derived with special account of confrontation of homogeneous grain strains and a linear smoothing homogenization results. It is important to note that for the special case of isotropic grains inside the RVE new material constants for polycrystal elasticity apart from Lame constants are not necessary.
  Nonproportionality of strain path is connected to microsymmetry properties of the microinertia tensor. Such a conclusion seems interesting in light of the existence of couple stresses.
  An extensive analysis of local and nonlocal micro-evolution equations is made. Some of them are quasi-rate-independent. The transition from inelastic micro-behaviour to the corresponding meso-behaviour is performed. Although successfully compared to experimental results, meso-evolution equations based on Vakulenko's concept of thermodynamic time and tensor representation theory have to be connected more firmly to intergranular inelastic slip micromechanisms. Then better and more general constitutive equations for stress and moment stress would be possible.

There are many subjects dropped from this monograph due to brevity of exposition. To mention just a few of them: practical aspects of damage, fracture mechanics, configurational forces, thermo-ratchetting, plastic waves, localized instability, metal forming applications, neutron irradiation, etc. More explicitly, some of these problems treated by the theory of quasi-rate-independent viscoplasticity have been:

- The theory of viscoplasticity of irradiated steels has been reported by the author at IUTAM symposium in Krakow ([Mic91b]). Although very actual for problems of strength of reactor steel serving for building the first wall (like AISI 316H), it has not been included here to save space and due to its resemblance to viscoplasticity of ferromagnetics. For the same reason, all the other more practical damage issues like those treated in the paper by [Mic03] have not been included. Nevertheless, the geometric approach presented in the first section and thermo-inelastic memory concept of the second section are broad enough to describe properly different types of damage.

- Localized instability of stainless steel sheets with nonproportional dynamic stress histories over a wide range of plastic strain rates have been analyzed in the paper [KM-01]. It served for comparison of tensor representation approach with J_2 approach with the so-called universal flow curve.
- Thermo-ratchetting with 3D viscoplastic strain was an excellent problem to test applicability of PCR and MAM models ([MV-98]). Comparison with experiments has shown much better agreement of the latter.
- The Hopkinson bar apparatus consisting of two elastic long bars enclosing much smaller specimen was analyzed in [MB-02]. Unlike traditional viscoplasticity, dealing with J_2-based approach without thermodynamic time, where only elastic wave speeds are met, the quasi-rate-independent model permitted visco-plastic wave speeds in the obtained hyperbolic wave equation.

The first two issues are not included here whereas the last two are shortly reviewed in the next two chapters of the book.

Part III

Applications of the Theory

9 Plastic Wave Propagation in Hopkinson Bar

The focus of this chapter (cf. [MB-02]) is an analysis of the experimental Hopkinson bar technique when such a device consists of a short tensile or shearing specimen surrounded by two very long elastic bars (explained in detail in [AM-79]). Unlike commonly applied bypass analysis which attempts to draw conclusions from behaviour of elastic bars, an attempt is made to take into account real plastic waves inside the specimen with a few hundred reflections. A quasi-rate-independent tensor function model for AISI 316H calibrated in [Mic97] is applied in its simplest yet nonlinear version. Some special slightly perturbed elastic incident and reflected waves in elastic bars serve here to simulate starting solutions. The numerical results show a good agreement with experimentally observed homogeneous strain state throughout the specimen during the process. Lindholm's procedure for finding specimen stress and strain by such a bypass procedure is criticized. An iterative procedure aimed to improve such a procedure is briefly discussed.

In this chapter, we wish to revisit standard techniques for analysis of the Hopkinson bar testing technique taking into account plastic wave propagation inside the standard (extremely short) tension specimen as well as elastic waves propagating along the very long incident-reflected wave bar as well as the transmitted wave bar. The strains inside the specimen are large up to 60%. The evolution equation for plastic stretching tensor as calibrated in [Mic97] on the basis of experiments performed in the dynamic testing laboratory of JRC-Ispra, Italy ([AM-79, AMP91, AMM91]) is applied. The tests have been performed using classical tension specimen as well as *bicchierrino*

M.V. Mićunović, *Thermomechanics of Viscoplasticity: Fundamentals and Applications*, Advances in Mechanics and Mathematics 20, DOI: 10.1007/978-0-387-89490-4_9,
© Springer Science + Business Media, LLC 2009

shear specimen[1] made of austenitic stainless steel AISI 316H in the range of strain rates $[10^{-3}, 10^{3}]\mathrm{s}^{-1}$.

9.1 Brief preliminaries

Geometry of deformation has been discussed widely in previous chapters. Thus we will not repeat its details now. We apply again Kröner's decomposition rule ([Kro60]):

$$\mathbf{F} = \mathbf{F}_E \mathbf{F}_P \tag{9.1}$$

where $\mathbf{F}$ is the deformation gradient tensor, $\mathbf{F}_E$ the elastic distortion tensor and $\mathbf{F}_P$ the plastic distortion tensor mapping, respectively, $(\chi_0) \to (\chi_t)$, $(\nu_t) \to (\chi_t)$ and $(\chi_0) \to (\nu_t)$. The term "distortion" is used to underline the fact that $\mathbf{F}_E$ and $\mathbf{F}_P$ are incompatible, i.e., compatibility conditions applied to a metric tensor of (ν_t) are not satisfied (cf., e.g., [Mic74b]).

According to the assumption that the elastic strain is caused and escorted by the corresponding stress tensor, Hooke's law holds for the mapping $(\nu_t) \to (\chi_t)$ and it should be written in an invariant way connected to the intermediate referential configuration (ν_t). For this aim aside from the stress tensor present in (χ_t) configuration, called Cauchy stress (or "true" stress), we quote also the first and second Piola–Kirchhoff stress tensor ([TN-65]):

$$\mathbf{T}_R = \det(\mathbf{F})\,\mathbf{T}\mathbf{F}^{-T}, \quad \mathbf{S} = \det(\mathbf{F_E})\,\mathbf{F}_E^{-1}\mathbf{T}\mathbf{F}_E^{-T}, \tag{9.2}$$

respectively. The first is connected to (χ_t) and (χ_0) and often it is termed engineering stress, whereas the second is referred to as the natural state local configuration (ν_t). If elastic strain is much smaller than finite total strain, then Hooke's law reads

$$\mathbf{S} = \mathbb{D}\colon \mathbf{E}_E. \tag{9.3}$$

Fourth-rank tensor $\mathcal{H}$ consisting of material constants should depend in general on temperature as well as on principal invariants of the (traceless) plastic strain tensor:

$$\pi_2 = \mathrm{tr}\left\{\varepsilon_P^2\right\}, \quad \pi_3 = \mathrm{tr}\left\{\varepsilon_P^3\right\}., \tag{9.4}$$

where logarithmic plastic strain is equal to $2\,\varepsilon_P = \log(\mathbf{F}_P\mathbf{F}_P^T)$. Two basic tensor constituents of the subsequent evolution equations are the plastic strain tensor and the second Piola–Kirchhoff stress tensor defined above. Thus, the relevant tensor generators read (the subscript d is used to denote the deviatoric part of a second-rank tensor) ([TN-65, Mrk79])

$$\begin{gathered}\mathbf{H}_1 = \mathbf{S}_d,\;\; \mathbf{H}_2 = \left(\mathbf{S}_d^2\right)_d,\;\; \mathbf{H}_3 = \varepsilon_P,\;\; \mathbf{H}_4 = \left(\varepsilon_P^2\right)_d, \\ \mathbf{H}_5 = \left(\mathbf{S}_d\varepsilon_P + \mathbf{e}_P\mathbf{S}_d\right)_d,\;\; \mathbf{H}_6 = \left(\mathbf{S}_d\varepsilon_P^2 + \varepsilon_P^2\mathbf{S}_d\right)_d, \\ \mathbf{H}_7 = \left(\mathbf{S}_d^2\varepsilon_P + \mathbf{e}_P\mathbf{S}_d^2\right)_d,\end{gathered} \tag{9.5}$$

[1] It consists of two rigid cylinders connected by the gage part—a thin circular crown explained in detail in the fifth chapter.

while the corresponding principal and mixed invariants will also be necessary in the sequel:

$$
\begin{aligned}
s_2 = \operatorname{tr}\left(\mathbf{S}_d^2\right), \quad s_3 = \operatorname{tr}\left(\mathbf{S}_d^3\right) \quad \mu_1 = \operatorname{tr}\left(\mathbf{S}_d \boldsymbol{\varepsilon}_P\right), \\
\mu_2 = \operatorname{tr}\left(\mathbf{S}_d \boldsymbol{\varepsilon}_P^2\right), \quad \mu_3 = \operatorname{tr}\left(\mathbf{S}_d^2 \boldsymbol{\varepsilon}_P\right), \quad \mu_4 = \operatorname{tr}\left(\mathbf{S}_d^2 \boldsymbol{\varepsilon}_P^2\right), \\
\gamma \equiv \left\{s_2, s_3, \pi_2, \pi_3, \mu_1, \mu_2, \mu_3, \mu_4\right\}.
\end{aligned}
\tag{9.6}
$$

As usual, some of the above principal invariants are used here to denote intensities of the corresponding tensors

$$
S = \sqrt{s_2}, \quad \dot{\pi} = \left(\mathbf{D}_P : \mathbf{D}_P\right)^{1/2}, \quad \pi = \int_0^\tau \dot{\pi}\left(\tau'\right) \, d\tau'. \tag{9.7}
$$

In the terminology of experimental plasticity in slightly different form

$$
\sigma^{eq} = S\sqrt{\frac{3}{2}}, \quad \dot{\varepsilon}_P^{eq} = \dot{\pi}\sqrt{\frac{2}{3}}, \quad \varepsilon_P^{eq} = \pi\sqrt{\frac{2}{3}}, \tag{9.8}
$$

they are commonly termed *equivalent stress*, *equivalent plastic strain* (i.e., accumulated plastic strain) and *equivalent plastic strain rate.*

9.2 Evolution equations of the problem

As already mentioned in the fifth chapter, it has been known by experimentalists for a long time that initial yield stress depends on strain rate or on stress rate such that at higher stress rates initial yield stress is larger. On the other hand, Rabotnov in his book suggested that there exists the phenomenon of delayed yielding inherent in some metals and alloys, i.e., it means that stress exceeds its static value after a certain time interval called *delay time* elapses. According to such an assumption, in the paper [Mic97] the integral equation

$$
\pi\left(\tau\right) = \int_0^\tau J\left(\tau - \tau'\right) \frac{DS\left(\tau'\right)}{D\tau'} d\tau' \equiv \int_0^\tau \psi\left(\tau, \tau'\right) \, d\tau' \tag{9.9}
$$

was postulated and calibrated.

If plastic deformation commences at time τ^* such that initial stress time rate equals

$$
S\left(\tau^*\right) \equiv Y_0 \left(\left. \frac{DS\left(\tau\right)}{D\tau} \right|_{\tau=\tau^*} \right), \tag{9.10}
$$

then the initial yield stress depends on the initial time rate of stress. Accordingly, the kernel in the above integral equation should read

$$
J\left(\tau - \tau'\right) = \begin{cases} 0, & \tau < \tau^*, \\ \exp\left(-\mathcal{M}\right), & \tau \geq \tau^*. \end{cases} \tag{9.11}
$$

Applying this expression for kernel to the above integral equation, the following representation is acquired:

$$\dot{\pi}(\tau) = J(0)\,\frac{DS(\tau)}{D\tau} = \begin{cases} 0, & \tau < \tau^*, \\ \exp(-\mathcal{M})\sqrt{2/3}\;\dot{\sigma}^{eq}(\tau), & \tau \geq \tau^*. \end{cases} \tag{9.12}$$

The integral appearing in (9.9) is the Riemann integral. Indeed, it is not difficult to show that it is uniformly bounded on $[0,\tau]$. On the other hand, a linear relationship between $D\pi/D\tau$ and $DS/D\tau$ was found in [Mic97] in the form

$$\frac{D\pi}{D\tau} = \exp(-\mathcal{M})\,\frac{DS}{D\tau} \tag{9.13}$$

with a material constant $\mathcal{M}$ holding both for tension and shear of AISI 316H such that ([Mic97])

$$\mathcal{M}_{tension} \approx \mathcal{M}_{shear} \approx \mathcal{M} = 6.8645. \tag{9.14}$$

The agreement of these values with the corresponding values obtained in the case of tension as well as shear is considerable, i.e., discrepancy between $\mathcal{M}_{tension}$ and $\mathcal{M}_{shear}$ amounts to approximately 0.3% for a very large range of strain rates $D\pi D\tau \in \left[10^{-3}, 10^{3}\right]\,\mathrm{s}^{-1}$. Thus it is expected that $\mathcal{M}$ is a material constant for AISI 316H.[2]

If the triggering value of the invariant S (cf. $(9.7)_1$) where plasticity onset occurs is denoted by Y, then the simplest nonlinear dependence on the initial stress rate could be given by the equation

$$Y = Y_0 + Y_1\left(\left.\frac{DS(\tau)}{D\tau}\right|_{\tau=\tau^*}\right)^m. \tag{9.15}$$

Its statical value Y_0 depends on the accumulated plastic strain accounting in such a way for the strain hardening effect. The other two quantities appearing above, namely, Y_1 and m, are taken to be constants giving rise to the simplest means of nonlinear stress rate hardening.

It has been shown by experiments on AISI 316H (by means of traditional tension specimen, *bicchierino*-type specimen as well as cruciform specimen) that the plastic stretching tensor is not perpendicular to yield surface (cf. [AMM91]). Taking such evidence into account, some constitutive models have been compared and calibrated in the paper [Mic97].[3] Since such deviation from normality is not large, as relatively simple and yet general enough to be concordant with experiments, the normality model introduced by Rice in [Ric71] based on a loading function normality and generalized to tensor functions is accepted here. Similar evolution equation was derived by Ziegler from the principle of least irreversible force. Such an equation reads

---

[2] Instead of $\mathcal{M}$ the value of this material constant may be more conveniently expressed (for $\tau \geq \tau^*$) by means of the integral kernel $J(0) \approx 1.044 \times 10^{-3}\mathrm{MPa}^{-1}$ allowing for explicit dimensions.

[3] In the fifth chapter, calibration of tensor representation type models with larger number of material constants has been made. Anyway, for the subsequent analysis this, the simplest, nonlinear model is sufficient.

$$\mathbf{D}_P = \Lambda \left(\frac{\partial \Omega}{\partial \mathbf{S}} \right)_d . \tag{9.16}$$

This relationship has been made explicit in the papers [Mic96b, Mic01] in such a way to include dependence of the loading function on stress tensor and plastic strain as representatives of Rice's PIR (pattern of internal rearrangements). More precisely,

$$\Omega = \hat{\Omega}\left(\mathbf{S}, \varepsilon_P\right) = \tilde{\Omega}\left(\gamma\right) \tag{9.17}$$

and stress derivatives of this function necessarily lead to the tensor generators (9.5).

On the other hand, the above consideration about time delay of plastic yielding and dependence of the yield stress on stress rate allow further specialization of (9.16) by means of

$$\Lambda = \frac{DS}{D\tau} J\left(0\right) \phi\left(\pi\right). \tag{9.18}$$

The simplest evolution equation nonlinear in stress tensor then reads (cf. (9.5))

$$\mathbf{D}_P = \frac{DS}{D\tau} \exp\left(-\mathcal{M}\right) \eta\left(S - Y\right) \phi\left(\pi\right) \left(c_1 \mathbf{S}_d + c_2 \left(\mathbf{S}_d^2\right)_d\right), \tag{9.19}$$

where $\eta\left(S - Y\right) = 1$ for $S > Y$ and $\eta\left(S - Y\right) = 0$ otherwise (the Heaviside function) and the strain function $\phi\left(\pi\right)$ might be either unity or some function aimed to take into account strain hardening such as

$$\phi\left(\pi\right) = \pi^{\lambda}$$

where, obviously for $\lambda \neq 0$, we have a nonlinear π-dependence. Such a model with only four material constants $\{\mathcal{M}, c_1, c_2, \lambda\}$ was calibrated in [Mic97] permitting a high correlation coefficient 0.9683 for tension and shear in the very large strain rate range $D\pi/D\tau \in \left[10^{-3}, 10^{3}\right]$ s^{-1}. The evolution equation (9.19) may be written as follows:

$$\mathbf{D}_P = \frac{DS}{D\tau} \exp\left(-\mathcal{M}\right) \eta\left(S - Y\right) \phi\left(\pi\right) \sum_{\alpha=1}^{7} \Gamma_\alpha\left(\gamma\right) \mathbf{H}_\alpha, \tag{9.20}$$

where the loading function

$$\Omega = \frac{1}{2} c_1 s_2 + \frac{1}{3} c_2 s_3$$

leads to $\Gamma_1 = c_1, \Gamma_2 = c_2, \Gamma_\alpha = 0\,(\alpha > 2)$ while the tensor generators $\mathbf{H}_\alpha\,(\alpha = 1, \ldots, 7)$ are shown above by (9.5). Such a model could be termed *quasi-rate-independent*. This means that if we multiply (9.19) by $d\tau$, then this equation becomes incremental. However, it should be taken into account that Y depends on $DS/D\tau$ which means that time rates influence the plastic stretching tensor.

More general rate-dependent model in its simplest form might be given by ([Mic97])

$$\mathbf{D}_P = \eta\left(S - Y\right)\phi\left(\pi\right)\sum_{\alpha=1}^{7}\left(\Gamma_\alpha\left(\gamma\right)\frac{DS}{D\tau}\exp\left(-\mathcal{M}\right) + \Gamma_\alpha^{\#}\left(\gamma\right)\right)\mathbf{H}_\alpha, \tag{9.21}$$

where, additionally, $\Gamma_1^{\#} = c_3, \Gamma_2^{\#} = c_4, \Gamma_\alpha^{\#} = 0\,(\alpha > 2)$. Necessarily, by experimental evidence, the constants c_3, c_4 must be much smaller than c_1, c_2 (cf. (9.13)). Its advantage with respect to (9.20) is that for slow processes it covers the case when stress rate vanishes while inelastic creep strain rate is different from zero. In our considerations dealing with high strain rates it is not so important to take these additional terms into account.

9.3 Longitudinal plastic waves

Consider inelastic deformation in an isotropic straight cylindrical bar with circular cross-section whose longitudinal material coordinate is $\zeta \in [0, L]$ and the other material coordinates ξ, η are also Cartesian. It is assumed in the sequel that during all the considered time interval $\tau \in [0, T]$ deplanation of cross-sections is negligible. Thus, all material points belonging initially to a normal cross-section belong to the same section during all of the motion. Therefore, $\zeta = \text{const}$ stands for a cross-section with such fixed material points. Moreover, it is assumed that shears are also negligible. Then deformation gradient tensor and plastic distortion tensor have the following forms:

$$\mathbf{F} = \begin{Bmatrix} 1+\omega & 0 & 0 \\ 0 & 1+\omega & 0 \\ 0 & 0 & 1+\varepsilon \end{Bmatrix}, \tag{9.22}$$

$$\mathbf{F}_P = \begin{Bmatrix} \left(1+\varepsilon_P\right)^{-1/2} & 0 & 0 \\ 0 & \left(1+\varepsilon_P\right)^{-1/2} & 0 \\ 0 & 0 & 1+\varepsilon_P \end{Bmatrix}.$$

Logarithmic plastic strain tensor and plastic stretching are then obtained as follows:

$$\boldsymbol{\varepsilon}_P = \sqrt{\frac{3}{2}}\ln\left(1+\varepsilon_P\right)\mathbf{N}, \quad \text{where} \quad \mathbf{N} = \sqrt{\frac{1}{6}}\begin{Bmatrix} -1 & 0 & 0 \\ 0 & -1 & 0 \\ 0 & 0 & 2 \end{Bmatrix}, \tag{9.23}$$

such that

$$\mathbf{D}_P = \dot{\pi}\,\mathbf{N} \quad \text{with} \quad \dot{\pi} \equiv \left\|\mathbf{D}_P\right\| = \left(\mathbf{D}_P : \mathbf{D}_P\right)^{1/2} = \sqrt{\frac{3}{2}}\frac{D}{D\tau}\ln\left(1+\varepsilon_P\right). \tag{9.24}$$

In such a case of special geometry and strain conditions we have $\mathbf{D}_P = D\varepsilon_P/D\tau$. Of course, such a relationship would not hold in the general case. The unit tensor $\mathbf{N}$ with the properties $\|\mathbf{N}\| = (\mathbf{N}:\mathbf{N})^{1/2} = 1$ is here introduced for convenience.

If only longitudinal Cauchy stress component $T_{33} \equiv \sigma$ differs from zero, then from Hooke's law written with respect to (ν_t) configuration, i.e.,

$$\mathbf{E}_E = \frac{1}{E}\left((1+\nu)\ \mathbf{S} - \nu\,\mathbf{1}\ tr\mathbf{S}\right) \tag{9.25}$$

$(E, \nu$ are elastic constants for isotropic body) we get the only nonzero components of Piola–Kirchhoff tensors $(9.2)_2$ in the form

$$S_{33} = E\,E_{E33} = \sigma\,(1+\omega)^2\,\frac{(1+\varepsilon_P)^2}{1+\varepsilon},$$
$$T_{R33} = \sigma\,(1+\omega)^2 = E\,E_{E33}\frac{1+\varepsilon}{(1+\varepsilon_P)^2},$$

where

$$E_{E33} = \frac{1}{2}\left(\left(\frac{1+\varepsilon}{1+\varepsilon_P}\right)^2 - 1\right) \equiv \frac{1}{2}\left((1+\varepsilon_E)^2 - 1\right)$$

is the corresponding longitudinal elastic strain component being very small for steels $(|E_{E33}| \ll 1)$. On the other hand, from $S_{11} = S_{22} = 0$ we get the lateral total stretch by means of

$$(1+\omega)^2 = \frac{1-2\nu E_{E33}}{1+\varepsilon_P}.$$

The equation of balance of linear momentum written with respect to the undeformed reference configuration (χ_0) reads

$$\frac{\partial}{\partial\zeta}T_{R33} = \rho_0\frac{Dv_3}{D\tau}, \tag{9.26}$$

where ρ_0 is the mass density in the undeformed (χ_0) configuration, v_3 is the longitudinal component of the velocity vector in spatial coordinates with respect to deformed configuration (χ_t) and T_{R33} is given above.

In order to complete field equations of the problem the geometric relation

$$\frac{\partial v_3}{\partial\zeta} = \frac{D\varepsilon}{D\tau}, \tag{9.27}$$

is necessary. Let us introduce nondimensional time t and nondimensional longitudinal material coordinate Z by means of the substitutions

$$Z = \frac{\zeta}{L},\ t = \frac{\tau}{T},\ \text{such that } Z \in [0,1],\ t \in [0,1] \text{ and } V = \frac{T}{L}v_3 \tag{9.28}$$

is the corresponding nondimensional velocity.

Now, the balance law (9.26) by means of (9.3) and (9.28) may be transformed into

$$\frac{DV}{Dt} = \frac{c_0^2}{2} \frac{\partial}{\partial Z} \left(\frac{1 + \varepsilon_E}{1 + \varepsilon_P} \left((1 + \varepsilon_E)^2 - 1 \right) \right), \tag{9.29}$$

where

$$c_0^2 = \frac{E}{\rho_0} \left(\frac{T}{L} \right)^2 \tag{9.30}$$

is the *nondimensional elastic wave speed* of the linearized wave equation. Indeed, in the elastic range, $\varepsilon_P = 0$, such that (9.29) together with the nondimensional equation

$$\frac{\partial V}{\partial Z} = \frac{D\varepsilon}{Dt}, \tag{9.31}$$

obtained from the geometric relation (9.27), give the elastic wave equation. However, if plastic strain rate does not vanish, then an additional equation is necessary. Such an equation is (9.21) rewritten in its nondimensional form. Therefore, equations (9.29)–(9.31) and such a transformed (9.21) are collected into the following set of nonlinear partial differential equations of first order:

$$\frac{\partial \mathcal{U}}{\partial t} + \mathcal{A}(\mathcal{U}) \frac{\partial \mathcal{U}}{\partial Z} = \mathcal{B}^{\#}(\mathcal{U}), \tag{9.32}$$

where

$$\mathcal{U} = \begin{Bmatrix} V \\ \varepsilon \\ \varepsilon_P \end{Bmatrix}, \quad \mathcal{B}^{\#}(\mathcal{U}) = \begin{Bmatrix} 0 \\ 0 \\ \eta T b^{\#} \end{Bmatrix},$$
$$\mathcal{A}(\mathcal{U}) = \begin{Bmatrix} 0 & -c_0^2\, a_{12} & -c_0^2\, a_{13} \\ -1 & 0 & 0 \\ -\eta\, a_{31} & 0 & 0 \end{Bmatrix}. \tag{9.33}$$

In the above expressions for the matrix $\mathcal{A}$ and the column-vector $\mathcal{B}^{\#}$, scalar functions are introduced:

$$2(1 + \varepsilon_P)^2 a_{12} = 3(1 + \varepsilon_E)^2 - 1,$$
$$(1 + \varepsilon_P)^2 a_{13} = -(1 + \varepsilon_E)\left(2(1 + \varepsilon_E)^2 - 1\right),$$
$$a_{31} = \frac{(1 + \varepsilon_E)\mathcal{D}}{1 - (1 + \varepsilon_E)^2 \mathcal{D}\eta}, \quad \mathcal{D} = \sqrt{\frac{2}{3}} \exp(-\mathcal{M})\, E\, \pi^{\lambda} \left(C_1\, s + C_2\, s^2\right), \tag{9.34}$$
$$b^{\#} = \sqrt{\frac{2}{3}}\,(1 + \varepsilon_P)\, \frac{\left(C_3\, s + C_4\, s^2\right) \pi^{\lambda}}{1 - (1 + \varepsilon_E)^2\, \mathcal{D}\, \eta}.$$

Herein the Heaviside function is denoted by $\eta = \eta(S - Y)$ and the nondimensional second Piola–Kirchhoff stress

$$s = \frac{S_{33}}{Y_0|_{\varepsilon_P = 0}} = \frac{1}{Y_0|_{\varepsilon_P = 0}} E E_{E33} \tag{9.35}$$

is scaled by means of the initial yield stress at the boundary of the original (virgin) elastic range such that the reduced material constants (cf. (9.20) and (9.21)) are introduced as follows:

$$C_\alpha = \frac{c_\alpha}{Y_0|_{\varepsilon_P=0}}, \ \alpha \in \{1,3\} \ \text{ and } \ C_\beta = \frac{c_\beta}{\sqrt{6}\, Y_0|_{\varepsilon_P=0}}, \ \beta \in \{2,4\}. \tag{9.36}$$

Let us assume a solution of the homogeneous part of (9.32) in the form $\mathcal{U} = \mathcal{U}_0 \exp(Z - \lambda t)$, i.e., as a wave propagating along the Z-axis at a speed $\lambda \equiv c$. With such an assumption (9.32) becomes reduced to

$$(\mathcal{A} - \lambda 1)\,\mathcal{U}_0 = 0. \tag{9.37}$$

Since the solutions of the characteristic equation

$$\det(\mathcal{A} - \lambda 1) = -\lambda^3 + \lambda\, c_0^2 \,(a_{12} + \eta\, a_{13} a_{31}) = 0 \tag{9.38}$$

are real and different, i.e.,

$$\lambda_1 = 0, \ \ \lambda_{2/3} = \pm c_0 \,(a_{12} + \eta\, a_{13} a_{31})^{1/2}, \tag{9.39}$$

the wave equation is hyperbolic. It should be noted that one of the solutions vanishes.

Consider now more closely the initial elastic range characterized by $\varepsilon_P = \varepsilon_{P0} = 0$. The above solutions of (9.32) specialize for this very special case into a very simple expression for initial nonlinear wave speed:

$$c_0^{el} = c_0\, a_{12}^{1/2} = c_0 \left(1 + 3\varepsilon_E + \frac{3}{2}\varepsilon_E^2\right)^{1/2}. \tag{9.40}$$

Of course, in this special case elastic and total strains coincide. In a subsequent elastic range characterized by means of $\varepsilon_P = \varepsilon_{P0} = \text{const} \neq 0$ we would have plastic strain dependent *nonlinear elastic wave speed* as follows:

$$c^{el} = c_0\, a_{12}^{1/2} = \frac{c_0}{|1 + \varepsilon_{P0}|} \left(1 + 3\varepsilon_E + \frac{3}{2}\varepsilon_E^2\right)^{1/2}. \tag{9.41}$$

Taking into account that for steels $|\varepsilon_E| \ll 1$, we note that in a subsequent elastic range with advanced previous plastic strains the corresponding elastic wave speed is predicted to be considerably smaller than the elastic wave speed in the initial elastic range. Such a proposition could serve as a basis for an experimental check of validity of the constitutive model proposed and calibrated in [Mic97] and applied here. Concerning character of such a wave, we see from the derivative of the elastic wave speed, i.e.,

$$\frac{dc^{el}}{d\varepsilon} = \frac{3\,(1 + \varepsilon_E)\, c^{el}}{2\left(1 + 3\varepsilon_E + \frac{3}{2}\varepsilon_E^2\right)} > 0, \tag{9.42}$$

that an acceleration wave could be transformed into a shock wave if large elastic strains are possible. However, this cannot happen since a much slower plastic wave appears immediately after a yield surface crossing.

Indeed, if $S > Y$, then plastic strain changes with time such that $\eta = \eta(S - Y) = 1$ and *plastic wave speed* c may be expressed by the following expression (cf. (9.34) and (9.39)):

$$\begin{aligned} c &= c^{el}\left(1 + \frac{a_{13}a_{31}}{a_{12}}\,\eta\right)^{1/2} \\ &= c^{el}\left(1 - \eta\frac{2\,(1+\varepsilon_E)\left(2\,(1+\varepsilon_E)^2 - 1\right)}{3\,(1+\varepsilon_E)^2 - 1}\,\frac{(1+\varepsilon_E)\,\mathcal{D}}{1 - (1+\varepsilon_E)^2\,\mathcal{D}\eta}\right)^{1/2} \end{aligned} \tag{9.43}$$

Let us note that $c < c^{el}$. For advanced plastic strains we may even neglect elastic strain in the above relationship which leads to

$$c \approx c^{el}\left(\frac{1 - 2\mathcal{D}\eta}{1 - \mathcal{D}\eta}\right)^{1/2}.$$

Thus, for a very long rod excited from one of its ends the plastic wave front is always delayed behind an elastic precursor wave traveling with the elastic speed c^{el} characteristic for the elastic range to which state of material at that instant belongs.

It is worthy of note that the special case of the above approximate relation when $\mathcal{D} \approx 0.5$ leads to vanishing of the plastic wave speed and this should give rise to a localization onset according to [Per88].

Let us introduce now left $l_{(\alpha)}$, $\alpha \in \{1,2,3\}$, and right $r_{(\beta)}$, $\beta \in \{1,2,3\}$, (column-type) eigenvectors of (9.37), i.e.,

$$l^T_{(\alpha)}\left(\mathcal{A}(\mathcal{U}) - \lambda_{(\alpha)}1\right) = 0 \qquad \text{and} \qquad \left(\mathcal{A}(\mathcal{U}) - \lambda_{(\beta)}1\right) r_{(\beta)} = 0, \tag{9.44}$$

which are orthogonal to each other, i.e., $l^T_{(\alpha)} r_{(\beta)} = 0$ if $\alpha \neq \beta$. They form the matrices

$$\left\{\begin{matrix} l^T_{(1)} \\ l^T_{(2)} \\ l^T_{(3)} \end{matrix}\right\} = \left\{\begin{matrix} 0 & -\eta\; a_{31} & 1 \\ 1 & -c_0^2\; a_{12}/c & -c_0^2\; a_{13}/c \\ 1 & c_0^2\; a_{12}/c & c_0^2\; a_{13}/c \end{matrix}\right\}, \tag{9.45}$$

$$\left\{ r_{(1)}\; r_{(2)}\; r_{(3)} \right\} = \left\{\begin{matrix} 0 & 1/2 & 1/2 \\ -\left(c_0/c\right)^2\; a_{13} & -1/2c & 1/2c \\ \left(c_0/c\right)^2\; a_{12} & -\eta\; a_{31}/2c & \eta\; a_{31}/2c \end{matrix}\right\}. \tag{9.46}$$

Suppose now that instead of the material coordinate Z and time t, new coordinates r and $s \equiv t$ are introduced by means of

$$r\,(Z,t) = \text{const.} \tag{9.47}$$

They are *characteristics* for the *loading acceleration wave* whose front $Z = \hat{Z}(t)$ moves with the speed (cf. (9.44))

$$\frac{d\hat{Z}}{dt} = \lambda_{(2)}, \quad \lambda_{(2)} = c, \quad c = -\frac{\partial r/\partial t}{\partial r/\partial Z}, \tag{9.48}$$

such that

$$\left(\mathcal{A}(\mathcal{U}) - \lambda_{(\alpha)} 1\right)\left[\frac{\partial \mathcal{U}}{\partial r}\right] = 0 \Rightarrow \left[\frac{\partial V}{\partial r}\right] = -c\left[\frac{\partial \varepsilon}{\partial r}\right], \quad \left[\frac{\partial \varepsilon_P}{\partial r}\right] = \eta\, a_{31}\left[\frac{\partial \varepsilon}{\partial r}\right] \tag{9.49}$$

holds. In the above relationships $[\partial \mathcal{U}/\partial r]$ denotes the jump of $\partial \mathcal{U}/\partial r$ passing from one side of the characteristic (9.47) to its opposite side.

Let us transform the wave equation (9.32) introducing new independent variables, i.e., $\{r, t\}$ instead of $\{Z, t\}$, and multiply such a transformed equation from the left side by the corresponding left eigen-vector $l_{(2)}^T$. In such a way we obtain the so-called *interior equation*

$$l_{(2)}^T \frac{\partial \mathcal{U}}{\partial t} = l_{(2)}^T \mathcal{B}^{\#}(\mathcal{U}), \tag{9.50}$$

which holds along each characteristic (9.47) governing change of the solution vector $\mathcal{U}$ along it. Obviously, the solution vector $\mathcal{U}$ is constant along a characteristic for the quasi-rate-independent model (9.20). In other words, such a wave is said to be *simple* (cf. [Now78], page 145). For the more general rate-dependent model (9.21) a change of the solution vector along a characteristic is very small since constants $\Gamma_\alpha^{\#} = \{c_3, c_4\}$ are much smaller than $\Gamma_\alpha = \{c_1, c_2\}$.

Let us derive an equation governing spatial and temporal changes of $\lceil \partial \mathcal{U}/\partial r \rceil$. First, transforming the wave equation from $\{Z, t\}$ to $\{r, t\}$ and differentiating such a transformed equation by r we get

$$\left(\mathcal{A}(\mathcal{U}) - c\, 1\right)\frac{\partial^2 \mathcal{U}}{\partial r^2}\frac{\partial r}{\partial Z} + \frac{\partial \mathcal{U}^T}{\partial r^2}\frac{\partial}{\partial \mathcal{U}}\left(\mathcal{A}(\mathcal{U}) - c\, 1\right)\frac{\partial \mathcal{U}}{\partial r}\frac{\partial r}{\partial Z} + \frac{\partial^2 \mathcal{U}}{\partial r\, \partial t} = \frac{\partial \mathcal{B}^{\#}}{\partial \mathcal{U}}\frac{\partial \mathcal{U}}{\partial r}. \tag{9.51}$$

If this equation is multiplied form the left by $l_{(2)}^T$ and the orthogonality of $l_{(\alpha)}$ and $r_{(\beta)}$ is remembered, then after taking jumps of all the terms the above equation becomes

$$\frac{\partial}{\partial t}\left\lceil\frac{\partial \varepsilon}{\partial r}\right\rceil + \mu_1\left\lceil\frac{\partial \varepsilon}{\partial r}\right\rceil + \mu_2\left\lceil\frac{\partial \varepsilon}{\partial r}\right\rceil^2 = 0, \tag{9.52}$$

which is the required evolution equation commonly called *amplitude equation.* If it is solved, then jumps$\lceil \partial V/\partial r \rceil$ and $\lceil \partial \varepsilon_P/\partial r \rceil$ are easily found from (9.49). The coefficients of the amplitude equation are obtained after a tedious calculation in the following form:

$$\mu_1 = \frac{1}{2c}\left(\frac{\partial c}{\partial t} + \eta\, \frac{c_0^2}{c} a_{13}\, \frac{\partial a_{31}}{\partial t}\right) + \mu_1^{\#} + m^T \frac{\partial \mathcal{U}^+}{\partial r}, \tag{9.53}$$

with notations

$$\mu_1^{\#} = -\frac{c_0^2}{2c^2}\eta\ T\ a_{13}\left(\frac{\partial b^{\#}}{\partial \varepsilon} + a_{31}\frac{\partial b^{\#}}{\partial \varepsilon_P}\right),\quad m^T \equiv \{\mu_{1V}\ \mu_{1\varepsilon}\ \mu_{1\varepsilon_P}\},$$

$$\mu_{1V} = \frac{\partial c}{\partial \varepsilon} + \eta\ a_{31}\frac{\partial c}{\partial \varepsilon_P} - \eta\ \frac{c_0^2}{c}a_{13}\left(\frac{\partial a_{31}}{\partial \varepsilon} + a_{31}\frac{\partial a_{31}}{\partial \varepsilon_P}\right),$$

$$\begin{aligned}\mu_{1\varepsilon} =& -\frac{\partial c}{\partial \varepsilon} + \frac{c_0^2}{c}\left(\frac{\partial a_{12}}{\partial \varepsilon} + \eta a_{31}\frac{\partial a_{13}}{\partial \varepsilon} - \eta a_{13}\frac{\partial a_{31}}{\partial \varepsilon}\right)\\ &- \frac{c_0^2}{c}\left(\frac{\partial a_{12}}{\partial \varepsilon} + \eta a_{31}\frac{\partial a_{12}}{\partial \varepsilon_P} - \frac{1}{c}\left(\frac{\partial c}{\partial \varepsilon} + \eta a_{31}\frac{\partial c}{\partial \varepsilon_P}\right)\right),\end{aligned}$$

$$\begin{aligned}\mu_{1\varepsilon_P} =& -\eta a_{31}\frac{\partial c}{\partial \varepsilon_P} + \eta\frac{c_0^2}{c}a_{31}\left(\frac{\partial a_{12}}{\partial \varepsilon_P} + a_{31}\frac{\partial a_{13}}{\partial \varepsilon_P} - a_{13}\frac{\partial a_{31}}{\partial \varepsilon_P}\right)\\ &+ \frac{c_0^2}{2c}\left(\frac{\partial a_{13}}{\partial \varepsilon} + \eta a_{31}\frac{\partial a_{13}}{\partial \varepsilon_P} - \frac{1}{c}a_{13}\left(\frac{\partial c}{\partial \varepsilon} + \eta a_{31}\frac{\partial c}{\partial \varepsilon_P}\right)\right),\end{aligned}$$

as well as

$$\begin{aligned}\mu_2 =& -\frac{\partial c}{\partial \varepsilon} - \eta a_{31}\frac{\partial c}{\partial \varepsilon_P} + \frac{c_0^2}{2c}\left(\frac{\partial a_{12}}{\partial \varepsilon} + \eta a_{31}\frac{\partial a_{13}}{\partial \varepsilon_P}\right)\\ &+ \frac{c_0^2}{2c}\eta a_{31}\left(\frac{\partial a_{13}}{\partial \varepsilon} + \eta a_{31}\frac{\partial a_{13}}{\partial \varepsilon_P}\right) - \frac{c_0^2}{2c}\eta a_{13}\left(\frac{\partial a_{31}}{\partial \varepsilon} + \eta a_{31}\frac{\partial a_{31}}{\partial \varepsilon_P}\right).\end{aligned} \tag{9.54}$$

The term $\mu_1^{\#}$ is shown separately in order to show that it vanishes for the quasi-rate-independent model (9.20). If a loading wave enters into an undisturbed region, we have $\partial \mathcal{U}^+/\partial r = 0$ so that the amplitude equation becomes significantly simplified. However, for such an assumption a caution must be made since in front of a plastic wave there exists the corresponding elastic precursor wave.

Finally, let us remark that if the indirect wave with the speed $\lambda_{(3)} = -c$ is considered, then for such a wave $\mathcal{U} = \mathcal{U}_0 \exp(Z + ct)$ and analogous calculations with the corresponding left eigenvector $l_{(3)}^T$ (cf. (9.39)) would give a new amplitude equation the same in form to (9.52) but with other coefficients (9.53) and (9.54).

9.4 Numerical simulation of a Hopkinson bar

9.4.1 A solution algorithm and its accuracy

Consider now a Hopkinson bar as an experimental apparatus consisting of two very long and thick cylindrical coaxial elastic bars with a cylindrical very thin and short viscoplastic cylindrical specimen (cf. [AM-79]). The left bar is preloaded by a constant elastic tensile strain on a major part of its length such that the rest part of the left bar is initially immobilized by a clamp which suddenly becomes broken at the beginning of the wave motion (cf. [AM-79]). Let their material coordinates as well as time be normalized in the way shown in the previous section, i.e., by

$$t = \frac{\tau}{T}, \quad t \in [0,1], \quad Z_k = \frac{\zeta_k}{L_k}, \quad Z_k \in [0,1], \; k \in \{1,2,3\}, \tag{9.55}$$

such that their nondimensional linearized elastic wave speeds are

$$(c_{0k}^{el})^2 = \frac{E_k}{\rho_{0k}} \left(\frac{T}{L_k} \right)^2, \; k \in \{1,2,3\}, \tag{9.56}$$

where the indices $1, 3$ stand for elastic bars whereas the index 2 serves to denote the tension specimen.

Boundary conditions among the bars and the specimen must include equality of contact normal forces leading to corresponding relationships connecting their first Piola–Kirchhoff stresses. Taking into account values of the nondimensional material and temporal coordinates t and Z_k, boundary conditions in stresses read

$$T_{R22}(0,t) = T_{R11}(1,t)\frac{A_{01}}{A_{02}} \quad \text{and} \quad T_{R33}(0,t) = T_{R22}(1,t)\frac{A_{02}}{A_{03}}, \tag{9.57}$$

where A_{0k}, $k \in \{1,2,3\}$, are areas of undeformed cross-sections. Similarly, boundary conditions for nondimensional velocities have to take the following form:

$$V_2(0,t) = V_1(1,t)\frac{L_1}{L_2} \quad \text{and} \quad V_3(0,t) = V_2(1,t)\frac{L_2}{L_3}. \tag{9.58}$$

For a numerical solution of the wave equations of the type (9.32) the following numerical method ([Bal83]) is applied here. Time and material coordinate are discretized as follows:

$t \in [0,1]$,	$\Rightarrow$ temporal index $K \in [1,M]$, $\Delta t = (M-1)^{-1}$
$Z \in [0,1]$,	$\Rightarrow$ spatial index $J \in [1,N]$, $\Delta Z = (N-1)^{-1}$

Implicit integration in the quasi-rate-independent case (cf. (9.19)) is shown by the following scheme. Its initialization at time step $K+1$ is determined by the last iteration values of the previous step, i.e.,

$$\varepsilon^{(i)}(J, K+1) = \varepsilon(J,K)$$

with $i = 1$ at initial iteration position and $\mathcal{U}^{(0)}(J, K+1) = \mathcal{U}(J,K)$.

$$\varepsilon^{(i)}(J,K+1)$$

$$\Downarrow$$

$$\varepsilon_P^{(i)}(J,K+1)=\varepsilon_P(J,K)+ \eta^{(i-1)}(J,K+1)\,a_{31}^{(i-1)}(J,K+1)\left(\varepsilon^{(i)}(J,K+1)-\varepsilon(J,K)\right)$$

$$\Downarrow$$

$$V^{(i)}(J,K+1)=V(J,K)+c_0^2\,a_{12}^{(i)}(J,K+1)\,\partial_Z\varepsilon^{(i)}(J,K+1)\,\Delta t+ c_0^2\,a_{13}^{(i)}(J,K+1)\,\partial_Z\varepsilon_P^{(i)}(J,K+1)\,\Delta t$$

$$\Downarrow$$

$$\partial_Z V^{(i)}(J,K+1)=0.5\,\left(V^{(i)}(J+1,K+1)-V^{(i)}(J-1,K+1)\right)/\Delta Z$$

$$\Downarrow$$

$$\varepsilon^{(i+1)}(J,K+1)=\varepsilon(J,K)+\partial_Z V^{(i)}(J,K+1)\,\Delta t$$

$$\Downarrow$$

$$ABS\left(\varepsilon^{(i+1)}(J,K+1)-\varepsilon^{(i)}(J,K+1)\right)<TOL$$

$\Downarrow$ YES $\Downarrow$ $\quad$ $\Downarrow$ NO $\Downarrow$

$K+1 \Leftarrow K$ $\quad$ $i+1 \Leftarrow i$

From the above algorithm we are able to derive its order of accuracy following the procedure explained in [Kuk92b]. To do this we recall that (9.32) is here approximated by

$$\frac{1}{\Delta t}\left(\mathcal{U}_J^{K+1}-\mathcal{U}_J^{K}\right)+\frac{1}{2\Delta Z}\mathcal{A}\left(\mathcal{U}_J^{K+1}\right)\left(\mathcal{U}_{J+1}^{K+1}-\mathcal{U}_{J-1}^{K}\right)=\mathcal{B}^{\#}\left(\mathcal{U}_J^{K+1}\right). \quad (9.59)$$

Now, taking into account that the development of

$$\mathcal{U}_{J+\beta}^{K+\alpha}\equiv\mathcal{U}\left(J\Delta Z+\beta\Delta Z,K\Delta t+\alpha\Delta t\right) \quad (9.60)$$

into power series and replacement of the obtained expression into (9.59) gives approximation order of $O(\Delta t+\Delta Z^2)$, we see that the algorithm is linear in time but it is of second order in material coordinate.

On the other hand, von Neumann stability analysis (cf., e.g., [PFT90]) requires that $\xi < 1$ in the following solution

$$\mathcal{U}_J^K = \xi^K \exp(i\chi J \Delta Z)\mathcal{U}_0 \tag{9.61}$$

of (9.59). Replacing (9.61) into (9.59) we arrive at the following form of the characteristic equation (9.38):

$$\det\left(\mathcal{A} - i\frac{\Delta Z}{\Delta t}\frac{\xi - 1}{\xi}\frac{1}{\sin(\chi\Delta Z)}\mathcal{I}\right) = 0, \tag{9.62}$$

such that its solutions permit the following inequality:

$$\Delta t \geq \left|1 - \frac{1}{\xi}\right| \frac{\Delta Z}{|c|} \sim \frac{\alpha_{BK}\Delta Z}{c^{el}}, \tag{9.63}$$

where plastic wave speed c is determined by (9.39). Therefore, the proposed procedure is unconditionally stable permitting unbounded time increments. In the paper [BT-78] the value $\alpha_{BK} = 5^{1/2}$ is suggested as convenient.

Practically, for meeting some accuracy requirement choosing $\xi \in [0.9, 1)$ we may reduce Δt as much as necessary. For instance, if the specimen is divided into 100 elements, then a convenient nondimensional time interval could be $\Delta t \sim 10^{-4}$ for the above established accuracy.

9.4.2 Appropriate boundary conditions

A very delicate point in this numerical routine is initialization due to the fact that geometrical changes in the apparatus are abrupt with large values of A_{01}/A_{02} as well as of L_1/L_2. Moreover, the length of the specimen is more than one hundred times smaller than the lengths of elastic bars. This means that only at the beginning of plastic wave motion plastic waves might be clearly recognized whereas during numerous subsequent reflections the state of specimen strain becomes practically homogeneous.

Thus, a more realistic initialization simulating background wave-type space-time values of velocity and strain is needed. Otherwise, a disturbance at the end of the specimen becomes numerically "frozen" not propagating at all along the specimen with the growth of time. In this chapter we proceed in the following way.

For the time being, suppose that a very small disturbance of the type

$$\varepsilon_1(0,t) = \varepsilon_0\, \eta\left(-Z_1 + \Theta L_1\right), \qquad \text{with} \qquad \Theta = \text{const} < 1 \tag{9.64}$$

is imposed to the left (so-called *incident-reflected* bar), such that the corresponding induced strain in the specimen stays inside its initial elastic range. Here $\eta(Z) = 1$, for $Z > 0$ and $\eta(Z) = 0$ otherwise, while the magnitude of ε_0 is chosen to be small enough to provoke only linear elastic wave inside the specimen due to approximately constant value of a_{12} for the wave speed c_{02}^{el} in

(9.40). Then after P reflections incident and reflected stresses and velocities in the bars as well as the specimen (under the assumption of linearity of the elastic wave equation) would have the following forms:

$$\sigma_{1I} = \frac{E_1}{2}\,\varepsilon_0\,\eta\left(t + \frac{-Z_1 + \Theta L_1}{c_{01}}\right) + \frac{E_1}{2}\,\varepsilon_0\,\eta\left(-t + \frac{-Z_1 + \Theta L_1}{c_{01}}\right), \tag{9.65}$$

$$\begin{aligned}\sigma_{1R} = &-\frac{E_1}{2}\varepsilon_0\,\frac{r_{12}-1}{r_{12}+1}\eta\left(t + \frac{Z_1 - (1+\Theta)\,L_1}{c_{01}}\right) - 2E_1\varepsilon_0\,\frac{r_{12}}{(r_{12}+1)^2}\\ &\times \frac{r_{23}-1}{r_{23}+1}\sum_{\alpha=1}^{P-1}(-r_{123})^{\alpha-1}\,\eta\left(t + \frac{Z_1 - (1+\Theta)\,L_1}{c_{01}} - \frac{2\alpha L_2}{c_{02}}\right),\end{aligned} \tag{9.66}$$

$$\begin{aligned}\sigma_{2I} = &\,E_2\,\varepsilon_0\,\frac{c_{01}L_1}{c_{02}L_2}\frac{r_{12}}{r_{12}+1}\\ &\times \sum_{\alpha=1}^{P-1}(-r_{123})^{\alpha-1}\,\eta\left(t + \frac{Z_1 - (1+\Theta)\,L_1}{c_{01}} - \frac{2\alpha L_2}{c_{02}}\right),\end{aligned} \tag{9.67}$$

$$\begin{aligned}\sigma_{2R} = &\,E_2\,\varepsilon_0\,\frac{c_{01}L_1}{c_{02}L_2}\frac{r_{12}}{r_{12}+1}\frac{r_{23}-1}{r_{23}+1}\\ &\times \sum_{\alpha=1}^{P-1}(-r_{123})^{\alpha-1}\,\eta\left(t + \frac{Z_1 - (1+\Theta)\,L_1}{c_{01}} - \frac{2\alpha L_2}{c_{02}}\right),\end{aligned} \tag{9.68}$$

$$\begin{aligned}\sigma_{3I} = &\,2E_3\,\varepsilon_0\,\frac{c_{01}L_1}{c_{03}L_3}\frac{r_{12}}{r_{12}+1}\frac{r_{23}}{r_{23}+1}\\ &\times \sum_{\alpha=1}^{P-1}(-r_{123})^{\alpha-1}\,\eta\left(t - \frac{Z_3}{c_{03}} - \frac{\Theta L_1}{c_{01}} - \frac{(1+2\alpha)\,L_2}{c_{02}}\right),\end{aligned} \tag{9.69}$$

with the following notations based on elastic impedances[4]:

$$r_{12} = \frac{E_1 A_1}{E_2 A_2}\frac{c_{02}L_2}{c_{01}L_1}, \quad r_{23} = \frac{E_2 A_2}{E_3 A_3}\frac{c_{03}L_3}{c_{02}L_2}, \quad r_{123} = \frac{r_{12}-1}{r_{12}+1}\frac{r_{23}-1}{r_{23}+1}.$$

The corresponding velocities would have the values

[4] At first sight, a special case of (9.67) when elastic impedance $r_{12} = 1$ leads to $r_{123} = 0$ such that σ_{2I} disappears which, obviously, is nonsense. Such a conclusion comes from the above compact notation. In fact, when $r_{123} \to 0$, then for $\alpha = 1$ we have $\lim_{r_{123}\to 0} r_{123}^0 = 1$ such that

$$\sigma_{2I} = 0.5E_2\,\varepsilon_0\,\frac{c_{01}L_1}{c_{02}L_2}\eta\left(t + \frac{Z_1 - (1+\Theta)\,L_1}{c_{01}} - \frac{2L_2}{c_{02}}\right).$$

Similar holds true for σ_{2R} and σ_{3I} given by the next two formulae, (9.68) and (9.69).

$$V_{1I} = \frac{1}{2}\, c_{01}\varepsilon_0\, \eta\left(t + \frac{-Z_1 + \Theta L_1}{c_{01}}\right) - \frac{1}{2}\, c_{01}\varepsilon_0\, \eta\left(-t + \frac{-Z_1 + \Theta L_1}{c_{01}}\right),$$

$$V_{1R} = \frac{c_{01}}{E_1}\sigma_{1R},\ V_{2I} = -\frac{c_{02}}{E_2}\sigma_{2I},\ V_{2R} = \frac{c_{02}}{E_2}\sigma_{2R},\ V_{3I} = -\frac{c_{03}}{E_3}\sigma_{3I}\ . \tag{9.70}$$

Due to the assumed linearity of the elastic wave equation (which is fulfilled for very small elastic strains) the additivity

$$\sigma_k = \sigma_{kI} + \sigma_{kR}\ ,\quad V_k = V_{kI} + V_{kR} \tag{4.12}$$

would hold.

9.4.3 Results of plastic waves inside the specimen

Let us imagine that the initial strain of the left bar, ε_0, in (9.64)–(9.69) is now augmented enough to cause plastic straining of the specimen and that after $P = 2$ stresses and strains in elastic bars are kept unchanged. In other words, this means that during the first two reflections inside the specimen it stays inside the elastic range. Then these formulae with $P = 2$ will serve as an input into the numerical routine shown above. With this type of initialization, plastic strain of the specimen, being calculated by the proposed algorithm as a function of time and its material longitudinal coordinate, is depicted in Fig. 9.1.

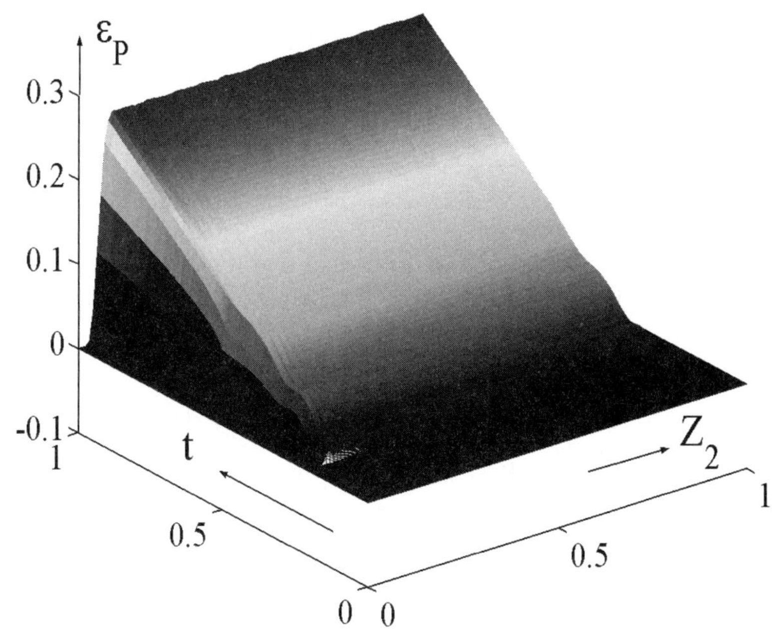

Fig. 9.1. Plastic wave inside the specimen as a function of space and time

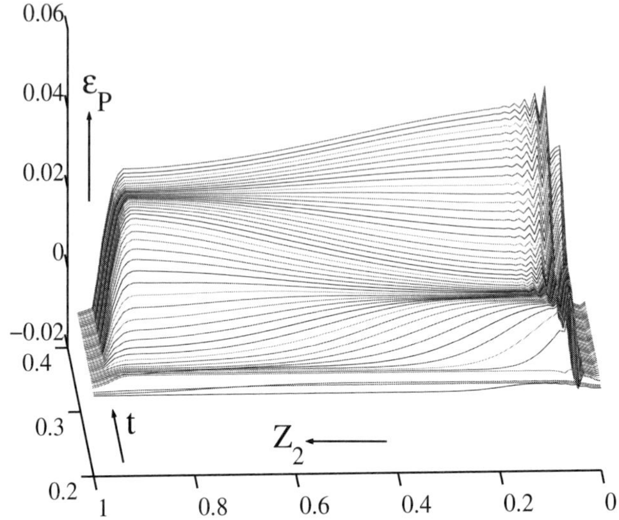

Fig. 9.2. Initial transition interval of plastic wave

In order to underline that at the initial time interval we have to deal with inhomogeneous distributions along the specimen whereas at advanced strains we have almost constant strain along its gage field, we have constructed Figs. 9.2 and 9.3. From the whole history, the two characteristic regions are here chosen for presentation: initial transition time interval and only the last segment of the subsequent steady time interval.

The considered example was attained by means of the following data: $T = 0.001$ s, $A_{01}/A_{02} = 25$, $L_1/L_2 = 250$, $E_1 = 210$ GPa, $E_2 = 190$ GPa, $A_{03} = A_{01}$, $L_3 = L_1$, $E_3 = E_1$, $\varepsilon_0 = 0.0016$, while ρ_0 is the mass density of steel. Taking into account the above accuracy analysis the specimen was divided into 100 equally spaced elements. The initial time increment was taken to be slightly smaller than the corresponding Courant value ([Kuk92b]). The geometric transition from elastic bars to the specimen was assumed to be gradual with rounded corners change of radius in order to diminish stress concentration ([AM-79]) such that the gage part of the specimen has a twofold smaller radius than their mounting ends. The initial as well as a yield stress at a nonzero plastic strain are taken, respectively, to be[5]

$$\frac{Y_0}{a} = 1 + b\left(\ln\left[\sqrt{\frac{3}{2}}E\frac{\partial\varepsilon}{\partial t}\right]_{t_0=0}\right)^c, \quad Y = Y_0 + ad\pi^e. \tag{9.71}$$

[5] The meaning of normalizing constant a is that it is used to denote the initial yield stress at zero plastic strain and equivalent stress rate equal to 1 MPa/s. The other constants appearing in (9.71) except Young's modulus as well as "evolution" constants c_1 and c_2 are nondimensional.

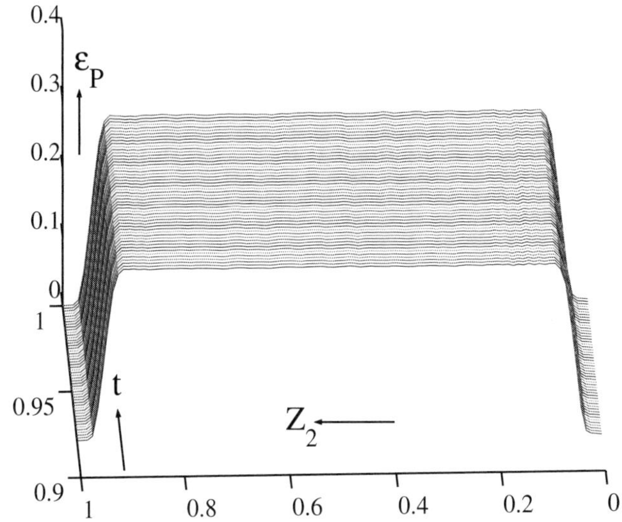

Fig. 9.3. Ending steady interval of plastic wave

These as well as material constants appearing in (9.19) are taken from [Mic97]:

$$a = 251.2\,\mathrm{MPa},\ b = 0.015,\ c = 1.44,\ d = 17.23,\ e = 0.5,$$
$$c_1 = 1.095\ \mathrm{MPa}^{-1},\ c_2 = -0.244\,\mathrm{MPa}^{-2},\ \lambda = 0.223.$$

It is worthy of note that unlike (9.15) the triggering relationship (9.71) for plasticity commencement at diverse loading–unloading paths takes into account combined strain–strain rate hardening. Thus, for strain controlled experiments the curves $Y_0(\pi)$ are not parallel when the strain rate is varied (cf. [AM-79, AMP91]).

9.4.4 A discussion about Lindholm's procedure

At the end of this section, let us consider carefully the standard determination of stress–strain state inside the specimen by means of measurements made on the elastic bars only. Suppose that two strain gages, SG1 and SG3, are situated symmetrically at the same distance from the specimen, i.e., $\vartheta L_1 = \vartheta L_3$, where $\vartheta < \Theta < 1$. In other words, SG1 has a position between the fixing clamp on the left (incident-reflected) bar and the left end of the specimen. Let the elastic waves in the left and the right elastic bar be

$$u_1 = f_1\left(t - \frac{Z_1}{c_{01}}\right) + g_1\left(t + \frac{Z_1}{c_{01}}\right),\quad u_3 = f_3\left(t - \frac{Z_3}{c_{03}}\right)$$

where f_1 is the *incident* wave, g_1 the *reflected* wave and f_3 the *transmitted* wave. Neglecting the length of the specimen we may take that time delays at SG1 and SG3 are approximately the same and equal to

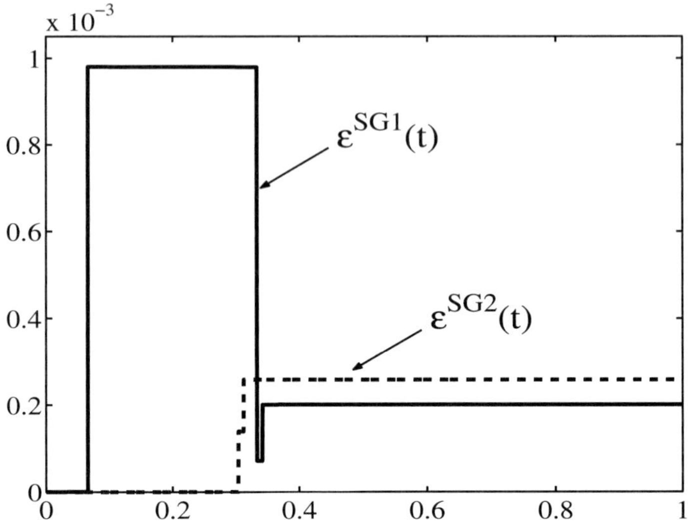

Fig. 9.4. Incident, reflected and transmitted strains from left (SG1) and right (SG2) strain gages as functions of nondimensional time

$$\Delta t = \vartheta L_1/c_{01} = \vartheta L_3/c_{03}.$$

Let us denote the incident, reflected and transmitted strains by means of $\varepsilon_I(Z_1,t) = \partial f_1(\cdot)/\partial Z_1, \varepsilon_R(Z_1,t) = \partial g_1(\cdot)/\partial Z_1$ and $\varepsilon_T(Z_3,t) = \partial f_3(\cdot)/\partial Z_3$, respectively. Then we have

$$\begin{aligned}
\varepsilon_1(1,t) &= \varepsilon_I^{SG1}(t-\Delta t) + \varepsilon_R^{SG1}(t+\Delta t) \equiv \varepsilon_I(t) + \varepsilon_R(t),\\
\varepsilon_3(0,t) &= \varepsilon_T^{SG3}(t+\Delta t) \equiv \varepsilon_T(t),
\end{aligned}$$

where ε with superscripts SG1 and SG3 show readings on the strain gages. For only two prescribed reflections caused by the augmented input ($P = 2$ in formulae (9.66)–(9.69)) as well as stresses and strains subsequently kept at fixed values we would get the following picture for incident, reflected and transmitted strain histories at the strain gages. Having such readings as inputs Lindholm proposed the approximate formulae for the Cauchy stress and presumably homogeneous linear strain (cf. [Lin67]) as follows:

$$\begin{aligned}
\sigma_2(0.5\,,t) &\approx 0.5\,(\sigma_2(0,t) + \sigma_2(1,t))\\
&= 0.5\,E_2\,(\,A_{01}(\varepsilon_I(t) + \varepsilon_R(t)) + A_{03}\,\varepsilon_T(t))/A_{02},
\end{aligned} \tag{9.72}$$

$$\varepsilon_2(0.5\,,t) = \frac{1}{L_2}\int_0^t \left(L_1 c_{01}\left(\varepsilon_I\left(t'\right) - \varepsilon_R\left(t'\right)\right) - L_3 c_{03}\,\varepsilon_T\left(t'\right)\right) dt'. \tag{9.73}$$

Comparing the above two functions of time with the corresponding values calculated by the applied numerical routine gives rise to Fig. 9.5, which shows

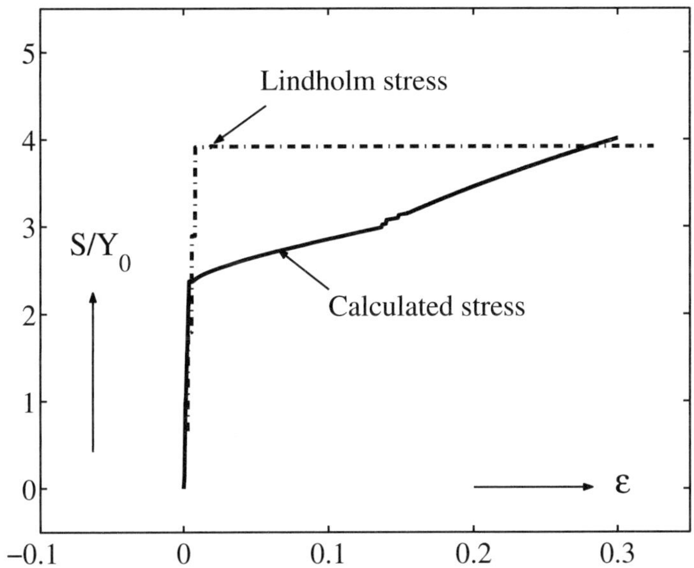

Fig. 9.5. Check of Lindholm's approximate formulae

that Lindholm's approach should be applied with caution having in mind that the corresponding error is considerably high. Similar conclusion, but from some other considerations, has been drawn by Wu and Gorham in [WG-97].[6]

9.5 Concluding remarks

At the end of this chapter we can draw the following conclusions:

- It has been previously shown that the so-called universal flow curve and associate flow rule based on the yield function relating only scalars like equivalent stress and equivalent plastic strain failed to describe simultaneously tension and shear even in the range of only small strain rates (compare for instance [Mic92b, Mic97]). Although commonly used for its simplicity, such an equation is intrinsically scalar since it can describe successfully only tension test up to large strains. The simplest yet approx-

---

[6] Comparing the last four figures we might conclude that (9.72) should be modified inserting a scaling factor:

$$\begin{aligned} \sigma_2\,(0.5\,,t) &\approx 0.5\,k_{\text{Lind}}\,\left(\sigma_2\,(0,t)+\sigma_2\,(1,t)\right) \\ &= 0.5\,k_{\text{Lind}}\,E_2\,\left(\,A_{01}\,(\varepsilon_I\,(t)+\varepsilon_R\,(t))+A_{03}\,\varepsilon_T\,(t)\right)/A_{02}, \end{aligned} \tag{9.72a}$$

where correction factor $k_{\text{Lind}} \leq 1$ is smaller in the initial range of small strains. On the basis of Fig. 9.5 the second formula (9.73) seems to be more exact.

imately correct approach is to combine loading function orthogonality with tensor functions.

- At present, it may be concluded that the standard Lindholm's approach to Hopkinson bar analysis does not give satisfactory answers to assumed homogeneous stress and strain states until failure. Instead, despite numerical difficulties met at time and space normalization the approach which accounts explicitly for plastic waves has obvious advantages.
- It is important to underline that the flat horizontal line in Fig. 9.5 does not follow either from the applied numerical scheme or constitutive assumptions. On the contrary, boundary conditions at both ends of the specimen are assumed to fulfill (9.64)–(9.69) dictating a fixed form of strains and stresses at the left and right end of the specimen. Then the determination of stress and strain by application of (9.72) and (9.73) necessarily leads to such a line. The point is that in such a procedure which eliminates plastic waves and reflections the specimen is considered as a "black box."
- However, it must be taken into account that the ingenious Lindholm's assumption has to be accepted at the beginning of a test analysis. Then, an interactive procedure should be applied to improve agreement between theory and test, especially at the initial transition range of inelastic strains.

Acknowledgment. Research done in this chapter, accomplished with Professor Angel Baltov, was motivated and made possible by experimental setups and techniques developed by our friends Dr. C. Albertini and the late Dr. M. Montagnani to whom our gratitude must be addressed. Very valuable contributions to this chapter were made by Professors V. Kukudzanov and W. Kosinski. Mr. D. Stanic made considerable effort in practical computer solving of the finite difference problem.

10

Ratchetting Phenomenon at Low Strain Rates for AISI 316H Stainless Steel

This chapter[1] deals with 3D viscoplastic strain of a rectangular block made of AISI 316H austenitic stainless steel. One of its sides is loaded by constant normal stress whereas two lateral side surfaces are acted upon by harmonically variable shear stress. It is experimentally observed that such temporal variation induces progressive but saturated increase of axial strain in the direction of tension stress components. The strain rate is of the order of $0.001\,\mathrm{s}^{-1}$. The problem is treated by two constitutive models:

a) Perzyna–Chaboche's model with incorporated evolution equations for back stress and equivalent flow stress (its eight material constants are taken from [Ele91]) and

b) the model explained in [Mic96a], based on tensor representations where plastic stretching is second-order polynomial in stress and linear in plastic strain (having six material constants). Comparison with experiments has shown superiority of the second model.

10.1 Introduction

Steel mantel of nuclear reactors composed of austenitic stainless steels is exposed during its exploitation to time-dependent stress–strain histories.

As a consequence, there are three typical types of response:

[1] A substantial part of this chapter has been published in [MV-98].

M.V. Mićunović, *Thermomechanics of Viscoplasticity: Fundamentals and Applications*, Advances in Mechanics and Mathematics 20, DOI: 10.1007/978-0-387-89490-4_10,
© Springer Science + Business Media, LLC 2009

- Incremental collapse is characterized by an increase of the mean stress curve up to fracture. Deviations around this curve are steady (i.e., with slightly changing amplitude) changing with unperturbed frequency.
- Low-cycle fatigue takes place with a constant mean stress and the other features are the same as in the previous case of behaviour.
- Elastic vibrations (*shakedown*) appearing as a consequence of the corresponding decrease of amplitudes of plastic strain vibrations. Hence, behaviour of the material body is such that stress tensor enters the elastic region, i.e., interior of the yield surface in the stress space.

The case of plastic saturation when stress frequency in "universal" flow curve diagram (i.e., equivalent Mises stress versus accumulated plastic strain) increases is of special interest. Such behaviour is called *ratchetting.* It has been shown to be the case at multiaxial stress–strain histories especially at nuclear reactors where such histories usually appear.

The paper deals with the problem of constant normal stress and a harmonically variable shear stress. Ratchetting predictions of the models presented in the papers [Per71, CR-83] and [Mic96a] are compared. Relevant stress amplitudes and shear stress frequency are chosen in such a way to be comparable with ratchetting experiments reported in [LRC79]. The equivalent plastic strain rate is of the order of magnitude of 10^{-3} s^{-1}, which corresponds to low strain rates.

10.2 Preliminary considerations and problem statement

For finite elastoplastic strains it is commonly accepted that aside from undeformed configuration (χ_0) and deformed current configuration (χ_t) there exists a local reference configuration of natural state elements (ν_t). Then, Kröner's decomposition rule ([Kro60, Sto62]) holds in the following form:

$$\mathbf{F}_P := \mathbf{F}_E^{-1}\mathbf{F}, \tag{10.1}$$

where
$\mathbf{F}$ is the deformation gradient tensor,
$\mathbf{F}_E$ is the elastic distortion tensor, and
$\mathbf{F}_P$ is the plastic distortion tensor,
determined by the mappings $(\chi_0) \to (\chi_t)$, $\chi_n \to (\chi_t)$ and $(\chi_0) \to (\nu_t)$, respectively. Let us apply the polar decomposition theorem on the plastic distortion tensor by means of

$$\mathbf{F}_P = \mathbf{R}_P\mathbf{U}_P = \mathbf{V}_P\mathbf{R}_P,$$

where $\mathbf{R}_P^{-1} = \mathbf{R}_P^T$ holds for the plastic rotation tensor.

Then as an invariant measure of plastic strain the Hill's logarithmic tensor

$$\varepsilon_P = \ln \mathbf{V}_P = 0.5 \ln \left(\mathbf{F}_P\mathbf{F}_P^T\right) \tag{10.2}$$

is chosen. Its principal advantage lies in the fact that it is a deviatoric tensor. In other words, its three principal invariants read

$$\pi_1 = \mathrm{tr}\boldsymbol{\varepsilon}_P = 0, \qquad \pi_2 = \mathrm{tr}\boldsymbol{\varepsilon}_P^2 \neq 0, \qquad \pi_3 = \mathrm{tr}\boldsymbol{\varepsilon}_P^3 \neq 0, \tag{10.3}$$

if plastic volume change is neglected. Then plastic stretching tensor equals

$$\mathbf{D}_P = \frac{1}{2}\frac{d}{dt}\left\{\begin{array}{ccc} 2\ln(1+e_P) & 0 & 0 \\ 0 & -\ln(1+e_P) & \gamma_P \\ 0 & \gamma_P & -\ln(1+e_P) \end{array}\right\} \tag{10.4}$$

determined by the symmetric part of plastic "velocity gradient" tensor

$$\mathbf{L}_P = \dot{\mathbf{F}}_P \mathbf{F}_P^{-1}.$$

As usual the superimposed dot stands for material time derivative.

Consider now a problem of three-dimensional straining of a rectangular parallelepiped. Two of its opposite sides are acted upon by constant normal stresses whereas on two lateral sides two harmonically variable shear stresses act. The loading scheme is shown in Fig. 10.1.

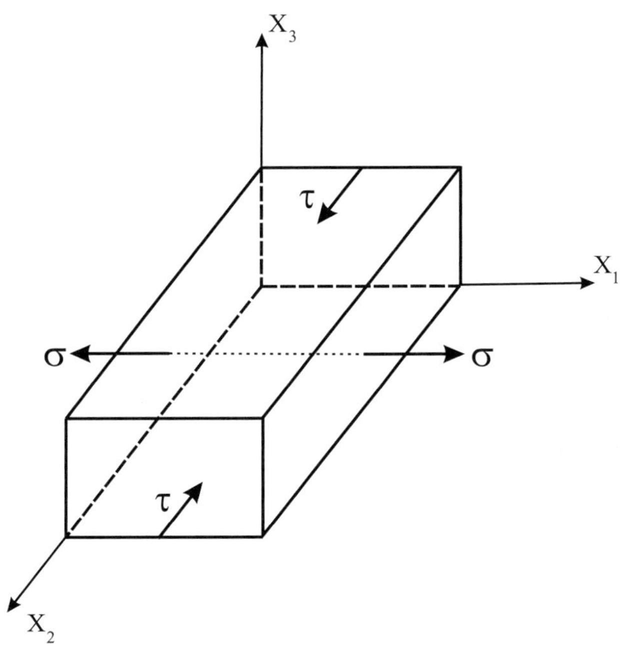

Fig. 10.1. Loading of the considered body

Cauchy stress tensor and Piola–Kirchhoff stress tensor now read

$$\mathbf{T} = \begin{Bmatrix} \sigma & 0 & 0 \\ 0 & 0 & \tau \\ 0 & \tau & 0 \end{Bmatrix}, \qquad \mathbf{S} = \mathbf{F}_E^{-1}\mathbf{T}\mathbf{F}_E^{-T} \approx \mathbf{T}, \tag{10.5}$$

where

$$\sigma = \text{const}, \qquad \tau = \tau_0 \sin(\omega t) .$$

It should by underlined that the second of the above tensors is defined with respect to local reference configuration (ν_t) and that for small elastic strains

$$\|\mathbf{F}_E\| \approx 1$$

if elastic rotations are also very small. On the other hand, the plastic distortion tensor equals

$$\mathbf{F}_P = \begin{Bmatrix} 1 & 0 & 0 \\ 0 & 1 & \gamma_P \\ 0 & 0 & 1 \end{Bmatrix} \begin{Bmatrix} \chi_P^2 & 0 & 0 \\ 0 & \chi_P^{-1} & 0 \\ 0 & 0 & \chi_P^{-1} \end{Bmatrix} = \begin{Bmatrix} \chi_P^2 & 0 & 0 \\ 0 & \chi_P^{-1} & \gamma_P \chi_P^{-1} \\ 0 & 0 & \chi_P^{-1} \end{Bmatrix}, \tag{10.6}$$

where notation $\chi_P^2 \equiv 1 + e_P$ is applied. For a given stress history $\mathbf{T}(t)$ a response of the material body determined by plastic strain history is looked for. We apply the additional condition on stress history

$$\frac{d}{dt} \|\mathbf{T}(t)\| \leq 16\,\text{MPa/s}$$

in order to keep strain rates in the low range of the order of magnitude $\|\mathbf{D}_P\| \leq 10^{-3}\,\text{s}^{-1}$.

10.3 Model of Perzyna–Chaboche

For a description of a viscoplastic body Chaboche ([CR-83]) has generalized Perzyna's model ([Per71]) to account for an evolution of the residual stress. His evolution equations might be written as follows:

$$\dot{\varepsilon}_P = \left\langle \frac{\mathcal{F} - D}{k} \right\rangle^n \frac{\mathbf{T}_d - \mathbf{B}_d}{\mathcal{F}}, \qquad \mathcal{F} = \|\mathbf{T}_d - \mathbf{B}_d\|, \tag{10.7}$$

$$\dot{\mathbf{B}}_d = c\,(A\mathbf{D}_P - \mathbf{B}_d \dot{p}) - \Gamma \|\mathbf{B}_d\|^{m-1} \mathbf{B}_d, \tag{10.8}$$

$$\dot{p} = \sqrt{\frac{2}{3}} \|\mathbf{D}_P\|, \tag{10.9}$$

$$\dot{D} = b\,(Q - D)\,\dot{p}, \tag{10.10}$$

where

$\dot{\varepsilon}_P \approx \mathbf{D}_P$ - the plastic strain rate (approximately equal to plastic stretching for small strains range),

$\mathbf{B}$ - the residual stress (back stress tensor),

D - a static equivalent flow stress,

p - the accumulated plastic strain scalar and

$\mathcal{A}_c = \{k, n, c, A, \Gamma, m, b, Q\}$ - a set of material constants to be determined from experiments.

Here the traditional notations for a second tensor intensity as well as its deviatoric part are employed:

$$\|\mathbf{A}\| = \left(\mathrm{tr}\mathbf{A}^2\right)^{1/2} \equiv (\mathbf{A} : \mathbf{A})^{1/2}, \qquad \mathbf{A}_d = \mathbf{A} - \frac{1}{3}\mathbf{1}\mathrm{tr}\mathbf{A},$$

while plastic loading indicator function is determined by means of $\langle x \rangle = 0$ if $x \geq 0$ or 0 if $x < 0$.

An identification of material constants has been made in [Ele91] on the basis of experiments for standard cyclic tension-compression test as well as cyclic torsion test. Their values for AISI 316H as reported in [Ele91] read

$$\mathcal{A}_c = \left\{68.38,\ 5.8,\ 65,\ 113.33,\ 8.7 \cdot 10^{-10},\ 1.3,\ 8.8,\ 220.45\right\}. \tag{10.11}$$

For the problem presented in the previous section, deviatoric residual stress (cf. also [Ele91]) equals

$$\mathbf{B}_d = \frac{1}{3}\left\{\begin{array}{ccc} 2B_L & 0 & 0 \\ 0 & -B_L & 3B_T \\ 0 & 3B_T & -B_L \end{array}\right\}, \tag{10.12}$$

such that

$$\mathcal{F} = \sqrt{\frac{2}{3}}\sqrt{(\sigma - B_L)^2 + 3\,(\tau - B_T)^2}. \tag{10.13}$$

We are going to consider a relatively small cycle number. This makes it possible to neglect the material constant Γ. Such an assumption simplifies the set (10.7)–(10.10) transforming it into

$$\dot{\varepsilon}_P = \frac{2}{3}\left\langle\frac{\mathcal{F} - D}{k}\right\rangle^n \frac{\sigma - B_L}{\mathcal{F}}, \qquad \dot{\gamma}_P = 2\left\langle\frac{\mathcal{F} - D}{k}\right\rangle^n \frac{\tau - B_T}{\mathcal{F}}, \tag{10.14}$$

$$\dot{B}_L = c\left(\frac{3}{2}A\dot{\varepsilon}_P - \dot{p}B_L\right), \qquad \dot{B}_T = c\left(\frac{1}{2}A\dot{\gamma}_P - \dot{p}B_T\right), \tag{10.15}$$

$$\dot{p} = \sqrt{\dot{\varepsilon}_P^2 + \frac{1}{3}\dot{\gamma}_P^2}, \tag{10.16}$$

$$\dot{D} = b\,(Q - D)\,\dot{p}, \tag{10.17}$$

where we introduced the notation $\varepsilon_P \equiv \ln(1 + e_P) \approx e_P$ such that $\dot{\varepsilon}_P \approx \dot{e}_P$.

It is worthy of note that Perzyna–Chaboche's model is developed for small strains only. Taking this fact into account we may simplify logarithmic strain tensor (10.2) into the following form:

$$\boldsymbol{\varepsilon}_P = \ln \mathbf{V}_P \approx \frac{1}{2} \left\{ \begin{array}{ccc} 2e_P & 0 & 0 \\ 0 & -e_P & \gamma_P \\ 0 & \gamma_P & -e_P \end{array} \right\} . \tag{10.18}$$

The evolution equations (10.14)–(10.17) have been integrated numerically for a simulated stress controlled test with normal stress $\sigma = 245\,\mathrm{MPa}$, shear stress amplitude $\tau_0 = 75\,\mathrm{MPa}$ and shear stress frequency $\omega = 0.5\,\mathrm{rad/s}$ for ten stress cycles. Such special choice of history parameters is made in order to acquire results comparable with experimental data reported in [LRC79]. These experiments have been made with AISI 316H at a constant tension of 250 MPa, maximal shear strain of 2.6% and amplitude of shear strain equal to 0.17%. Number of experimental cycles was ten.

Results of integration of evolution equations (10.14)–(10.17) with material constants (10.11) and $\Gamma = 0$ are presented in Fig. 10.2. Aside from time-plastic strain component plots the so-called universal flow curve (i.e., equivalent stress versus accumulated plastic strain) as well as phase portrait of plastic strain components are given. The densification feature characteristic for ratchetting has been observed in the "universal" flow curve.

10.4 MAM model with loading function based normality

Following the exposition given in the fifth chapter, let us take the tensor representation approach with loading function based normality. Rice's model [Ric71] by itself is based on normality of plastic strain rate tensor on a loading function Ω taking account of microstructural rearrangements during plastic straining. Translated to finite strains and with Ziegler's modification derived from the notion of least irreversible force, the evolution equation reads

$$\mathbf{D}_P = \Lambda \frac{\partial \Omega}{\partial \mathbf{S}}, \tag{10.19}$$

where the microstructural rearrangements are supposed to be completely determined by the plastic strain tensor given by (10.2).

An essential generalization of this model was given in [Mic96a]. In this paper the loading function was assumed to depend on the set

$$\Omega = \Omega(\gamma), \qquad \gamma \equiv (S_2, S_3, \mu_1, \mu_3) \tag{10.20}$$

of proper and mixed invariants of stress and plastic strain tensors

$$S_2 = \mathrm{tr}\mathbf{S}_d^2, \quad S_3 = \mathrm{tr}\mathbf{S}_d^3, \quad \mu_1 = \mathrm{tr}\{\mathbf{S}_d \boldsymbol{\varepsilon}_P\}, \quad \mu_3 = \mathrm{tr}\{\mathbf{S}_d^2 \boldsymbol{\varepsilon}_P\}. \tag{10.21}$$

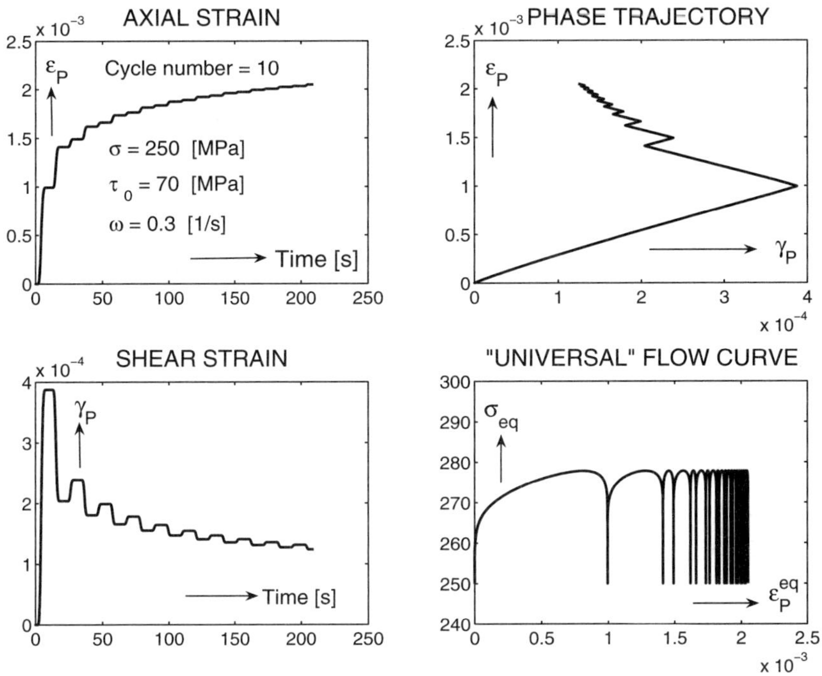

Fig. 10.2. Ratchetting under stress control simulated by Perzyna–Chaboche model

If we take for plastic stretching a second-order approximation in stress and linear approximation in plastic strain, we arrive at the formula

$$\begin{aligned}\Lambda^{-1}\,\mathbf{D}_P &= (A_1 + A_3\mu_1)\,\mathbf{S}_d + A_2\left(\mathbf{S}_d^2\right)_d \\ &\quad + \frac{1}{2}A_3 S_2 \boldsymbol{\varepsilon}_P + A_4\left(\mathbf{S}_d\boldsymbol{\varepsilon}_P + \boldsymbol{\varepsilon}_P\mathbf{S}_d\right)_d,\end{aligned} \tag{10.22}$$

where again subscript d denotes the deviatoric part of a second-rank tensor. Let us introduce the *overstress tensor* (cf. (2.21)) by means of

$$\delta\mathbf{S} \equiv \mathbf{S} - \mathbf{S}^*. \tag{10.23}$$

Moreover, we need to know a direction of $\mathbf{S}^*$. Let us assume that dynamic stress $\mathbf{S}$ as well as static stress $\mathbf{S}^*$ have the same direction. This is justified by the hypothetical nature of the static stress, namely, it is just a projection of the dynamic stress tensor to the yield surface in the stress space. Hence

$$\mathbf{S}_d = \sqrt{S_2}\,\mathbf{M}, \qquad \mathbf{S}_d^* = \sqrt{S_2^*}\,\mathbf{M}, \tag{10.24}$$

where a unit tensor $\mathbf{M}$ with the property $\mathrm{tr}\mathbf{M}^2 = 1$ is introduced. Let us denote the initial static yield stress by Y_0, i.e., $Y_0^2 = S_{2_0}^*$ and a relative equivalent overstress by means of

$$\Delta = h_{eq} - h^*_{eq} \equiv \sqrt{\frac{S_2}{S_{2_0}}} - \sqrt{\frac{S^*_2}{S^*_{2_0}}}. \tag{10.25}$$

The scalar coefficient Λ is obtained from the condition that plastic stretching vanishes when overstress equals zero. Then the evolution equation takes its final form:

$$\begin{aligned} \mathbf{D}_P = \Delta \left[A_1 \mathbf{S}_d + A_4 \left(\mathbf{S}_d \varepsilon_P + \varepsilon_P \mathbf{S}_d \right)_d \right] \\ + \Delta^2 \left[A_2 \left(\mathbf{S}_d^2 \right)_d + A_3 \left(\mu_1 \mathbf{S}_d + \frac{1}{2} S_2 \varepsilon_P \right) \right] \end{aligned} \tag{10.26}$$

On the basis of experimental data for tension and shear of AISI 316H ([AMP91]) material constants of this model are here found to be

$$\begin{aligned} \mathcal{A}_c &= \{A_1,\ A_2,\ A_3,\ A_4,\ \lambda_1,\ \lambda_2\} \\ &= \{0.01181,\ 0.01033,\ -0.008012,\ -0.01978,\ 20,\ 0.2\} \end{aligned} \tag{10.27}$$

by means of the Nelder–Mead method of analyzing chi-square functional. The two new constants λ_1 and λ_2 appearing in (10.27) arrive from the usual approximation of static equivalent flow stress:

$$h^*_{eq} = (1 + \lambda_1 p)^{\lambda_2} \tag{10.28}$$

where p is the accumulated plastic strain. The governing set of differential equations is obtained from (10.26) by substituting constants (10.27) into it. Its integration for the given stress history produced the results shown in Fig. 10.3.

10.5 Comparisons and concluding remarks

In the papers [LRC79, HW-87, BM-87, MSS85] the authors have investigated ratchetting behaviour in combined tension–torsion tests. On the other hand, the paper [RK-88] deals with tension tests only. For comparison purposes the paper [LRC79] is especially convenient since its loading programme was such that after 10 cycles the maximal normal strain was 2.6% at the corresponding tension stress of 250 MPa and a prescribed harmonically changed shear strain whose amplitude was 0.17%. Of course, the specimens tested were made of the stainless steel AISI 316H.

Examining the situation of the end point of the phase trajectory calculated by the Perzyna–Chaboche model (Fig. 10.2.) we see that at these conditions the maximal normal strain has the value of 0.18% which means that the predicted strain is approximately 15 times smaller than the corresponding experimentally acquired strain.

On the other hand, the phase trajectory depicted in Fig. 10.3 gives the maximal tension strain equal to 2.4%. Therefore the relative error amounts to 8%.

A short conclusion could be formulated as follows:

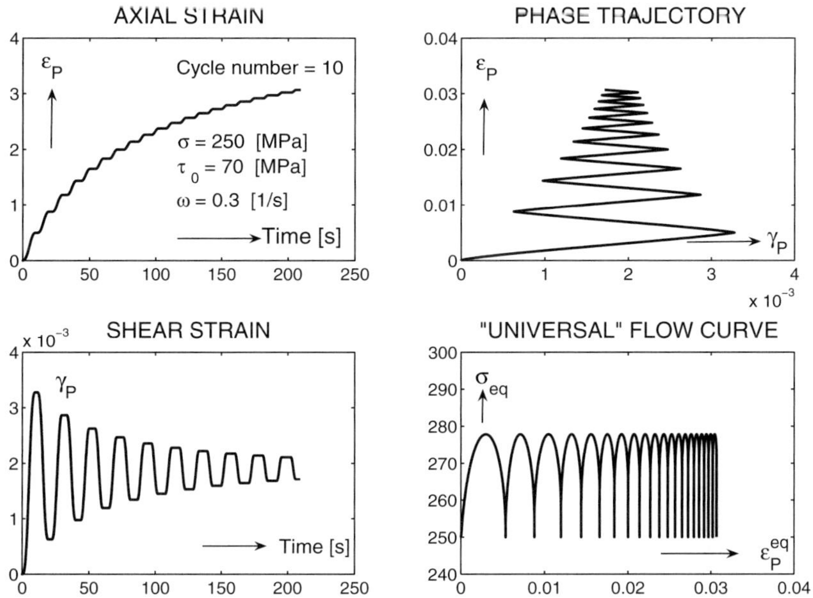

Fig. 10.3. Ratchetting with stress control for the overstress MAM model with generalized Rice–Ziegler normality

- Although handicapped by the absence of compression data which are essential for a good cyclic behaviour prediction, the tensor representation model [Mic96a] has been shown to cover multiaxial variable stress–strain histories in a surprisingly good way.
- An eventual inclusion of the residual stress tensor into (10.26) would even improve the predicting abilities of the tensor representation based viscoplasticity models. This depends on the available experimental data bank and not on theoretical considerations.
- Both models for the given number of stress cycles do not forecast a commencement of elastic strain vibrations, i.e., the shakedown behaviour (cf. also [Kng82]). However, the model of Perzyna–Chaboche aside from poor predictions gives results very near these vibrations.

11

Stress and Strain Measures for Orthotropic Metals at Large Nonproportional Plastic Strain Histories

This chapter reexamines the definitions of equivalent Mises stress and equivalent plastic strain rate for anisotropic materials. Such an issue is of great importance if plastic strain induced anisotropy of initially isotropic materials (for instance, car body sheets) is considered. It is shown that for nonproportional strain histories with alternating simple tension and simple shear intervals, the traditional Hill's definition for orthotropic materials contains variable material functions instead of three material constants. It is demonstrated that tensor function representation based on plastic strain and stress tensors is worthwhile to be considered even in the framework of classical concepts.

11.1 Generalities

For each constitutive theory clear stress and strain measures are necessary. It is commonly accepted that aside from undeformed configuration (χ_0) and instant deformed configuration (χ_t) an intermediate local reference configuration (ν_t) is introduced. Then Kröner's decomposition rule holds:

$$\mathbf{F} = \mathbf{F}^E \mathbf{F}^P \tag{11.1}$$

where $\mathbf{F}$ is the deformation gradient tensor, $\mathbf{F}^E$ the elastic distortion tensor and $\mathbf{F}^P$ the plastic distortion tensor mapping, respectively, $(\chi_0) \rightarrow (\chi_t), (\nu_t) \rightarrow (\chi_t)$ and $(\chi_0) \rightarrow (\nu_t)$. Concerning plastic strain measures, two of

M.V. Mićunović, *Thermomechanics of Viscoplasticity: Fundamentals and Applications*, Advances in Mechanics and Mathematics 20, DOI: 10.1007/978-0-387-89490-4_11,

© Springer Science + Business Media, LLC 2009

them could be convenient for us—both of them related to the intermediate reference configuration. First is the logarithmic plastic strain

$$\boldsymbol{\varepsilon}^P = \frac{1}{2}\ln(\mathbf{F}^P\mathbf{F}^{PT}) \tag{11.2}$$

where the superscript T denotes transpose. Its advantage is that it is traceless exactly for large plastic strains

$$\varepsilon_{11}^P + \varepsilon_{22}^P + \varepsilon_{33}^P = 0 \tag{11.3}$$

if plastic volume change is negligible, i.e., if $\det \mathbf{F}^P = 1$. However, its shortcoming is that for nondiagonal $\mathbf{F}^P$ (which always appears in the case of shear) its determination is tedious since we have to find proper directions of the symmetric tensor $\mathbf{F}^P\mathbf{F}^{PT}$, determine logarithms of its eigenvalues and then transform such obtained diagonal matrix to original coordinate directions.

Another choice is the Eulerian plastic strain tensor

$$\mathbf{e}^P = \frac{1}{2}\left(\mathbf{F}^P\mathbf{F}^{PT} - \mathbf{1}\right) \tag{11.4}$$

which is not traceless but its determination is straightforward. Of course $\boldsymbol{\varepsilon}^P = 0 \Leftrightarrow \mathbf{e}^P = 0$ or if one of them is equal to zero, then the other also must be zero. This is easily seen from the relationship $\boldsymbol{\varepsilon}^P = \ln(\mathbf{1}+2\,\mathbf{e}^P)/2$. It is important to note that due to the plastic incompressibility only five out of six components of $\boldsymbol{\varepsilon}^P$ are independent. This allows us to use only the deviatoric part of $\mathbf{e}^P$, i.e.,

$$\mathbf{e}_d^P = \mathbf{e}^P - \mathbf{1}\,\mathrm{tr}\{\mathbf{e}^P\}/3 \qquad (e_{d,11}^P + e_{d,22}^P + e_{d,33}^P = 0). \tag{11.5}$$

Apart from plastic strains, as another strain measure the Lagrangian elastic strain will be used:

$$\mathbf{E}^E = (\mathbf{F}^{ET}\mathbf{F}^E - \mathbf{1})/2. \tag{11.6}$$

All these strain measures are referred to vectorial base vectors of (ν_t). Another tensor connected also to the configuration (ν_t), being of importance for the following considerations, is the *plastic stretching tensor*:

$$\mathbf{D}^P = \frac{1}{2}[\dot{\mathbf{F}}^P(\mathbf{F}^P)^{-1} + (\mathbf{F}^P)^{-T}(\dot{\mathbf{F}}^P)^T]. \tag{11.7}$$

where the superposed dot stands for material differentiation. For incremental plasticity theories it is often convenient to introduce the plastic strain increment tensor by the following identity:

$$d\tilde{\boldsymbol{\varepsilon}}^P \equiv \mathbf{D}^P dt, \tag{11.8}$$

where dt is infinitesimal time increment.

Remark 11.1 (On plastic strain rate tensors).

It is often preferred to use either $\dot{\boldsymbol{\varepsilon}}^P$ or $\dot{\mathbf{e}}^P$ instead of $\mathbf{D}^P$. These two tensors may be called plastic strain rate tensors. However, $\mathrm{tr}\mathbf{D}^P = 0$ for arbitrarily large strains if condition (11.4) is satisfied. On the other hand traces, $\mathrm{tr}\dot{\boldsymbol{\varepsilon}}^P \approx 0$ as well as $\mathrm{tr}\dot{\mathbf{e}}^P \approx 0$ only for small strains. Indeed, representing $\boldsymbol{\varepsilon}^P$ by its eigenvectors and eigenvalues

$$\boldsymbol{\varepsilon}^P = \sum_{\alpha=1}^{3} \varepsilon_\alpha^P \mathbf{n}_\alpha \otimes \mathbf{n}_\alpha \tag{11.9}$$

we see that although $\dot{\varepsilon}_1^P + \dot{\varepsilon}_2^P + \dot{\varepsilon}_3^P = 0$ we have $\mathrm{tr}\dot{\boldsymbol{\varepsilon}}^P \neq 0$ since eigenvectors are time-dependent. In the special case of pure normal strains when shears are equal to zero $\mathbf{n}_\alpha = \mathrm{const}(\alpha = 1, 2, 3)$ and in this special case $\dot{\boldsymbol{\varepsilon}}^P = \mathbf{D}^P$. However, such an equality does not hold for shear strains.

According to the discussion that the elastic strain is caused by the corresponding stress tensor, Hooke's law (compare (3.21)) holds for the mapping $(\nu_t) \rightarrow (\chi_t)$ and it should be written in an invariant way connected to the intermediate referential configuration (ν_t). For this aim aside from stress tensor $\boldsymbol{\sigma}$ present in (χ_t) configuration, called *Cauchy stress* (or *true stress*), we quote also *first* and *second Piola–Kirchhoff stress tensor*

$$\boldsymbol{\sigma}^{eng} = (\det \mathbf{F})\boldsymbol{\sigma}\mathbf{F}^{-T}, \quad \mathbf{S} = (\mathbf{F}^E)^{-1}\boldsymbol{\sigma}(\mathbf{F}^E)^{-T}, \tag{11.10}$$

respectively. The first is connected to (χ_t) and (χ_0), its matrix is not symmetric in general and often it is called engineering stress, whereas the second is referred to natural state local configuration (ν_t). If elastic strain is much smaller than finite total strain, i.e., $2\,\|\mathbf{E}^E\| \ll \|\mathbf{F}^T\mathbf{F} - \mathbf{1}\|$, then Hooke's law reads

$$\mathbf{S} = \mathbb{D}\colon \mathbf{E}^E \quad \text{or} \quad \text{equivalently} \quad \mathbf{E}^E = \mathbb{D}^{-1}\colon \mathbf{S}. \tag{11.11}$$

Fourth-rank tensor of material constants $\mathbb{D}$ depends in general on temperature and damage.

If elastic strains are small (for metals they are of the order of 10^{-3}), then difference between true stress and the second Piola–Kirchhoff stress becomes negligible.

The following principal invariants (being the same in all coordinate frames) will also be necessary in the sequel:

$$s_2 = \mathrm{tr}\{\mathbf{S}_d^2\}, \quad \bar{\sigma} \equiv (3s_2/2)^{1/2}, \quad \dot{\bar{\varepsilon}}^P = (2\mathbf{D}^P : \mathbf{D}^P/3)^{1/2}. \tag{11.12}$$

Here again subscript d is used to denote the deviatoric part of a second-rank tensor: $\mathbf{S}_d \equiv \mathbf{S} - \mathbf{1}\,\mathrm{tr}\,\{\mathbf{S}\}/3$. Special attention should be paid here to $\bar{\sigma}$ and $\dot{\bar{\varepsilon}}^P$. For isotropic materials they are called *equivalent stress* and *equivalent plastic strain rate*, respectively.

Remark 11.2 (On plastic strain rate tensors).

Suppose that we have another reference configuration $(\chi_0^{'})$ *such that* $\mathbf{F}_{\text{prev}}$ maps $(\chi_0^{'}) \to (\chi_0)$. In this case let $\mathbf{F}_{rel}, \mathbf{F}_{rel}^E, \mathbf{F}_{rel}^P$ denote the tensors defined above, mapping, respectively, $(\chi_0) \to (\chi_t), (\nu_t) \to (\chi_t)$ and $(\chi_0) \to (\nu_t)$. Let, moreover, previous elastic, i.e., residual, strain be negligible such that $\mathbf{F}_{\text{prev}} \approx \mathbf{F}_{\text{prev}}^P$. Then total deformation gradient tensor $\boldsymbol{F}$ which maps $(\chi_0^{'}) \to (\chi_t)$ equals

$$\mathbf{F} = \mathbf{F}_{rel}\mathbf{F}_{\text{prev}} \quad \textit{and} \quad \mathbf{F}^E = \mathbf{F}_{rel}^E, \quad \mathbf{F}^P = \mathbf{F}_{rel}^P \mathbf{F}_{\text{prev}}. \tag{11.13}$$

In accordance with (11.7) relative plastic stretching reads

$$\mathbf{D}_{rel}^P = \left[\dot{\mathbf{F}}_{rel}^P (\mathbf{F}_{rel}^P)^{-1} + (\mathbf{F}_{rel}^P)^{-T} (\dot{\mathbf{F}}_{rel}^P)^T\right] /2. \tag{11.14}$$

Due to the fact that $\mathbf{F}_{\text{prev}} = \text{const}$ we have the equality

$$\mathbf{D}_{\text{rel}}^P = \mathbf{D}^P. \tag{11.15}$$

Finally, it should be noted that $\boldsymbol{\sigma}^{eng}$ may be defined by means of $\mathbf{F}_{rel}$ instead of $\mathbf{F}$.

11.2 Previous tension in rolling direction

Consider a rolled plate with the rolling direction along the coordinate axis X_1 and coordinate perpendicular to the sheet by X_3. Then the previous deformation gradient tensor reads

$$\mathbf{F}_{\text{prev}} = \begin{Bmatrix} 1 + \varepsilon_{E\text{prev}} & 0 & 0 \\ 0 & 1 + \varepsilon_{2\text{prev}} & 0 \\ 0 & 0 & 1 + \varepsilon_{3\text{prev}} \end{Bmatrix}. \tag{11.16}$$

Here $\varepsilon_{E\text{prev}} = \Delta L'/L_0'$ is the previous engineering strain measured only in the rolling direction whereas $\varepsilon_{3\text{prev}}$ is the change of thickness and $\varepsilon_{2\text{prev}}$ is the change of width. Neglecting previous elastic strain and assuming that plastic incompressibility holds, we obtain finally

$$\varepsilon_{2\text{prev}} = (1 + \varepsilon_{E\text{prev}})^{-\xi} - 1, \quad \varepsilon_{3\text{prev}} = (1 + \varepsilon_{E\text{prev}})^{-1+\xi} - 1 \tag{11.17}$$

where for arbitrary scalar ξ we have $\det \mathbf{F}_{\text{prev}} = 1$ which was desired by the above assumptions. In the case of proportional straining path $\xi = \text{const}$. Taking into account these formulae for the *plastic deformation velocity tensor* we have

$$\mathbf{L}_{\text{prev}}^P \equiv \dot{\mathbf{F}}_{\text{prev}} \mathbf{F}_{\text{prev}}^{-1} \approx \frac{\dot{\varepsilon}_{E\text{prev}}}{1 + \varepsilon_{E\text{prev}}} \begin{Bmatrix} 1 & 0 & 0 \\ 0 & -\xi & 0 \\ 0 & 0 & -1 + \xi \end{Bmatrix} \equiv \frac{\dot{\varepsilon}_{E\text{prev}}}{1 + \varepsilon_{E\text{prev}}} \mathbf{A}_{\text{prev}}. \tag{11.18}$$

where $\mathbf{A}_{\text{prev}}$ is a constant tensor describing its direction. Its symmetric part is the *plastic stretching tensor* (cf. (11.7)):

$$\mathbf{D}^P_{\text{prev}} = \frac{1}{2}\left(\mathbf{L}^P_{\text{prev}} + \mathbf{L}^{PT}_{\text{prev}}\right) = \mathbf{A}_{\text{prev}} \frac{d}{dt}\left[\ln(1+\varepsilon_{E\text{prev}})\right] \tag{11.19}$$

whose second invariant leads to the standard definition of *equivalent plastic strain rate*:

$$\dot{\bar{\varepsilon}}^P_{\text{prev}} = \left[\frac{2}{3}\mathbf{D}^P_{\text{prev}} : \mathbf{D}^P_{\text{prev}}\right]^{1/2} = \frac{2}{\sqrt{3}}\sqrt{1-\xi+\xi^2}\frac{d}{dt}\left[\ln(1+\varepsilon_{E\text{prev}})\right]. \tag{11.20}$$

It must be noted, however, that in the frame of classical theory governed by Levi-Mises equations such a definition holds true only for isotropic materials. For orthotropic materials such framework is extended here. If (11.20) is accepted, then the accumulated plastic strain amounts to

$$\bar{\varepsilon}^P_{\text{prev}} = (2/\sqrt{3})\sqrt{1-\xi+\xi^2}\ln\left(1+\varepsilon_{E\text{prev}}\right). \tag{11.21}$$

In the sequel this will be considered as an input into subsequent tension or shear experiments.

The previous plastic strain tensor and its deviator (with the notation: $\Psi \equiv (1+\varepsilon_{\text{prev}})^2$) are

$$\begin{aligned}\mathbf{e}_{\text{prev}} &= \frac{1}{2}(\mathbf{F}_{\text{prev}}\mathbf{F}^T_{\text{prev}} - \mathbf{1}) \\ &= \frac{1}{2}\begin{Bmatrix} -1+\Psi & 0 & 0 \\ 0 & -1+\Psi^{-\xi} & 0 \\ 0 & 0 & -1+\Psi^{-1+\xi}\end{Bmatrix},\end{aligned} \tag{11.22}$$

$$6\,\mathbf{e}_{d,\text{prev}} = \tag{11.23}$$

$$\begin{Bmatrix} 2\Psi - \Psi^{-\xi} - \Psi^{-1+\xi} & 0 & 0 \\ 0 & -\Psi + 2\Psi^{-\xi} - \Psi^{-1+\xi} & 0 \\ 0 & 0 & -\Psi - \Psi^{-\xi} + 2\Psi^{-1+\xi}\end{Bmatrix}.$$

On the other hand, the logarithmic plastic strain tensor following (11.2) reads (cf. also (11.18))

$$\varepsilon_{\text{prev}} = \ln(\mathbf{F}_{\text{prev}}\mathbf{F}^T_{\text{prev}})/2 = \ln\left(1+\varepsilon_{E\text{prev}}\right)\,\mathbf{A}_{\text{prev}}. \tag{11.24}$$

Remark 11.3 (In-plane previous straining).

If previous straining is in-plane, then $\xi = 0$ which simplifies the above relationships such that (11.21) and (11.18) are reduced to

$$\bar{\varepsilon}^P_{\text{prev}} = 2\ \ln\left(1 + \varepsilon_{E\text{prev}}\right)/\sqrt{3} \quad \textit{and} \quad \mathbf{A}_{\text{prev}} = \begin{Bmatrix} 1\,0\,0 \\ 0\,0\,0 \\ 0\,0\,-1 \end{Bmatrix}.$$

Of course, the above relations will require modification for orthotropic materials. The corresponding simplification of (11.22) gives in-plane strain state, i.e.,

$$\mathbf{e}_{\text{prev}} = \frac{1}{2}\begin{Bmatrix} -1+\Psi & 0\,0 \\ 0 & 0\,0 \\ 0 & 0\,-1+\Psi^{-1} \end{Bmatrix}.$$

11.3 Subsequent in-plane tension in arbitrary direction

Suppose that the straining process depicted in Fig. 11.1 has two steps. Namely, after the previous straining along axis X_1' the subsequent uniaxial tension along x_1 axis takes place.

Let preferred directions in (χ_0) be given by "fibre" vectors $\mathbf{a}_{01}$ and $\mathbf{a}_{02}$. If uniaxial stress is in the α-direction with respect to material axis X_1, then "fibre" vectors take new positions determined by angles β_1, β_2 with respect to spatial axes x_1, x_2. For convenience such obtained "fibre" vectors $\mathbf{a}_1$ and $\mathbf{a}_2$ are taken to be unit. Also, all the matrices (constituting tensor components) in this and next section will be written in spatial coordinate frame x_k $(k \in \{1,2,3\})$. First, relative deformation gradient tensor reads

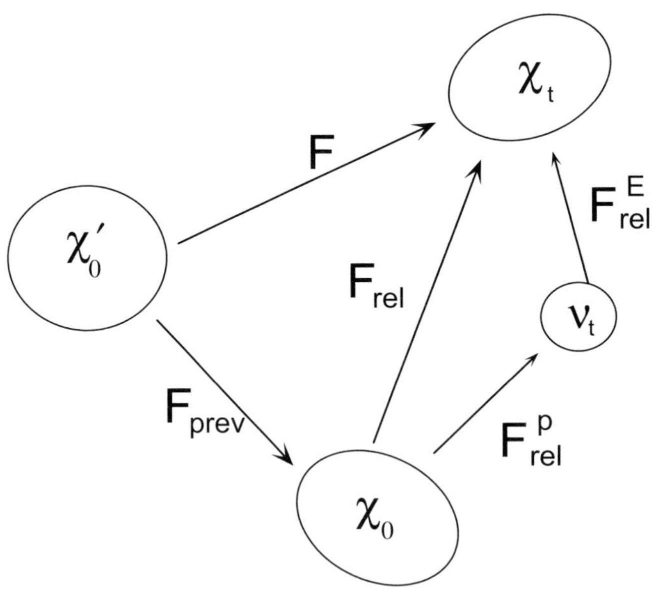

Fig. 11.1. Decomposition of deformation gradients

$$\mathbf{F}_{\text{rel}} = \begin{Bmatrix} 1+p & 0 & 0 \\ 0 & 1+p_2 & 0 \\ 0 & 0 & \dfrac{1}{(1+p)(1+p_2)} \end{Bmatrix} \begin{Bmatrix} 1 & \gamma & 0 \\ 0 & 1 & 0 \\ 0 & 0 & 1 \end{Bmatrix}$$
$$= \begin{Bmatrix} 1+p & \gamma(1+p) & 0 \\ 0 & 1+p_2 & 0 \\ 0 & 0 & \dfrac{1}{(1+p)(1+p_2)} \end{Bmatrix}. \tag{11.25}$$

if elastic strain is neglected, i.e.,

$$\mathbf{F}_{rel} \approx \mathbf{F}^P_{rel} \tag{11.26}$$

and plastic incompressibility ($\det \mathbf{F}^P = 1$) holds. Then unit vectors $\mathbf{a}_1$ and $\mathbf{a}_2$ read

$$\mathbf{a}_1 = \mathbf{F}_{rel} \cdot \mathbf{a}_{01} = \begin{Bmatrix} \cos\beta_1 \\ -\sin\beta_1 \\ 0 \end{Bmatrix},$$
$$\cos\beta_1 = \frac{(1+p)(\cos\alpha - \gamma\sin\alpha)}{\sqrt{(1+p)^2(\cos\alpha-\gamma\sin\alpha)^2 + (1+p_2)^2\sin^2\alpha}}, \tag{11.27}$$

$$\mathbf{a}_2 = \mathbf{F}_{rel} \cdot \mathbf{a}_{02} = \begin{Bmatrix} \cos\beta_2 \\ -\sin\beta_2 \\ 0 \end{Bmatrix},$$
$$\sin\beta_2 = \frac{(1+p)(\sin\alpha - \gamma\cos\alpha)}{\sqrt{(1+p)^2(\sin\alpha+\gamma\cos\alpha)^2 + (1+p_2)^2\cos^2\alpha}}. \tag{11.28}$$

It should be noted that usually during a testing only the strain along the tension direction is measured, i.e.,

$$p \approx \varDelta L_0/L_0, \tag{11.29}$$

whereas p_2 and γ have not been measured. Moreover the two following fibre dyadic deviators with the corresponding fibre stress invariants are

$$\mathbf{A}_1 = (\mathbf{a}_1 \otimes \mathbf{a}_1)_d = \frac{1}{3}\begin{Bmatrix} 3\cos^2\beta_1 - 1 & -3\sin\beta_1\cos\beta_1 & 0 \\ -3\sin\beta_1\cos\beta_1 & 3\sin^2\beta_1 - 1 & 0 \\ 0 & 0 & -1 \end{Bmatrix},$$

$$A_{\sigma_1} = \boldsymbol{\sigma} : \mathbf{A}_1 = \frac{1}{3}(3\cos^2\beta_1 - 1)\,\sigma,$$

$$\mathbf{A}_2 = (\mathbf{a}_2 \otimes \mathbf{a}_2)_d = \frac{1}{3}\begin{Bmatrix} 3\sin^2\beta_2 - 1 & 3\sin\beta_2\cos\beta_2 & 0 \\ 3\sin\beta_2\cos\beta_2 & 3\cos^2\beta_2 - 1 & 0 \\ 0 & 0 & -1 \end{Bmatrix},$$

$$A_{\sigma_2} = \boldsymbol{\sigma} : \mathbf{A}_2 = \frac{1}{3}(3\sin^2\beta_2 - 1)\,\sigma.$$

Of course, here $\bar{\sigma} = \sigma$ because the stress tensor as well as its deviator read

$$\boldsymbol{\sigma} = \sigma \begin{Bmatrix} 1\,0\,0 \\ 0\,0\,0 \\ 0\,0\,0 \end{Bmatrix}, \quad \boldsymbol{\sigma}_d = \frac{\sigma}{3} \begin{Bmatrix} 2\;0 & 0 \\ 0\;-1\;0 \\ 0\;0 & -1 \end{Bmatrix}. \tag{11.30}$$

The corresponding engineering stress tensor is determined by (cf. (11.10))

$$\boldsymbol{\sigma}^{eng} = \frac{\sigma}{1+p} \begin{Bmatrix} 1\,0\,0 \\ 0\,0\,0 \\ 0\,0\,0 \end{Bmatrix}, \quad \sigma^{eng} = \frac{\sigma}{1+p} \quad \text{with} \quad \sigma^{eng} = \frac{F}{A_0}. \tag{11.31}$$

Now the plastic stretching tensor (11.7) (with tacitly introduced notations $\pi \equiv \ln(1+p)$ and $\xi \equiv \ln(1+p_2) - \ln(1+p)$) is equal to

$$\mathbf{D}^P = \begin{Bmatrix} \dot{\pi} & \frac{1}{2}\dot{\gamma}\; e^{-x} & 0 \\ \frac{1}{2}\dot{\gamma}\; e^{-x} & \dot{\pi} + \dot{\xi} & 0 \\ 0 & 0 & -\,2\dot{\pi}\,-\,\dot{\xi} \end{Bmatrix}. \tag{11.32}$$

The related invariant showing the magnitude of (11.32) according to list (11.12) equals

$$\dot{\bar{\varepsilon}}^P = \sqrt{4\;(\dot{\pi}^2 + \dot{\pi}\;\dot{\xi} + \;\dot{\xi}^2/3) + \dot{\gamma}^2 e^{-2\xi}/3}, \quad d\bar{\varepsilon}^P = \dot{\bar{\varepsilon}}^P dt. \tag{11.33}$$

Another two invariants related to the fibre directions are defined by means of

$$A_{d\varepsilon_1} = \mathbf{A}_1 \,:\, d\tilde{\boldsymbol{\varepsilon}}^P = d\pi + d\xi\;\sin^2\beta_1 - d\gamma\; e^{-\xi}\sin\beta_1\cos\beta_1, \tag{11.34}$$

$$A_{d\varepsilon_2} = \mathbf{A}_2 \,:\, d\tilde{\boldsymbol{\varepsilon}}^P = d\pi + d\xi\;\cos^2\beta_2 + d\gamma\; e^{-\xi}\sin\beta_2\cos\beta_2. \tag{11.35}$$

a) Rolling direction. In this case $\alpha = 0$, $\cos\beta_1 = 1$, $\sin\beta_2 = \gamma\left(\gamma^2 + e^{2\xi}\right)^{-1/2}$, $d\bar{\sigma}$ and $d\overline{\varepsilon^P}$ remain unchanged, such that

$$\mathbf{A}_1 = \frac{1}{3}\begin{Bmatrix} 2\;0 & 0 \\ 0\;-1\;0 \\ 0\;0 & -1 \end{Bmatrix},$$

$$\mathbf{A}_2 = \frac{1}{3(\gamma^2 + e^{2\xi})}\begin{Bmatrix} 2\gamma^2 - e^{2\xi} & 3\,\gamma\, e^{\xi} & 0 \\ 3\,\gamma\, e^{\xi} & 2e^{2\xi} - \gamma^2 & 0 \\ 0 & 0 & -\gamma^2 - e^{2\xi} \end{Bmatrix},$$

$$A_{\sigma_1} = \tfrac{2}{3}\sigma, \; A_{\sigma_2} = \frac{\sigma}{3}\frac{2\gamma^2 - e^{2\xi}}{\gamma^2 + e^{2\xi}},$$

$$A_{d\varepsilon_1} = d\pi, \; A_{d\varepsilon_2} = \frac{1}{\gamma^2 + e^{2\xi}}\left[(\gamma^2 + e^{2\xi})d\pi + e^{2\xi}d\xi + \gamma d\gamma\right].$$

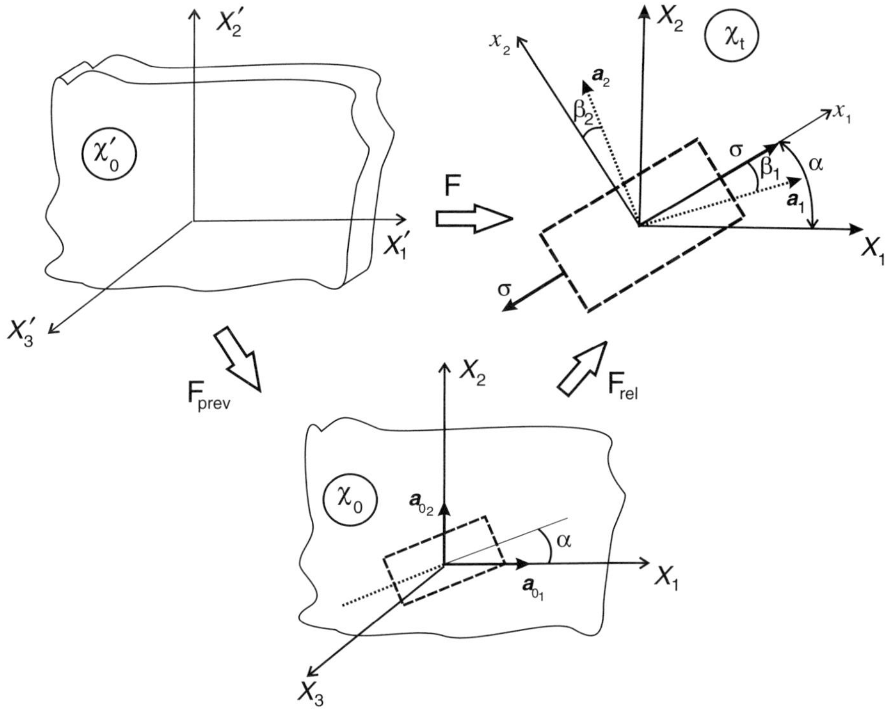

Fig. 11.2. Subsequent in-plane tension in an α-direction

b) Direction perpendicular to rolling. According to Fig. 11.2 we have $\alpha = \pi/2$, $\sin\beta_2 = 1$, $\sin\beta_1 = -\gamma\left(\gamma^2 + e^{2\xi}\right)^{-1/2}$, leading to the following simplified expressions:

$$\mathbf{A}_1 = \frac{1}{3(\gamma^2 + e^{2\xi})} \begin{Bmatrix} 2\gamma^2 - e^{2\xi} & 3\,\gamma\, e^{\xi} & 0 \\ 3\,\gamma\, e^{\xi} & 2e^{2\xi} - \gamma^2 & 0 \\ 0 & 0 & -\gamma^2 - e^{2\xi} \end{Bmatrix},$$

$$\mathbf{A}_2 = \frac{1}{3} \begin{Bmatrix} 2 & 0 & 0 \\ 0 & -1 & 0 \\ 0 & 0 & -1 \end{Bmatrix},$$

$$A_{\sigma_1} = \tfrac{\sigma}{3}\tfrac{2\gamma^2 - e^{2\xi}}{\gamma^2 + e^{2\xi}}, \qquad A_{\sigma_2} = \frac{2}{3}\sigma,$$

$$A_{d\varepsilon_1} = \tfrac{1}{\gamma^2 + e^{2\xi}}\left[(\gamma^2 + e^{2\xi})d\pi + e^{2\xi}d\xi + \gamma d\gamma\right], \; A_{d\varepsilon_2} = d\pi.$$

c) Equibiaxial direction. This special case corresponds to $\sin\beta_1 = e^{\xi}\left[(1-\gamma)^2 + e^{2\xi}\right]^{-1/2}$, $\alpha = \pi/4$, $\sin\beta_2 = (1+\gamma)\left[(1+\gamma)^2 + e^{2\xi}\right]^{-1/2}$ and

$$\mathbf{A}_1 = \frac{1}{3[(1-\gamma)^2 + e^{2\xi}]} \begin{Bmatrix} 2(1-\gamma)^2 - e^{2\xi} & -3(1-\gamma)e^{\xi} & 0 \\ -3(1-\gamma)e^{\xi} & 2e^{2\xi} - (1-\gamma)^2 & 0 \\ 0 & 0 & -(1-\gamma)^2 - e^{2\xi} \end{Bmatrix},$$

$$\mathbf{A}_2 = \frac{1}{3[(1+\gamma)^2 + e^{2\xi}]} \begin{Bmatrix} 2(1+\gamma)^2 - e^{2\xi} & 3(1+\gamma)e^{\xi} & 0 \\ 3(1+\gamma)e^{\xi} & 2e^{2\xi} - (1+\gamma)^2 & 0 \\ 0 & 0 & -(1+\gamma)^2 - e^{2\xi} \end{Bmatrix},$$

$$A_{\sigma_1} = \frac{\sigma}{3}\,\frac{2(1-\gamma)^2 - e^{2\xi}}{(1-\gamma)^2 + e^{2\xi}},$$

$$A_{d\varepsilon_1} = \frac{1}{(1-\gamma)^2 + e^{2\xi}} \left\{ \left[(1-\gamma)^2 + e^{2\xi}\right] d\pi + e^{2\xi} d\xi - (1-\gamma) d\gamma \right\},$$

$$A_{\sigma_2} = \frac{\sigma}{3}\,\frac{2(1+\gamma)^2 - e^{2\xi}}{(1+\gamma)^2 + e^{2\xi}},$$

$$A_{d\varepsilon_2} = \frac{1}{(1+\gamma)^2 + e^{2\xi}} \left\{ \left[(1+\gamma)^2 + e^{2\xi}\right] d\pi + e^{2\xi} d\xi + (1+\gamma) d\gamma \right\}.$$

However, it must be noted that in all the special cases

$$\bar{\sigma} = \sigma, \quad d\bar{\varepsilon}^P = \frac{1}{\sqrt{3}} \left(4\,(3\,d\pi^2 + 3\,d\pi\,d\xi + d\xi^2) + e^{-2\xi} d\gamma^2 \right)^{1/2} \tag{11.36}$$

irrespective of stress direction.

11.4 Subsequent shear in arbitrary direction

Let us consider another two-step straining process composed of previous straining in rolling direction followed by shear in arbitrary direction in sheet plane—Fig. 11.3.

In the previous section it was assumed that a tensile stress causes not only normal strains but shearing strains as well. Similarly, it is natural to assume here that shear stress τ has as a consequence also normal strains. Then, formulae (11.25)–(11.28) as well as (11.32)–(11.35) remain unchanged.

However, stress tensor and stress invariants must be analyzed carefully. Suppose that we know, as in a general case, only the reference configuration (χ_0) and that the body in it has the shape of a rectangular block with thickness $\mathcal{H}$. If a subsequent shear happens perpendicular to the thickness direction with a displacement equal to x, then the measured strain amounts to

$$\gamma = \frac{x}{\mathcal{H}}, \tag{11.37}$$

whereas the Cauchy (i.e., true) stress tensor and its magnitude have the forms

$$\boldsymbol{\sigma} = \tau \begin{Bmatrix} 0 & 1 & 0 \\ 1 & 0 & 0 \\ 0 & 0 & 0 \end{Bmatrix} = \boldsymbol{\sigma}_d, \quad \tau = \frac{F/2}{A} = \frac{F}{2b\delta}. \tag{11.38}$$

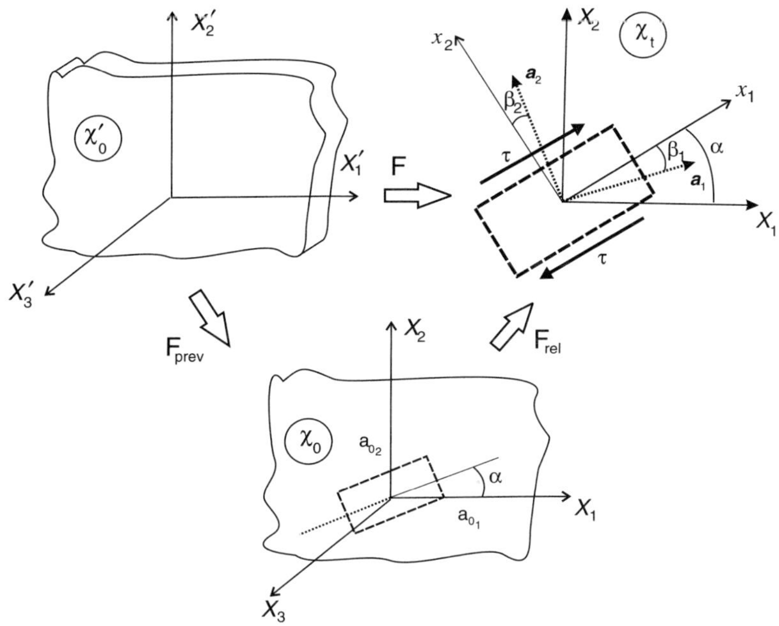

Fig. 11.3. Previous tension followed by an arbitrary α-shear

The fibre stress invariants and the stress invariant $\bar{\sigma}$ (cf. (1.15)) are

$$A_{\sigma_1} = \boldsymbol{\sigma} : \mathbf{A}_1 = -\tau \sin 2\beta_1 \qquad \text{and} \qquad A_{\sigma_2} = \boldsymbol{\sigma} : \mathbf{A}_2 = \tau \sin 2\beta_2, \quad (11.39)$$

$$\bar{\sigma} = \left\{3\,\mathrm{tr}\boldsymbol{\sigma}_d^2/2\right\}^{1/2} = \tau\sqrt{3}. \quad (11.40)$$

a) Rolling direction. For this case, $\mathbf{A}_1$ and $\mathbf{A}_2$ are the same as for subsequent tension, while fibre stress invariants are

$$A_{\sigma_1} = \boldsymbol{\sigma} : \mathbf{A}_1 = 0, \quad A_{\sigma_2} = \boldsymbol{\sigma} : \mathbf{A}_2 = 2\tau\gamma e^{\xi}\left(\gamma^2 + e^{2\xi}\right)^{-1}. \quad (11.41)$$

b) Direction perpendicular to rolling. Again $\mathbf{A}_1$ and $\mathbf{A}_2$ remain the same, but fibre stress invariants now read

$$A_{\sigma_2} = \boldsymbol{\sigma} : \mathbf{A}_2 = 0, \qquad A_{\sigma_1} = \boldsymbol{\sigma} : \mathbf{A}_1 = 2\tau\gamma e^{\xi}\left(\gamma^2 + e^{2\xi}\right)^{-1}. \quad (11.42)$$

c) Equibiaxial direction. For the direction $\alpha = \pi/4$, comparison with the case of subsequent tension shows that only fibre stress invariants are modified as follows:

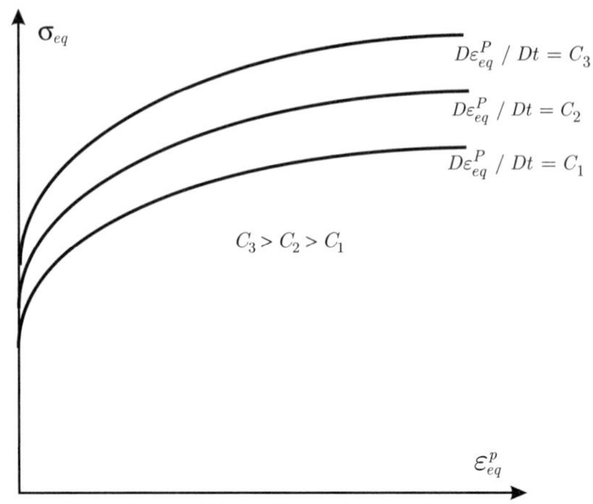

Fig. 11.4. Notion of universal flow curves

$$
\begin{aligned}
A_{\sigma_1} &= \boldsymbol{\sigma} : \mathbf{A}_1 = -2\tau \frac{(1-\gamma)e^{\xi}}{(1-\gamma)^2 + e^{2\xi}}, \\
A_{\sigma_2} &= \boldsymbol{\sigma} : \mathbf{A}_2 = 2\tau \frac{(1+\gamma)e^{\xi}}{(1+\gamma)^2 + e^{2\xi}}.
\end{aligned}
\tag{11.43}
$$

After all these cases, we may conclude that all the stress, strain and strain rate measures indispensable for a formulation of constitutive equations.

11.5 Yield function in the case of orthotropy

In this chapter we consider only models belonging to the class based on the notion of "universal" flow curve. The family of curves ([Per71])

$$
f\left(\sigma_{eq}, \varepsilon_{eq}^{P}, \dot{\varepsilon}_{eq}^{P}\right) = \sigma_{eq} - \Phi\left(\sigma_{eq}, \varepsilon_{eq}^{P}, \dot{\varepsilon}_{eq}^{P}\right) \begin{cases} = 0 \\ < 0 \end{cases} \tag{11.44}
$$

defines for $f = 0$ viscoplastic (i.e., rate dependent) deformation whereas $\sigma_{eq} - \Phi\left(\varepsilon_{eq}^{P}, 0\right) < 0$ defines elastic deformation. In a more traditional understanding the constant values of $\dot{\varepsilon}_{eq}^{P} = c_1, c_2, \ldots, c_n$ define family of universal flow curves (cf. Fig. 11.4). The word *universal* means that such curves should hold for all multiaxial strain states. It means that straining history can be accounted for only by scalar functionals of the form

$$
\sigma_{eq}(t) = \underset{\tau=0}{\overset{t}{\mathfrak{F}}} \left[\varepsilon_{eq}^{P}(\tau), \dot{\varepsilon}_{eq}^{P}(\tau)\right], \tag{11.45}
$$

where presence of $\dot{\varepsilon}_{eq}^{P}(\tau), \varepsilon_{eq}^{P}(\tau)(0 < \tau < t)$ takes into account plastic strain rate as well as plastic strain history. Now, associate flow rule formulated by

Prager and Hochenemser (and then applied by Drucker, Perzyna and others) is based on normality of plastic stretching tensor $\mathbf{D}^P$ on the yield surface $f = \text{const}$. Therefore such a postulate gives

$$\mathbf{D}^P = \Lambda \frac{\partial f}{\partial \boldsymbol{\sigma}}. \tag{11.46}$$

It must be noted that although seemingly tensorial (11.46) is completely specified by scalars listed in (11.44). Traditionally, all these scalars are determined by tension experiments and then applied to arbitrary multiaxial strain states. In the sequel we will see that such a procedure is wrong.

Remark 11.4 (Fault of "universal" flow rules).
Experiments do not support the notion of universal flow curves (cf. [AMM91]). Thus, yield function not only must include σ_{eq} and ε^P_{eq} but must be of the form

$$f\left(\boldsymbol{\sigma}_d, \varepsilon^P_d, \mathbf{D}^P\right) \begin{cases} = 0 \\ < 0 \end{cases} \tag{11.47}$$

where tensors (i.e., their selected proper and mixed invariants) instead of scalars for multiaxial stress states are to be taken into account.

Despite strong criticism mentioned in the previous note we proceed here with the approach based on universal flow curves. For orthotropic materials whose anisotropy directions are given by vectors $\mathbf{a}_1, \mathbf{a}_2 \in \xi$ the stress invariants aligned to such directions have the following form:

$$A_{\sigma_1} = \mathbf{a}_1 \cdot \boldsymbol{\sigma}_d \cdot \mathbf{a}_1 \equiv \boldsymbol{\sigma}_d : \mathbf{A}_1, \qquad A_{\sigma_2} = \mathbf{a}_2 \cdot \boldsymbol{\sigma}_d \cdot \mathbf{a}_2 \equiv \sigma_d : \mathbf{A}_2. \tag{11.48}$$

Let us introduce a function of the mentioned stress invariants

$$\sigma^2_{eq} = \frac{3}{2}\left\{ \mathrm{tr}\tilde{\boldsymbol{\sigma}}^2_d + \frac{3}{2b+2c-5}\left[(2-b)A^2_{\sigma_1} + (2-c)A^2_{\sigma_2}\right]\right\} \tag{11.49}$$

as an equivalent stress for orthotropic materials. This serves the yield function (cf. (11.44)):

$$2f = \frac{2\sigma^2_{eq}}{3h(\varepsilon^P_{eq,rel})} - 1 \begin{cases} = 0 \\ < 0 \end{cases}, \quad \Phi = \sqrt{3\,h/2}, \tag{11.50}$$

Moreover the following notation is suitable for incremental strains:

$$\mathbf{D}^P dt \equiv d\tilde{\boldsymbol{\varepsilon}}^P, \tag{11.51}$$

$$\begin{aligned} A_{d\varepsilon_1} &= \mathbf{a}_1 \cdot d\tilde{\boldsymbol{\varepsilon}}^P \cdot \mathbf{a}_1 \equiv \mathbf{A}_1 : d\tilde{\boldsymbol{\varepsilon}}^P, \\ A_{d\varepsilon_2} &= \mathbf{a}_2 \cdot d\tilde{\boldsymbol{\varepsilon}}^P \cdot \mathbf{a}_2 \equiv \mathbf{A}_2 : d\,\tilde{\boldsymbol{\varepsilon}}^P. \end{aligned} \tag{11.52}$$

Here $d\tilde{\boldsymbol{\varepsilon}}^P$ is the increment of plastic strain tensor induced by $\mathbf{D}^P$ and $A_{d\varepsilon_1}$, $A_{d\varepsilon_2}$ are its invariant scalar parts aligned with anisotropy directions (i.e., fibre directions). According to [Hil50] equivalent plastic strain rate is defined by means of equality of plastic powers such that

$$\sigma_{eq} d\varepsilon_{eq}^P = \boldsymbol{\sigma}_d : d\tilde{\boldsymbol{\varepsilon}}^P. \tag{11.53}$$

The *associate flow rule* may be shown also in the form (derived from (11.46) by means of (11.51))

$$d\tilde{\boldsymbol{\varepsilon}}^P = d\lambda \, \frac{\partial f}{\partial \boldsymbol{\sigma}}, \tag{11.54}$$

where it is easily found from (11.53) that

$$d\lambda = \sigma_{eq} \, d\varepsilon_{eq}^P. \tag{11.55}$$

Substituting (11.51) into (11.55) we arrive at evolution equations

$$d\,\tilde{\boldsymbol{\varepsilon}}^P = \frac{3}{2}\frac{d\varepsilon_{eq}^P}{\sigma_{eq}} \left\{ \boldsymbol{\sigma}_d + \frac{3}{2b+2c-5} \left[(2-b) A_{\sigma_1} \mathbf{A}_1 + (2-c) A_{\sigma_2} \mathbf{A}_2 \right] \right\}, \tag{11.56}$$

which in the case of isotropy are called Levi-Mises equations. Dyadic deviators $\mathbf{A}_1$ and $\mathbf{A}_2$ are given in the third section of this chapter. Making use of (11.56) we get a definition of the *equivalent plastic strain*:

$$\begin{aligned} \left(d\varepsilon_{eq}^P\right)^2 = \frac{2}{3}\frac{1}{bc-1} & \left[(b-2)(2b-1) A_{d\varepsilon 1}^2 + (c-2)(2c-1) A_{d\varepsilon 2}^2 \right. \\ & \left. - 2(b-2)(c-2) A_{d\varepsilon 1}^2 A_{d\varepsilon 2}^2 \right] + \frac{2}{3} tr\{d\,\tilde{\varepsilon}^P\}^2, \end{aligned} \tag{11.57}$$

where $\left(d\bar{\varepsilon}^P \right)^2 = (2/3) \, \mathrm{tr} \, d\{\tilde{\varepsilon}^{\mathrm{P}}\}^2$.

The meaning of material constants is found from Levi-Mises equations when either only σ_1 or σ_2 are different from zero as follows:

$$\begin{aligned} b \equiv 1 + r_2 = 1 + \left. \frac{d\varepsilon_{33}^P}{d\varepsilon_{11}^P} \right|_{\sigma_1 = 0, \tau = 0}, \\ c \equiv 1 + r_1 = 1 + \left. \frac{d\varepsilon_{33}^P}{d\varepsilon_{22}^P} \right|_{\sigma_2 = 0, \tau = 0}. \end{aligned} \tag{11.58}$$

In the special case of *isotropy* $r_1 = r_2 = 1$ such that $b = c = 2$ and $\sigma_{eq} = \bar{\sigma}$, $d\varepsilon_{eq}^P = d\bar{\varepsilon}^P$.

Hill in his book [Hil50] gave the definition of equivalent stress for *orthotropic* materials (with different properties in all three characteristic directions) by means of

$$
\begin{aligned}
2f &= \frac{2}{3}\frac{\sigma_{eq}^2}{h} - 1 \\
&= \frac{2}{3h}\left[F(\sigma_2-\sigma_3)^2 + G(\sigma_3-\sigma_1)^2 + H(\sigma_1-\sigma_2)^2\right] - 1.
\end{aligned}
\tag{11.59}
$$

For a sheet with in-plane preferred directions, $\mathbf{a}_1$ and $\mathbf{a}_2$, comparison of (11.58) with (11.50) gives

$$
F = \frac{b-1}{2b+2c-5}, \qquad G = \frac{c-1}{2b+2c-5}, \qquad H = \frac{1}{2b+2c-5}. \tag{11.60}
$$

In the sequel the special cases of loading are confronted with the above evolution equations. Special attention is paid to possible simplifications of Levi-Mises equations for diverse loading histories. In order to make comparison with isotropy easier, let us introduce the following notation:

$$
\sigma_{eq} \;=\; k_\sigma\,\bar{\sigma}. \tag{11.61}
$$

In the special case of isotropic materials the above-introduced scaling factor is $k_\sigma = 1$.

11.5.1 Subsequent arbitrary tension

Let the loading correspond to Fig. 11.2. Then (11.25)–(11.35) hold for arbitrary loading direction determined by an angle $\alpha \in [0, \pi/2]$. Equivalent stress and equivalent plastic strain increment are equal to

$$
\begin{aligned}
\sigma_{eq}^2 &= \sigma^2\left\{1 + \frac{1}{2(2b+2c-5)}\left[(2-b)(3\cos^2\beta_1 - 1)^2\right.\right. \\
&\quad \left.\left. + (2-c)(3\sin^2\beta_2 - 1)^2\right]\right\}, \\
\left(d\varepsilon_{eq}^P\right)^2 &= \frac{2}{3}\frac{1}{bc-1}\left[(b-2)(2b-1)A_{d\varepsilon 1}^2 + (c-2)(2c-1)A_{d\varepsilon 2}^2\right. \\
&\quad \left. -2(b-2)(c-2)A_{d\varepsilon 1}^2 A_{d\varepsilon 2}^2\right] + (d\bar{\varepsilon}^P)^2, \\
\bar{\sigma} &= \sigma, \quad (d\bar{\varepsilon}^P)^2 = 4\left(d\pi^2 + d\pi\, d\xi + \frac{1}{3}d\xi^2\right) + \frac{1}{3}\; e^{-2\xi}\, d\gamma^2,
\end{aligned}
\tag{11.62}
$$

with scalars

$$
\begin{aligned}
A_{d\varepsilon_1} &= d\pi + \sin^2\beta_1\; d\xi \;-\; e^{-\xi}\sin\beta_1\;\cos\beta_1\; d\gamma, \\
A_{d\varepsilon_2} &= d\pi + \cos^2\beta_2\; d\xi \;+\; e^{-\xi}\sin\beta_2\;\cos\beta_2\; d\gamma.
\end{aligned}
$$

Consider, now the special cases of $\alpha = 0$, $\quad \alpha = \pi/2$ and $\alpha = \pi/4$. In all these cases Levi-Mises equations will be written explicitly.

a) Rolling direction. In the case when $\alpha = 0$, the equivalent stress and Levi-Mises equations (by means of (11.56)) allow the following expressions:

$$\sigma_{eq}^2 = \frac{\sigma^2}{2(2b+2c-5)}\left[4c-2+(2-c)\left(\frac{2\gamma^2-e^{2\xi}}{\gamma^2+e^{2\xi}}\right)^2\right] \tag{11.63}$$

and

$$\begin{aligned} d\pi &= \frac{\kappa}{2(2b+2c-5)}\left[4c-2+(2-c)\left(\frac{2\gamma^2-e^{2\xi}}{\gamma^2+e^{2\xi}}\right)^2\right], \\ d\gamma &= \frac{\kappa}{2}\frac{2-c}{2b+2c-5}\frac{6e^{2\xi}(2\gamma^2-e^{2\xi})}{(\gamma^2+e^{2\xi})^2}\gamma, \\ d\pi + d\xi &= \frac{\kappa}{2(2b+2c-5)}\left[-2c+1+(2-c)\frac{(2\gamma^2-e^{2\xi})(2e^{2\xi}-\gamma^2)}{(\gamma^2+e^{2\xi})^2}\right], \end{aligned} \tag{11.64}$$

with the notation $\kappa \equiv \sigma\, d\varepsilon_{eq}^P/\sigma_{eq}$. The straining process begins always with vanishing shear. Thus, the trivial solution of the second of (11.64), namely, $\gamma = 0$, is admissible. Then, equations (11.64) for a proportional straining path allow the solution

$$\xi = -\frac{c+1}{c}\pi, \tag{11.65}$$

such that equivalent stress and equivalent strain increment are mostly simplified to

$$\sigma_{eq}^2 = \frac{3}{2}\frac{c}{2b+2c-5}\sigma^2, \quad (d\varepsilon_{eq}^P)^2 = \left(\frac{2}{3}\frac{2b+2c-5}{c}\right)d\pi^2. \tag{11.66}$$

Hence, (11.61) and (11.66) give the scaling equivalent stress factor:

$$k_\sigma = \left(\frac{3}{2}\frac{c}{2b+2c-5}\right)^{1/2} \equiv k_0. \tag{11.67}$$

b) Direction perpendicular to rolling. For $\alpha = \pi/2$, similar argument leads to

$$\sigma_{eq}^2 = \frac{\sigma^2}{2(2b+2c-5)}\left[4b-2+(2-b)\left(\frac{2\gamma^2-e^{2\xi}}{\gamma^2+e^{2\xi}}\right)^2\right] \tag{11.68}$$

and Levi-Mises differential equations read

$$\begin{aligned} d\pi &= \frac{\kappa}{2(2b+2c-5)}\left[4b-2+(2-b)\left(\frac{2\gamma^2-e^{2\xi}}{\gamma^2+e^{2\xi}}\right)^2\right], \\ d\gamma &= \frac{\kappa}{2}\frac{2-b}{2b+2c-5}\frac{6e^{2\xi}(2\gamma^2-e^{2\xi})}{(\gamma^2+e^{2\xi})^2}\gamma, \\ d\pi + d\xi &= \frac{\kappa}{2(2b+2c-5)}\left[-2b+1+(2-b)\frac{(2\gamma^2-e^{2\xi})(2e^{2\xi}-\gamma^2)}{(\gamma^2+e^{2\xi})^2}\right]. \end{aligned} \tag{11.69}$$

Again, the trivial solution for shear, i.e., $\gamma = 0$, may be accepted. Then, proportional straining path is determined by

$$\xi = -\frac{b+1}{b}\pi \tag{11.70}$$

such that equivalent stress, equivalent strain increment and the scaling factor read

$$\sigma_{eq}^2 = \frac{3}{2}\frac{b}{2b+2c-5}\sigma^2, \quad (d\varepsilon_{eq}^P)^2 = \frac{2}{3}\frac{2b+2c-5}{c}\,d\pi^2, \tag{11.71}$$

$$k_\sigma = \left(\frac{3}{2}\frac{b}{2b+2c-5}\right)^{1/2} \equiv k_{90}. \tag{11.72}$$

c) Equibiaxial direction. If $\alpha = \pi/4$, then with the following convenient notation:

$$\omega_0 \equiv \frac{2-c}{2b+2c-5}, \qquad \omega_{90} \equiv \frac{2-b}{2b+2c-5} \tag{11.73}$$

(both of these coefficients vanish for isotropic materials) Levi-Mises equations become

$$\begin{aligned} d\pi &= \frac{\kappa}{2}\left\{2 + \omega_0\left[\frac{2(1+\gamma)^2 - e^{2\xi}}{(1+\gamma)^2 + e^{2\xi}}\right]^2 + \omega_{90}\left[\frac{2(1-\gamma)^2 - e^{2\xi}}{(1-\gamma)^2 + e^{2\xi}}\right]^2\right\}, \\ d\xi &= \frac{3}{2}\kappa\left\{-1 + \omega_0\frac{[2(1+\gamma)^2 - e^{2\xi}][e^{2\xi} - (1+\gamma)^2]}{[(1+\gamma)^2 + e^{2\xi}]^2}\right. \\ &\quad \left. +\omega_{90}\frac{[2(1-\gamma)^2 - e^{2\xi}][e^{2\xi} - (1-\gamma)^2]}{[(1-\gamma)^2 + e^{2\xi}]^2}\right\}, \\ d\gamma &= 3\kappa e^{\xi}\left\{\omega_0\frac{(1+\gamma)[2(1+\gamma)^2 - e^{2\xi}]}{[(1+\gamma)^2 + e^{2\xi}]^2} - \omega_{90}\frac{(1-\gamma)[2(1-\gamma)^2 - e^{2\xi}]}{[(1-\gamma)^2 + e^{2\xi}]^2}\right\}, \end{aligned} \tag{11.74}$$

whereas the equivalent stress may be represented by the following expression:

$$\sigma_{eq}^2 = \sigma^2\left\{1 + \frac{1}{2}\omega_0\left[\frac{2(1+\gamma)^2 - e^{2\xi}}{(1+\gamma)^2 + e^{2\xi}}\right]^2 + \frac{1}{2}\omega_{90}\left[\frac{2(1-\gamma)^2 - e^{2\xi}}{(1-\gamma)^2 + e^{2\xi}}\right]^2\right\}. \tag{11.75}$$

For this loading path the trivial solution for shear, i.e., $\gamma = 0$, is possible only when $\omega_0 = 0, \quad \omega_{90} = 0$, which could hold only when isotropy takes place. The differential equations are nonlinear and can be integrated only numerically. In turn, this requires knowledge of ω_0 and ω_{90}, i.e., material constants b and c. However, the equivalent plastic strain rate and the scaling factor may be found more directly by making use of definition (11.53) and (11.61):

$$d\varepsilon_{eq}^P = \frac{1}{\sigma_{eq}}\sigma_d : d\,\tilde{\varepsilon}^P = \frac{\sigma}{\sigma_{eq}}\,d\pi \equiv \frac{1}{k_\sigma}d\pi, \tag{11.76}$$

$$k_\sigma^2 = 1 + \frac{1}{2}\omega_0\left[\frac{2(1+\gamma)^2 - e^{2\xi}}{(1+\gamma)^2 + e^{2\xi}}\right]^2 + \frac{1}{2}\omega_{90}\left[\frac{2(1-\gamma)^2 - e^{2\xi}}{(1-\gamma)^2 + e^{2\xi}}\right]^2 \equiv k_{45}^2. \tag{11.77}$$

Obviously, it is not constant and can be found explicitly only after solving Levi-Mises differential equations (11.74).

At the end of this section it is worthy of note that a defining of strain scaling factor like $d\varepsilon_{eq}^P \equiv k_\varepsilon \, d\bar{\varepsilon}^P$ is neither convenient nor necessary.

11.5.2 Subsequent arbitrary shear

When a rectangular plate specimen is cut out from a previously deformed plate such that the axis of the so-obtained specimen takes an angle α with respect to rolling direction and shearing forces are concordant with Fig. 11.3, then formulae (11.37)–(11.40) hold. However, the fibre stress invariants (11.41)–(11.43) must be calculated again. For instance, the equivalent stress reads

$$\sigma_{eq} = \bar{\sigma} \left(1 + \frac{3}{2}\omega_{90} \sin^2 2\beta_1 + \frac{3}{2}\omega_0 \sin^2 2\beta_2\right)^{1/2} . \tag{11.78}$$

Here $\bar{\sigma} = \tau\sqrt{3}$ is the equivalent stress of a hypothetical isotropic material. The expression for equivalent plastic strain increment invariants is the same as in (11.64).

Let us pay attention now to the cases where $\alpha = 0$, $\alpha = \pi/2$ as well as $\alpha = \pi/4$.

a) Rolling direction. For the special case of loading history with $\alpha = 0$, we have

$$\sigma_{eq} = \tau\sqrt{3} \left[1 + \frac{3}{2}\,\omega_0 \Big(\frac{2\gamma\, e^\xi}{\gamma^2 + e^{2\xi}}\Big)^2\right]^{1/2} \equiv \sqrt{3}\,\tau\, k_0, \tag{11.79}$$

while Levi-Mises equations have the form

$$\begin{aligned} d\pi &= \frac{3\tau\ d\varepsilon_{eq}^P}{\sigma_{eq}}\,\omega_0 \frac{(2\gamma^2 - e^{-2\xi})\ \gamma\ e^\xi}{(\gamma^2 + e^{2\xi})^2}, \\ d\gamma &= \frac{3\tau\ d\varepsilon_{eq}^P}{\sigma_{eq}} e^\xi \left[1 + \omega_0 \frac{6\gamma^2 e^{2\xi}}{(\gamma^2 + e^{2\xi})^2}\right], \\ d\pi + d\xi &= \frac{3\tau\ d\varepsilon_{eq}^P}{\sigma_{eq}}\,\omega_0 \frac{(2e^{2\xi} - \gamma^2)\ \gamma\ e^\xi}{(\gamma^2 + e^{2\xi})^2}. \end{aligned} \tag{11.80}$$

Indeed, unless $\omega_0 = 0$, which holds only for isotropic materials, a pure shear stress along rolling direction causes not only shearing but also normal strains as well. Again the identity (11.53) is helpful here to find equivalent strain increment:

$$d\varepsilon_{eq}^P = \frac{1}{\sigma_{eq}} \boldsymbol{\sigma}_d : d\,\tilde{\boldsymbol{\varepsilon}}^P = \frac{1}{\sigma_{eq}} \tau\ e^{-\xi}\ d\gamma\ \equiv \frac{1}{\sqrt{3}\,k_0} e^{-\xi}\ d\gamma. \tag{11.81}$$

b) Direction perpendicular to rolling. When $\alpha = \pi/2$, the equivalent stress reads

$$\sigma_{eq} = \sqrt{3}\,\tau\,k_{90}, \quad \text{where} \quad k_{90} = \left[1 + \frac{3}{2}\,\omega_{90}\left(\frac{2\gamma\,e^{\xi}}{\gamma^2 + e^{2\xi}}\right)^2\right]^{1/2}. \tag{5.38}$$

Levi-Mises equations in this loading case are quite similar to (11.80) such that

$$\begin{aligned}
d\pi &= \frac{3\tau\ d\varepsilon^P_{eq}}{\sigma_{eq}}\,\omega_{90}\frac{(2\gamma^2 - e^{-2\xi})\,\gamma\,e^{\xi}}{(\gamma^2 + e^{2\xi})^2},\\
d\gamma &= \frac{3\tau\ d\varepsilon^P_{eq}}{\sigma_{eq}}e^{\xi}\left[1 + \omega_{90}\frac{6\gamma^2 e^{2\xi}}{(\gamma^2 + e^{2\xi})^2}\right],\\
d\pi + d\xi &= \frac{3\tau\ d\varepsilon^P_{eq}}{\sigma_{eq}}\,\omega_{90}\frac{(2e^{2\xi} - \gamma^2)\,\gamma\,e^{\xi}}{(\gamma^2 + e^{2\xi})^2}.
\end{aligned} \tag{11.82}$$

Again application of the identity (11.53) gives rise to the equivalent strain increment:

$$d\varepsilon^P_{eq} = \frac{1}{\sigma_{eq}}\sigma_d : d\,\tilde{\varepsilon}^P = \frac{1}{\sigma_{eq}}\tau\,e^{-\xi}\,d\gamma \equiv \frac{1}{\sqrt{3}\,k_{90}}e^{-\xi}\,d\gamma \tag{11.83}$$

which is different from (11.81) because $\omega_{90} \neq \omega_0$ for orthotropic materials.

c) Equibiaxial direction. When $\alpha = \pi/4$, then the corresponding equivalent stress is

$$\begin{aligned}
\sigma^2_{eq} &= \bar{\sigma}^2\left\{1 + \frac{3}{2}\omega_0\left[\frac{2e^{\xi}(1+\gamma)}{(1+\gamma)^2 + e^{2\xi}}\right]^2 + \frac{3}{2}\omega_{90}\left[\frac{2e^{\xi}(1-\gamma)}{(1-\gamma)^2 + e^{2\xi}}\right]^2\right\}\\
&\equiv \bar{\sigma}^2\,k^2_{45},
\end{aligned} \tag{11.84}$$

with $\bar{\sigma} = \tau\sqrt{3}$. Finally, Levi-Mises equations have the following form:

$$\begin{aligned}
d\pi &= \frac{3\tau\,d\varepsilon^P_{eq}}{2\,\sigma_{eq}}\left\{-\,\omega_0\frac{(1-\gamma)e^{\xi}[2(1-\gamma)^2 - e^{2\xi}]}{[(1-\gamma)^2 + e^{2\xi}]^2}\right.\\
&\quad\left. +\,\omega_{90}\frac{(1+\gamma)e^{\xi}[2(1+\gamma)^2 - e^{2\xi}]}{[(1+\gamma)^2 + e^{2\xi}]^2}\right\},\\
d\pi + d\xi &= \frac{3\tau\,d\varepsilon^P_{eq}}{2\,\sigma_{eq}}\left\{-\,\omega_0\frac{(1-\gamma)e^{\xi}[2e^{2\xi} - (1-\gamma)^2]}{[(1-\gamma)^2 + e^{2\xi}]^2}\right.\\
&\quad\left. +\,\omega_{90}\frac{(1+\gamma)e^{\xi}[2e^{2\xi} - (1+\gamma)^2]}{[(1+\gamma)^2 + e^{2\xi}]^2}\right\},\\
d\gamma &= \frac{3\tau\ d\varepsilon^P_{eq}}{\sigma_{eq}}\left\{1 + \omega_0\frac{6\,e^{2\xi}[2(1-\gamma)^2 - e^{2\xi}]}{[(1-\gamma)^2 + e^{2\xi}]^2}\right.\\
&\quad\left. +\,\omega_{90}\frac{6\,e^{2\xi}[2(1+\gamma)^2 - e^{2\xi}]}{[(1+\gamma)^2 + e^{2\xi}]^2}\right\}.
\end{aligned} \tag{11.85}$$

The equivalent plastic strain increment is found in a similar way as above:

$$d\varepsilon_{eq}^{P} = \frac{1}{\sigma_{eq}} \boldsymbol{\sigma}_d : d\,\tilde{\varepsilon}^{P} = \frac{1}{\sigma_{eq}} \tau\, e^{-\xi}\, d\gamma \equiv \frac{1}{\sqrt{3}\,k_{45}} e^{-\xi}\, d\gamma \tag{11.86}$$

where $k_\sigma \equiv k_{45}$ is given by (11.84).

11.6 Some concluding remarks

The main subject of this chapter as stated in the introduction has been to show that even the classical theory of plasticity (as formulated in the famous book [Hil50]):

- cannot be applied correctly without application of tensor invariants and correct deformation picture,
- plastic strain induced anisotropy makes application of classical notions of equivalent plastic strain and equivalent stress tedious,
- tensor function representation is worthwhile to be considered even in the framework of classical concepts,
- limit strains (cf. [MK-67] for a review) must be considered by these functions (like in [MK-97]) especially for nonproportional strain histories.

Remarks on Applications of the Theory

The last three chapters are devoted to applications of the theory to plastic waves (in a tension-type split Hopkinson bar), ratchetting phenomenon as well as issue of metal forming of orthotropic sheets. These examples serve to enlighten advantages and applicability of the tensor representation approach whenever multiaxial stresses and strains appear. Some general remarks could be useful now.

- The so-called universal flow curve and associate flow rule based on the yield function relating only scalars like equivalent stress and equivalent plastic strain are completely useless when we want to describe simultaneously tension and shear even in the range of only small strain rates. It has multiaxial suit but scalar essence being able to describe successfully only tension test up to large strains. The simplest yet approximately correct approach is to combine loading function orthogonality with tensor functions.
- Since ratchetting is inherently multiaxial and nonproportional, it is logical that tensor representation gives best results even without residual stress explicitly taken into account.
- If an engineer in the field of metal forming wants to make a reliable forecast, either of diffuse or localized instability, then tensor representation approach is inevitable even if the classical theory of plasticity is applied. The so-called Hill's yield function for anisotropic materials cannot be applied correctly without application of tensor invariants and correct deformation picture. As stated in the previous chapter, tensor function representation

is worthwhile to be considered even in the framework of classical concepts. If we want to apply an advanced theory with plastic strain induced anisotropy and/or damage, then it is impossible to do anything without it.

If a careful and proper geometry of deformation is combined with irreversible thermodynamics, then we arrive at best results.

References

[Aer02] AERO, E. L.: Essentially nonlinear theory of microdeformations in medium with periodic structure, *Theor. Appl. Mech.*, **28-29**, 1–25 (2002)

[AM-79] ALBERTINI, C., MONTAGNANI, M.: Testing techniques in dynamic biaxial loading, *Inst. Phys. Conf. Ser.*, **47/1**, 21–34 (1979)

[Alb89] ALBERTINI, C., MONTAGNANI, M., MIĆUNOVIĆ, M.: Viscoplastic behaviour of AISI 316 H—multiaxial experimental results, In: Hadjian, A. H. (ed) *Transactions of SMIRT-10,* 31–36, American Association for SMIRT, Los Angeles (1989)

[AMM91] ALBERTINI, C., MONTAGNANI, M., MIĆUNOVIĆ, M.: Viscoplastic behaviour of AISI 316 H: multiaxial experimental results and preliminary numerical analysis, *Nucl. Eng. Des.*, **130**, 205–210 (1991a)

[AMP91] ALBERTINI, C., MONTAGNANI, M., PIZZINATO, E., RODIS, A.: Comparison of the equivalent flow curves in tension and shear at low and high strain rate for AISI 316 and ARMCO iron, In: Jono, M., Inoue, T. (eds) *Proc. of Mech. Behaviour of Mater. VI*, Pergamon, Kyoto (1991b)

[AGM91] ALBERTINI, C., GRIFFITHS, L. J., MONTAGNANI, M., RODIS A., MARIOTTI, P., PALUFFI, A., PAZIENZA, G.: Material characterization by an innovative biaxial shear experiment at very large strains and at very high strain rates, *J. Phys. IV*, **1/C3**, 435–440 (1991)

[AM-92] ALBERTINI, C., MONTAGNANI, M.: Private communication (1992)

[Alt70] ALTS, T.: Thermodynamics of thermoelastic bodies with kinematic constraints. Fiber reinforced materials, *Arch. Rational Mech. Anal.*, **61**, 253–289 (1976)

[Ant70] ANTHONY, K.-H.: Nonmetric connexions, quasidislocations and quasidisclinations. A contribution to the theory of nonmechanical stresses in crystals, In: Simmons, J. A., de Wit, R., Bullough, R. (eds) *Fundamental Aspects of Dislocation Theory, Vol. I, National Bureau of Standards Special Publication*, **317**, 637–649, Washington, D.C. (1970)

[Ant92] ANTHONY, K.-H.: Dislocation dynamics, plasticity, thermodynamics: a unification by means of Lagrangian formalism, In: Anthony, K.-H., Wagner H.-J. (eds) *Continuum Models and Discrete Systems CMDS-7, Material Science Forum*, **123–125**, 175–616, Trans Tech Publ, Aedermannsdorf (1992)

[Asa83] ASARO, R. J.: Micromechanics of crystals and polycrystals, *Adv. Appl. Mech.,* **23**, 1–115 (1983)

[Bal83] BALTOV, A.: Investigation of dynamic processes of straining and fracture of inelastic bodies (in Russian), *Bulgarian Acad. Theoretical and Applied Mechanics*, Sofia, **IV/1** (1983)

[BT-78] BALTOV, A., TODOROV S.: Application of the finite system method in the case of longitudinal bar oscillations, *J. Theor. Appl. Mech.*, Sofia, **3**, 56–73 (1978)

[BK-79] BATAILLE, J., KESTIN, J.: Irreversible processes and physical interpretation of rational thermodynamics, *J. Non-Equilib. Thermodyn.*, **11**, 229–258 (1979)

[BBC86] BEGHI, M. G., BOTTANI, C., CAGLIOTI, G.: Irreversible thermodynamics of metals under stress, *Res Mechanica*, **19**, 365–379 (1986)

[Bel65] BELL, J.: The experimental foundations of solid mechanics, In: Flügge, S. (ed) *Handbuch der Physik*, **VIa/1**, 1–813, Springer, Berlin (1973)

[Ber98] BERTRAM, A.: An alternative approach to finite plasticity based on material isomorphisms, *Int. J. Plasticity*, **52**, 353–374 (1998)

[BOZ91] BERTRAM, A., OLSCHEWSKI, J., ZELEWSKI, M., SIEVERT, R.: Anisotropic Creep Modeling for f.c.c. single crystals, In: Zyczkowski, M. (ed) *Proceedings of IUTAM: Creep in structures* , pp. 29–36, Springer, Berlin (1991)

[Bil60] BILBY, E.: Continuous distributions of dislocations, *Progress in Solid Mechanics*, **1**, 329–398 (1960)

[Bod87] BODNER, S. R.: Review of a unified elastic-viscoplastic theory, In: Miller, A. K. (ed) *Unified Constitutive Equations for Plastic Deformation and Creep of Engineering Alloys*, p. 273. Elsevier, Essex (1987)

[BS-77] BOEHLER, J. P., SAWCZUK, A.: On yielding of oriented solids, *Acta Mech.*, **27**, 185–206 (1977)

[Bor49] BORN, M.: *Natural Philosophy of Cause and Chance*, Oxford University Press, Oxford (1949)

[BCN82] BOTTANI, C. E., CAGLIOTI, G., NOVELLI, A., OSSI, P. M., ROSSITTO, F., SILVA, G.: Il limite elastoplastico: realtà fisica e attuali criteri di qualificazione, *La Metallurgia Italiana*, **74**, 530 (1982)

[BM-87] BROOKFIELD, D. J., MARETON, D. N.: The ratchetting behavior of stainless steel type 316 at ambient and elevated temperature, In: Wittman, F. H. (ed) *Proceedings of SMIRT-9*, pp 427–432, Balkema, Rotterdam (1987)

[Bud65] BUDIANSKY, B.: On the elastic moduli of some heterogeneous materials, *J. Mech. Phys.Solids*, **13/4**, 223–227 (1965)

[Cai92] CAILLETAUD, G., A micromechanical approach to inelastic behaviour of metals, *Int. J. Plasticity*, **8**, 55–73 (1992)

[Car09] CARATHEODORY, C.: Untersuchungen über die Grundlagen der Thermodynamik, *Math. Ann.*, **67**, 355–386 (1909)

[CLJ01] CASAS-VAZQUEZ, J., LEBON, G., JOU, D.: *Extended Irreversible Thermodynamics*, Springer, Berlin (2001)

[CK-80] CERNOCKY, E. P., KREMPL, E.: A theory of thermoviscoelasticity based on infinitesimal total strain, *Int. J. Solids Structures*, **16**, 723–741 (1980)

[CR-83] CHABOCHE, J. L., ROUSSELIER, G.: On the plastic and viscoplastic constitutive equations, *Trans. ASME, PVT*, **105**, 153–164 (1983)

[Chi87] CHIKADZUMI, S.: *Ferromagnetism Physics—Magnetic Features and Applications, Vol. II* (translated from Japanese to Russian), Nauka, Moscow (1987)

[Cla97] CLAUSEN, B.: Characterisation of polycrystal deformation, RISO-R-985, *RISO reports*, Roskilde, Denmark (1997)

[CN-60] COLEMAN, B. D., NOLL, W.: An approximation theorem for functionals with applications in continuum mechanics, *Arch. Rational Mech. Anal.*, **6**, 355–370 (1960)

[CN-61] COLEMAN, B. D., NOLL, W.: Foundations of linear viscoelasticity, *Rev. Mod. Phys.*, **33**, 239–249 (1961)

[CN-63] COLEMAN, B. D., NOLL, W.: The thermodynamics of elastic materials with heat conduction and viscosity, *Arch. Rational Mech. Anal.*, **13**, 167–178 (1963)

[Daf84] DAFALIAS, Y. F.: The plastic spin concept and a simple illustration of its role in finite plastic transformations, *Mechanics of Materials*, **3**, 223–233 (1984)

[Das86] DASHNER, P. A.: Invariance considerations in large strain elastoplasticity, *J. Appl. Mech.*, **53**, 55–60 (1986)

[DP-91] DAVINI, C., PARRY, G. P.: Invariants for defective crystals, In: Maugin, G. (ed) *Transactions of CMDS-6*, **2**, 102–109, Longman, Harlow (1991)

[Dru52] DRUCKER, D.: A more fundamental approach to plastic stress-strain relations, *Proc. 1st U.S. Nat. Congr. Appl. Mech.* Amer. Soc. Mech. Eng., 487–491 (1952)

[Eck48] ECKART, C.: The thermodynamics of irreversible processes IV. The theory of elasticity and anelasticity, *Phys. Rev.*, **73/4**, 373–382 (1948)

[ElA02] EL-AZAB, A.: On the development of a statistical mechanics framework for crystal plasticity, In: Maruszewski, B. T. (ed) *Int. Symp. Structured Media—in memory of E. Kröner*, 59–70, Poznan Univ. Technology Publ, Poznan (2002)

[Ele91] ELEICHE, A. M.: Inelastic shear deformation of stainless steels under quasi-static and impact sequential reverse torsion (final report), *Contract No. 2892-55-12 ED ISP ET, JRC CEC Ispra - Cairo University* (1991)

[EM-00] EPSTEIN, M., MAUGIN, G. A.: Thermomechanics of volumetric growth in uniform bodies, *Int. J. Plasticity,*, **16**, 951–978 (2000)

[EGD97] ERBER, T., GURALNICK, S. A., DESAI, R. D., KWOK W.: Piezomagnetism and fatigue, *J. Phys. D, Appl. Phys.*, **30**, 2818–2836 (1997)

[Eri60] ERICKSEN, J. L.: Tensor fields, In: Flügge, S. (ed) *Handbuch der Physik*, **III/1**, pp. 794–858, Springer, Berlin (1960)

[Esh57] ESHELBY, J. D.: The determination of the elastic field of an ellipsoidal inclusion, and related problems, *Proc. Roy. Soc.*, **241**, 376–396 (1957)

[FFP91] FÄHNLE, M., FURTHMÜHLER, J., PAWELLEK. R.: Continuum models of amorphous and polycrystalline ferromagnets: magnetostriction and internal stresses, In: Maugin, G. (ed) *Transactions of CMDS-6*, **2**, 120–127, Longman, Harlow (1991)

[Faz88] FAZIO, C. J.: A finite strain plasticity theory with endochronic anisotropic hardnening, In: Cailletaud et al. (eds) *Proceedings of MECAMAT*, III/185–III/195, Mecamat, Besancon (1988)

[FH-97] FLECK, N. A., HUTCHINSON, J. W.: Strain gradient plasticity, *Adv. Appl. Mech,* **33**, 295–361 (1997)

[Fom75] FOMIN, V. L.: *Continuum Mechanics for Engineers* (in Russian), St. Peterburg University Publ., St. Peterburg (1975)

[For02] FOREST, S.: Homogenization methods and the mechanics of generalized continua—part 2., *Theor. Appl. Mech.*, **28-29**, 113–143 (2002)

[Gam92] GAMBIN, W.: Refined analysis of elastic-plastic crystals, *Int. J. Solids Structures*, **29/16**, 2013–2021 (1992)

[GHA81] GIL SEVILANO, J.F., VAN HOUTTE, P., AERNOUDT,E.: Large strain work hardening and textures, In: Chalmers et al (eds) *Progress in Material Science*, Pergamon, Oxford (1981)

[GNS83] GERMAIN, P., NGUYEN, Q. S., SUQUET, P.: Continuum thermodynamics, *ASME J. Appl. Mech.,* **50**, 1010–1020 (1983)

[GFG03] GRILLO, A., FEDERICO,S., GIAQUINTA,G., HERZOG,W., LA ROSA, G.: Restoration of the symmetries broken by reversible growth in hyperelastic bodies, *Theoret. Appl. Mech.*, **30/4**, 311–331 (2003)

[Gya70] GYARMATI, I.: *Non-equilibrium Thermodynamics*, Springer, Berlin (1970)

[Hak87] HAKEN, H.: Thermodynamics—synergetics—life, *J. Non-Equilib. Thermodyn.,* **12**, 1–10 (1987)

[HW-87] HAN, S., WACK, B.: Torsion ratchet with mechanical hysteresis, In: Wittman, F. H. (ed) *Proceedings of SMIRT-9* , pp 421–426, Balkema, Rotterdam (1987)

[Hil50] HILL, R.: *The Mathematical Theory of Plasticity*, Clarendon, Oxford (1950)

[Hil65] HILL, R.: A self–consistent mechanics of composite materials, *J. Mech. Phys. Solids*, **13**, 213–222 (1965)

[Hil68] HILL, R.: On constitutive inequalities for simple materials, *J. Mech. Phys. Solids*, **16**, 229–242 (1968)

[Hol73] HOLSAPPLE, K. A.: A finite elastic-plastic theory and invariance requirements, *Acta Mech.,* **17**, 277–290 (1973)

[Hut77] HUTTER, K.: Review article—the foundations of thermodynamics, its basic postulates and implications. A review of modern thermodynamics, *Acta Mech.,* **27**, 1–54 (1977)

[Ili61] ILI'USHIN, A. A.:, On a postulate of plasticity, *J. Appl. Math. Mech. - Translation of PMM,* **25**, 746–750 (1961)

[IIS85] INOUE, T., IMATANI, S., SAHASHI, T.: Some remarks on the behavior of high temperature materials under biaxial stress state, In: Gittus, J. H., Krempl, E. (eds) *Proceedings of SMIRT 8,* **L 1/2**, pp. 7–14, North Holand: Amsterdam (1985)

[IK-50] IRVING, J. H, & KIRKWOOD, J. G.: The statistical mechanical theory of transport processes. IV. The equations of hydrodynamics, *J. Chem. Phys.,* **18/6**, 817–829 (1950)

[Kle75] KLEIBER, M.: Kinematics of deformation process in materials subjected to finite elastic–plastic strains, *Int. J. Eng. Sci.,* **13**, 513–525 (1975)

[Klp72] KLEPACZKO, J.: Some experimental investigations of the elastic-plastic wave propagation in bars, In: Sawczuk, A. (ed) *Foundations of Plasticity*, Noordhoff Publ., Groningen (1972)

[Kon58] KONDO, K.: *RAAG Memoirs*, **II(D)**, Gakujutsu, Tokyo (1958)

[Kng82] KONIG, J. A.: Shakedown analysis in structural design, In: Kosinski, W. (ed) *Mechanics of Inelastic Media and Structures*, pp. 133-142, PWN, Warszawa (1982)

[Kos86] KOSIŃSKI, W.: *Field Singularities and Wave Analysis in Continuum Mechanics*, PWN, Warsaw (1986)

[KG-04] KOWALCZYK, K., GAMBIN, W.: Model of plastic anisotropy evolution with texture-dependent yield surface, *Int. J. Plasticity,* **20**, 19–54 (2004)

[KO-04] KOWALCZYK-GAJEWSKA, K., OSTROWSKA-MACIEJEWSKA, J.: The influence of internal restrictions on the elastic properties of anisotropic materials, *Arch. Mech.*, **56**, 205–232 (2004)

[KS-59] KRÖNER, E., SEEGER, A.: Nicht-lineare Elastizitätstheorie der Versetzungen und Eigenspannungen, *Arch. Rational Mech. Anal.*, **3**, 97–119 (1959)

[Kro60] KRÖNER, E.: Allgemeine Kontinuumstheorie der Versetzungen und Eigenspannungen, *Arch. Rational Mech. Anal.*, **4**, 273–334 (1960)

[Kro70] KRÖNER, E.: Initial studies of a plasticity theory based upon statistical mechanics, In: Kanninen, M. Adler, W. Rosenfeld A., Jofee, R. (eds) *Inelastic Behavior of Solids*, 137–148, McGraw-Hill, New York (1970)

[Kro71] KRÖNER, E.: *Statistical Continuum Mechanics*, CISM Course **92**, Springer, Wien (1971)

[KA-75] KRÖNER, E., ANTHONY, K.-H.: Dislocations and disclinations in material structures: the basic topological concepts, *Annual Rev. Mater. Sci.*, **5**, 43–72 (1975)

[Kro80] KRÖNER, E.: Graded and perfect disorder in random media, *J. Eng. Mech. Division*, 889–914 (1980)

[Kro86] KRÖNER, E.: How general is thermodynamics? *J. Non-Equilib. Thermodyn.,* **11**, 277–293 (1986)

[KM-01] KUDRJAVCEVA, L., MIĆUNOVIĆ, M.: Some examples of texture evolution at polycrystal viscoplasticity, *Proceedings of YUMECH' 2001*, Belgrade (2001)

[Kuk92a] KUKUDZANOV, V. N.: A numerical method for solution of nonsteady elastoviscoplastic problems at large strains, In: Besdo, D., Stein, E. (eds) *Finite Inelastic Deformations - Theory and Applications*, IUTAM Proc., Springer, Berlin (1992)

[Kuk92b] KUKUDZANOV, V. N.: *Finite difference methods in solving solid mechanics problems*, (in Russian), MFTI Publications, Moscow (1992)

[Kun65] KUNIN, I. A.: Theory of dislocations, In: Schouten J. A. *Tensor Analysis for Physicists,* (Russian translation), Nauka, Moscow (1965)

[KS-71] KUNIN, I. A., SOSNINA, E. G.: Ellipsoidal inhomogeneity in an elastic medium, (in Russian), *Dokl. Akad. Nauk SSSR*, **199/3**, 571–574 (1971)

[KS-73] KUNIN, I. A., SOSNINA, E. G.: Stress concentration on an ellipsoidal inhomogeneity in an anisotropic elastic medium, (in Russian), *Appl. Math. Mech. PMM*, **37**, 306–315 (1973)

[Lam77] LAM, L.: Dissipation functions and conservation laws of molecular liquids and solids, *Z. Phys.*, **B 27**, 101–110 (1977)

[Lmb74] LAMBERMONT, J. H.: On the unification of the mathematical theory of plasticity and dislocation theory by using thermodynamics, *Int. J. Eng. Sci.,* **12**, 937–965 (1974)

[LRC79] LEBEY, J., ROCHE, R. L., COUSSERAU, P.: Experimental tests on ratchet of tabular specimens (torsion and tension tests on ratchet of tubular specimens (torsion and tension) use for material characterization, In: Sawczuk, A., Zudans, Z. (eds) *Proceedings of SMIRT-5, L5/3* North Holland, Amsterdam (1979)

[Lee69] LEE, E. H.: Elastic-plastic deformation at finite strain, *ASME J. Appl. Mech.,* **36**, 1–6 (1969)

[Lev76] LEVIN V. M.: On determination of elastic and thermoelastic constants of composite materials (in Russian), *Izv. AN SSSR Mekhanika Tverdogo Tela,* **6**, 137–145 (1976)

[Lev82] LEVIN V. M.: On thermoelastic stresses in composite media, (in Russian), *Appl. Math. Mech. PMM,* **46/3**, 502–506 (1982)

[Lin67] LINDHOLM, U. S.: Some experiments in dynamic plasticity under combined stres, In: Lindholm, U. S. (ed) *Symposium on the Mechanical Behavior of Metals under Dynamic Loading,* San Antonio, Texas (1967)

[Liu72] LIU, I.-S.: Method of Lagrange multipliers for exploitation of the energy principle, *Arch. Rational Mech. Analysis*, **46**, 131–148 (1972)

[LC-83] LLEBOT, J. E., CASAS-VAZQUEZ, J.: A thermodynamic approach to heat and electric conduction in solids, *Physica,* **121/3**, 552–562 (1983)

[LL-81] LUBARDA, V., LEE, E. H.: A correct definition of elastic and plastic deformation and its computational significance, *ASME J. Appl. Mech.,* **48**, 35–40 (1981)

[LM-00] LUBARDA, V. A., MARKENSCOFF, X.: Conservation integrals in couple stress elasticity, *J. Mech. Phys. Solids,* **48**, 553–564 (2000)

[Lub84] LUBLINER, J.: A maximum dissipation principle in generalized plasticity, *Acta Mech.,* **52**, 225–237 (1984)

[MA-95] MAKAR, J. M., ATHERTON, D. L.: Effects of isofield uniaxial cyclic stress on the magnetization of 2% Mn pipeline steel - behavior on minor hysteresis loops and small major hysteresis loops, *IEEE Trans. Magn.*, **31/3**, 2220–2227 (1995)

[Man64] MANDEL, J.: Contibution théorique a l'étude de l'écrouissage et des lois l'écoulement plastique, *Proc. 11th Int. Congr. Appl. Mech.*, 502–509 (1964)

[Man81] MANDEL, M.: Sur la definition de la vitesse de déformation élastique et sa relation avec la vitesse de constrainte, *Int. J. Solids Structures,* **17**, 873–878 (1981)

[MK-67] MARCINIAK, Z., KUCZYNSKY, K.: Limit strains in the processes of stretch-forming sheet metal, *Int. J. Mech. Sci.,* **9**, 609–620 (1967)

[Mar99] MARKOV, K.Z.: In: Markov, K. Z., Preziosi, L. (eds) *Heterogeneous Media: Modelling and Simulation*, 105–145. Birkhauser, Boston (1999)

[MM-89] MARUSZEWSKI, B., MIĆUNOVIĆ, M.: On neutron irradiation of an isotropic thermoplastic body, *Int. J. Eng. Sci.*, **27/8**, 955–965 (1989)

[MSC87] MAUGIN, G. A., SABIR, M., CHAMBON,P.: Coupled magnetomechanical hysteresis effects: applications to nondestructive testing, In: Yamamoto, Y., Miya, K. (eds) *Electromagnetic Interactions in De-*

formable Solids and Structures, 255–264, North Holland, Amsterdam (1987)

[Mau88] MAUGIN, G. A.: *Continuum Mechanics of Electromagnetic Solids*, North Holland, Amsterdam (1988)

[Mau92] MAUGIN G. A.: *The Thermomechanics of Plasticity and Fracture*, Cambridge University Press, Cambridge (1992)

[Mau95] MAUGIN, G. A.: Material forces: concepts and applications, *Appl. Mech. Rev.*, **48/5**, 213–245 (1995)

[Mei76] MEIXNER, J.: On the foundation of thermodynamics of processes, In: Kestin, J. (ed) *The Second Law of Thermodynamics*, 313–323, Dowden, Hutchinson & Ross, Stroudsburg (1976)

[Mic74a] MIĆUNOVIĆ, M.: *Geometric Theory of Thermoelasticity and Continuous Distribution of Dislocations*, IFTR Reports, Warsaw (1974a)

[Mic74b] MIĆUNOVIĆ, M.: A geometrical treatment of thermoelasticity of simple inhomogeneous bodies: I - geometric and kinematic relations, *Bull. Acad. Pol. Sci., Ser. Sci. Tech.*, **22**, 579-588 (1974b)

[Mic75] MIĆUNOVIĆ, M.: A geometrical treatment of thermoelasticity of simple inhomogeneous bodies: II (Constitutive equations), & III (Approximations), *Bull. Acad. Pol. Sci.- Ser. Sci. Techn.*, **22**, 633–641 & **23**, 89–97 (1975)

[Mic77] MIĆUNOVIĆ, M.: Statistical analysis of linear solid body with continuous distribution of screw dislocations, *Theor. Appl. Mech.*, **3**, 59–68 (1977)

[Mic79] MIĆUNOVIĆ, M.: Kinetic stress tensor in statistical theory of screw parallel dislocations, *Theor. Appl. Mech.*, **5**, 91–96 (1979)

[Mic85] MIĆUNOVIĆ, M.: A yield condition for anisotropic materials, In: Gittus J. H., Krempl, E. (eds) *Transactions of SMIRT-8,* **L5/4**, 243—210, North-Holland, Amsterdam (1985)

[Mic86a] MIĆUNOVIĆ, M.: A thermodynamic approach to viscoplasticity, In: Kounadis, A. et al (eds) *Proceedings of the 1st National Congress of Mechanics*, 364–376, HSTAM, Athens (1986a)

[Mic86b] MIĆUNOVIĆ, M.: A thermodynamic theory of rate independent thermoplasticity of fibre-reinforced materials, In: Loo, T. T., Sun, C. T. (eds) *Proceedings of International Symposium on Composite Materials and Structures - Beijing*, 94–101, Technomic, Lancaster (1986b)

[Mic87a] MIĆUNOVIĆ, M.: A thermodynamic viscoplasticity theory of ideal fibre-reinforced materials, In: Green, W. A., Mićunović, M. (eds), *Mechanical Behavior of Composites and Laminates*, 199–208, Elsevier, London (1987a)

[Mic87b] MIĆUNOVIĆ, M.: A thermodynamic model of ideal cyclic viscoplasticity, In: Wittmann, F. H. (ed) *Transactions of SMIRT-9,* **L**, 195–203, Balkema, Rotterdam (1987b)

[Mic88] MIĆUNOVIĆ, M., ALBERTINI, C., MONTAGNANI, M.: Viscoplastic material properties at non-proportional strain histories of AISI 316 H stainless steel, In: Cailletaud et al (eds), *Proceedings of MECAMAT*, pp. II/107–II/118, Mecamat, Besancon (1988)

[Mic91a] MIĆUNOVIĆ, M.: Materials with thermoplastic memory, In: Maugin, G. A. (ed) *Transactions of CMDS-6,* **2**, 187–195, Longman, Harlow (1991a)

[Mic91b] Mićunović, M.: A viscoplasticity theory of irradiated materials, In: Zyckowski M. (ed) *Creep in Structures - 4th IUTAM Symposium*, 139–146, Springer, Berlin (1991b)

[Mic92a] Mićunović, M.: On the thermal elastoplastic transition in viscoelasticity of metals, *Facta Universitatis, Nis,* **1/2**, 155–168 (1992a)

[Mic92b] Mićunović, M.: Normality rule from plastic work extremals?, In: Anthony, K.-H., Wagner, H.-J. (eds) *Continuum Models and Discrete Systems CMDS-7, Material Science Forum*, **123-125**, 609–616, Trans Tech Publ, Aedermannsdorf (1992b)

[Mic93] Mićunović, M.: Inelastic memory versus viscoplasticity of continuously damaged materials, *J. Mech. Behav. Mater.*, **4/4**, 331–341 (1993)

[Mic94] Mićunović, M.: On viscoplasticity of irradiated steels, *Theor. Appl. Mech.,* **20**, 167–176 (1994)

[Mic96a] Mićunović, M.: Multiaxial dynamic experiments and viscoplasticity of metals, *Scientific Review, Belgrade*, **16**, 1–22 (1996a)

[Mic96b] Mićunović, M.: Multiaxial dynamic experiments versus tensor function representation, In: Markov, K. Z. (ed) *Proc. of CMDS-8*, World Scientific, 564-572, (1996b)

[Mic97] Mićunović, M., Albertini, C., Montagnani, M.: High strain rate viscoplasticity of AISI 316 stainless steel from tension and shear experiments, In: Miljanic, P. (ed) *Solid Mechanics, Serbian Acad. Sci. Meetings - LXXXVII, Dept. Techn. Sci.*, **3**, 97–106 (1997)

[MK-97] Kudrjavceva, L., Mićunović, M.: Localization instability from advanced viscoplasticity models, *Proceedings of YUMECH' 1997*, Vrnjacka Banja (1997)

[MV-98] Mićunović, M., Veljković, D.: Ratchetting phenomenon at low strain rates for AISI 316H stainless steel, *Facta Universitatis*, **2/8**, 503–511 (1998)

[Mic01] Mićunović, M.: On viscoplasticity of ferromagnetics, *Theor. Appl. Mech.*, **26**, 107–126 (2001)

[Mic02a] Mićunović, M.: Some issues in polycrystal viscoplasticity of steels, In: Maruszewski, B. T. (ed) *Int. Symp. Structured Media - in memory of E. Kröner*, pp. 196–207, Poznan Univ. Technology Publ, Poznan (2002a)

[Mic02b] Mićunović, M.: Low order inelastic micromorphic polycrystals, *Theor. Appl. Mech.*, **28-29**, 235–259 (2002b)

[MB-02] Mićunović, M., Baltov, A.: Plastic wave propagation in Hopkinson bar—revisited, *Arch. Mech.*, **54/5-6**, 577–602 (2002c)

[Mic03] Mićunović, M., Albertini, C., Montagnani, M.: High strain rate thermo-inelasticity of damaged AISI 316 H, *Int. J. Damage Mech.*, **12**, 267–303 (2003)

[MicWWW] Mićunović, M.: Presentation at the Meeting of the Society of Engineering Science, Ann Arbor, USA (2003) http://staff.mfkg.kg.ac.yu/mmicunovic/AnnArborMM.pps

[Mic05] Mićunović, M.: Self-consistent method applied to quasi-rate independent polycrystals, *Phil. Magazine*, **85 (33-35)**, 4031–4054 (2005)

[Mic06] Mićunović, M.: Thermodynamical and self consistent approach to inelastic ferromagnetic polycrystals, *Arch. Mech.*, **58/4-5**, 393–430 (2006)

[MCA87] MOLINARI, A., CANOVA, R. G., AHZI, S.: A self–consistent approach to the large deformation polycrystal viscoplasticity, *Acta Metall.*, **35**, 2983 (1987)

[MF-90] MOTOGI, S., FUKUDA, T.: Modelling of ΔE effect in polycrystalline ferromagnets, In: Hsieh, R. K. T. (ed) *Mechanical Modellings of New Electromagnetic Materials*, 281–288, Elsevier, Amsterdam (1990)

[MSD73] MROZ, Z., SHRIVASTAVA, H. R., DUBEY, R. N.: A non-linear hardening model and its application to cyclic loading, *Acta Mech.*, **17**, 277–290 (1973)

[MBK93] MULLER, D., BERVEILLER, M., KRATOCHVIL, J.: Non-local hardenings in metals, In: Anthony, K.-H., Wagner, H.-J. (eds) *Continuum Models and Discrete Systems CMDS-7, Material Science Forum*, **123–125**, 195–204, Trans Tech Publ, Aedermannsdorf (1993)

[Mul71] MÜLLER, I.: The coldness, a universal function in thermoelastic bodies, *Arch. Rational Mech. Anal.*, **41**, 319–332 (1971)

[Mul73] MÜLLER, I.: From thermostatics to thermodynamics; discussion paper, In: Delgado Domingos, J. J., Nina, M. N. R., Whitelaw, J. H. (eds) *Foundations of Continuum Thermodynamics*, John Wiley & Sons, New York (1973)

[Mul85] MÜLLER, I.: *Thermodynamics*, Pitman, Boston (1985)

[Mur88] MURA, T.: *Micromechanics of Defects in Solids*, Martinus Nijhoff, Dordrecht (1988)

[Mrk79] MURAKAMI, S.: Tensor function approach to constitutive equations of inelasticity, In: Sawczuk, A., Zudans, Z. (eds) *Transactions of SMIRT-5*, **L 1/4**, 1–10, North-Holland, Amsterdam (1979)

[Mrk88] MURAKAMI, S.: Mechanical modelling of material damage, *ASME J. Appl. Mech.*, **55**, 280–286 (1988)

[MSS85] MURAKAMI, S., SANOMURA, Y., SAITO, K.: Formulation of cross hardening in creep and its effects on creep damage process of copper, In: Gitus, J. H., Krempl, E. (eds) *Proceedings of SMIRT-8*, 273–280, North Holland, Amsterdam (1985)

[Nae77] NAERLOVIC-VELJKOVIĆ, N.: Field equations for micropolar current and a heat conducting magnetically-saturated solid, *Arch. Mech.*, **29/4**, 625–630 (1977)

[Nag90] NAGHDI, P. M.: A critical review of the state of finite plasticity, *ZAMP*, **41**, 315–394 (1990)

[NTJ90] NEALE, K.W., TOTH, L. S., JONAS, J. J.: Large strain shear and torsion of rate-sensitive fcc polycrystals, *Int. J. Plasticity*, **6**, 45–61 (1990)

[Nem75] NEMAT-NASSER, S.: On nonequilibrium thermodynamics of continua, In: Nemat-Naser, S. (ed) *Mechanics Today*, **2**, 94–158, Pergamon, New York (1975)

[Nem79] NEMAT-NASSER, S.: Decomposition of strain measures and their rates in finite deformation elastoplasticity, *Int. J. Solids Structures*, **15**, 155–166 (1979)

[NY-81] NOVOZHILOV, YU.V., YAPPA, YU. A.: *Electrodynamics*, Mir Publishers, Moscow (1981)

[Now78] NOWACKI, W. K.: *Stress Waves in Non-Elastic Solids*, Pergamon, Oxford (1978)

[Ogd03] Ogden, R.: *Nonlinear Elasticity with Application to Material Modelling*, Warsaw: IFTR - AMAS Lecture Notes (2003)

[Ost02] Ostoja-Starzewski, M., Microstructural randomness versus representative volume element, *ASME J. Appl. Mech.,* **69**, 25–35 (2002)

[Owe70] Owen, D. R.: A mechanical theory of materials with elastic range, *Arch. Rational Mech. Anal.*, **37**, 85–110 (1970)

[Pec83] Pecherski, R. B.: Relations of microscopic observations to constitutive modelling for advanced deformations and fracture initiation of viscoplastic materials, *Arch. Mech.*, **35**, 257–277 (1983)

[Per71] Perzyna, P., Fundamental problems in viscoplasticity, *Adv. Appl. Mech.*, **11**, 313–387 (1971)

[Per88] Perzyna, P.: Temperature and rate dependent theory of plasticity of crystalline solids, *Revue Phys. Appl.*, **23**, 445–459 (1988)

[Pil88] Pilvin, P.: Identification des Parametres de Models de Comportement, In: Cailletaud et al. (eds), *Proceedings of MECAMAT*, II/155–II/164, Mecamat, Besancon (1988)

[PFT90] Press, W. H., Flannery, B. P., Teukolsky S. A., Vetterling, W. T.: *Numerical Recipes in C*, Cambridge University Press, Cambridge (1990)

[Pod94] Podio-Guidugli, P.: *Dodici lezioni su argomenti costitutivi in elasticita - XVIII Scuola estiva di fisica matematica*, Gruppo nazionale di fisica matematica, Roma (1994)

[Pod00] Podio-Guidugli, P.: *A Primer in Elasticity*, Kluwer, Dordrecht (2000)

[Pri55] Prigogine, I.: *Thermodynamics of Irreversible Processes*, Interscience Publishers, New York (1955)

[Rab80] Rabotnov, Yu. N. *Elements of Hereditary Solid Mechanics,* Mir Publishers, Moscow (1980)

[Ric71] Rice, J. R.: Inelastic constitutive relations for solids: an internal variable theory and its application to metal plasticity, *J. Mech. Phys. Solids*, **19**, 433–455 (1971)

[Riv02] Rivlin, R. S.: Material symmetry revisited, *GAMM-Mitteilungen*, **1/2**, 109–126 (2002)

[RK-88] Ruggles, M., Krempl, E.: Ratchetting behavior of type 304 stainless steel at room and elevated temperatures, In: Cailletaud et al (eds) *Proceedings of MECAMAT*, II/67–II/79, Mecamat, Besancon (1988)

[Ruu88] Ruuskanen, P.:, Magnetomechanical effect in polycrystalline iron, In: Kettunen, P. O., Lepistö, T. K., Lehtonen, M. E. (eds) *Strength of Metals and Alloys, (ICSMA 8)*, 1433–1438, Pergamon, Oxford (1988)

[Ryc83] Rychlewski, J.: *"CEIIINOSSSTTUV" Mathematical Structure of Elastic Bodies*, (in Russian), Technical Report, **217**, Inst. Probl. Mech. USSR Acad. Sci. Publ., Moscow (1983)

[Sch54] Schouten, J. A.: *Ricci-Calculus*, Springer, Berlin (1954)

[See88] Seeger, A.: Thermodynamics of open systems, self-organization and crystal plasticity, In: Kettunen, P. O., Lepistö, T. K., Lehtonen, M. E. (eds) *Strength of Metals and Alloys, (ICSMA 8)*, 463–468, Pergamon, Oxford (1988)

[She77] Shermergor, T. D.: *Elasticity Theory of Microheterogeneous Media* (in Russian), Nauka, Moscow (1977)

[ST-87] SIDOROFF, F., TEODOSIU C.: Microstructure and phenomenological models for metals, In: Gittus, J., Zarka, J., Nemat-Nasser S. (eds) *Large Deformations of Solids: Physical Basis and Math. Modelling*, pp. 163-186, Elsevier, London (1987)

[Spe71] SPENCER, A. J. M.: Theory of invariants, In: Eringen A. C. (ed) *Continuum Physics*, Academic Press, New York (1971)

[Sto62] STOJANOVIĆ, R.: On the reference-state problem in the non-linear elasticity theory of continua with dislocations, *Physica Status Solidi,* **2**, 566–575 (1962)

[Sto72] STOJANOVIC, R.: *Recent Development in the Theory of Polar Continua*, CISM Courses and Lectures, **27**, Springer, Wien (1972)

[Tay38] TAYLOR, G.: Plastic strain in metals, *J. Inst. Metals*, **62**, 307–324 (1938)

[Teo70] TEODOSIU, C., A dynamic theory of dislocations and its applications to the theory of the elastic-plastic continuum, In: Simmons, J. A., de Wit, R., Bullough, R. (eds) *Fundamental Aspects of Dislocation Theory, Vol. I, National Bureau of Standards Spec. Publ.*, **317**, 837–877, Washington, D.C. (1970)

[TN-65] TRUESDELL, C., NOLL, W.: The non-linear field theories of mechanics, In: Flügge, S. (ed) *Handbuch der Physik*, **III/3**, Springer, Berlin (1965)

[Tru72] TRUESDELL, C.: *A First Course in Rational Continuum Mechanics* , The Johns Hopkins University Publ., Baltimore (1972)

[Vak70] VAKULENKO, A. A.: Superposition in continuum rheology (in Russian), *Izv. AN SSSR Mekhanika Tverdogo Tela*, **1**, 69–74 (1970)

[Val71] VALANIS, K. C.: A theory of viscoplasticity without a yield surface, *Arch. Mech.*, **23**, 517–533 (1971)

[Ver84] VERHÁS, J.:, (1984). A thermodynamic approach to viscoelasticity and plasticity, *Acta Mech.*, **53**, 125–139 (1984)

[Woo81] WOODS, L. C.: On the local form of the second law of thermodynamics in continuum mechanics, *Quart. Appl. Math.*, April, 119–126 (1981)

[Woo82] WOODS, L.: Thermodynamic inequalities in continuum mechanics, *IMA J. Appl. Math.*, **29**, 221–246 (1982)

[WG-97] WU, X. J., GORHAM, D. A.: Stress equilibrium in the split Hopkinson pressure bar test, *J. Phys. IV*, **7/C3**, 91–96 (1997)

[Zar86] ZARKA, J.: In: Gittus, J. & Zarka J. (eds) *Modelling Small Deformations Of Polycrystals*, 19–40. Elsevier, London (1986)

[ZH-88] ZHOU, S. A., HSIEH, R. K. T.:, A theoretical model of eddy current nondestructive test for electromagnetoelastic materials, *Int. J. Eng. Sci.*, **26/1**, 12–26 (1988)

[Zie63] ZIEGLER, H.: Some extremum principles in irreversible thermodynamics with application to continuum mechanics, In: Sneddon, I. N., Hill, R. (eds) *Progress in Solid Mechanics*, **4**, 91 (1963)

[Zie73] ZIEGLER, H.: Orthogonality in continuum mechanics, (in Polish), In: Perzyna, P. (ed) *Thermoplasticity*, 73–128, Ossolineum, Warsaw (1973)

[Zor68] ZORSKI, H.: Statistical theory of dislocations, *Int. J. Eng. Sci.*, **4**, 959–974 (1968)

[Zrw74] ZORAWSKI, M., Private communication (1974)

Glossary of Notation

Symbol	Name and description	Place of first occurrence
Latin letters		
$(\mathbf{A} \otimes \mathbf{B})_{\alpha\beta\gamma\delta} \equiv A_{\alpha\beta} B_{\gamma\delta}$	Tensor product of two 2-tensors	(3.22), p.52
$(\mathbf{A} \, \Box \, \mathbf{B})_{\alpha\beta\gamma\delta} \equiv A_{\alpha\gamma} B_{\beta\delta}$	Tensor product of two 2-tensors with 1234 $\rightarrow$ 1324 permutation	(3.22), p.52
$(\mathbf{A} \overset{T}{\Box} \mathbf{B})_{\alpha\beta\gamma\delta} \equiv (\mathbf{A} \diamond \mathbf{B})_{\alpha\beta\gamma\delta} \equiv A_{\alpha\delta} B_{\beta\gamma}$	Tensor product of two 2-tensors with 1234 $\rightarrow$ 1432 permutation	(3.22), p.52
$\mathcal{A} \circ \mathcal{B} = \mathcal{A}^{abc} \mathcal{B}_{abc}$	Scalar product of two 3-tensors	(8.49), p.155
$\overset{(m)}{\mathcal{A}}(t) \equiv D_t^m \mathcal{A}$	m-th material time derivative of arbitrary tensor $\mathcal{A}$	(2.23), p.42
$\langle \mathcal{A} \rangle$	Volume average of arbitrary tensor $\mathcal{A}$	(1.86), p.28
$\dot{\mathcal{A}}(t) \equiv D_t \mathcal{A}$	First time derivative of arbitrary tensor $\mathcal{A}$	(2.24), p.42
$\mathcal{A}, \mathcal{J}_{\mathcal{A}}, \sigma_{\mathcal{A}}$	Arbitrary balanced tensor field, its flux and source density	(4.1), p.61
$\mathcal{A}', \mathcal{A}^*$	An RVE-tensor field, its value at some point $\mathbf{x}'$ and its fluctuation	(8.17), p.149
$\overline{\mathcal{A}} \equiv \mathcal{A}$	Average of an RVE-tensor field obtained by homogenization	(8.37), p.153

$\mathbf{A}_E$	True spatial dislocation density	(1.41), p.17
$\mathcal{A}_E$	Spatial inhomogeneity of elastic distortion	(1.48), p.18
$\mathcal{A}_{MAM}$	Viscoplastic constants of MAM model	(5.27), p.92
$\mathbf{A}_P$	Reference material dislocation density	(1.50), p.19
$\mathcal{A}_{PCR}$	Viscoplastic constants of modified Perzyna–Chaboche–Rabotnov model	(5.28), p.92
$\mathcal{B}$	A real crystal body	p.10
$\mathbf{B}, \beta_{eq}$	Back stress and equivalent back stress	(5.18), p 89
$\mathcal{B}_x, \mathcal{B}_y$	Elastic–plastic frontier in stress and strain spaces	p.39
$\mathbb{D}_\chi,\ (\mathbb{D}_\chi)^{abcd}$	Hooke's 4-tensor of elasticity constants in (χ_t) configuration	(1.88), p.28
$\mathbb{D},\ (\mathbb{D})^{\alpha\beta\gamma\delta}$	Hooke's 4-tensor of elasticity constants in (ν_t) configuration	(3.21), p.52
$\mathbb{D}^{\text{eff}}$	Hooke's 4-tensor of effective elasticity constants	(7.15), p.125
$\mathcal{D}_E$	True spatial disclination density	(1.48), p.18
$\mathbb{D}_\Lambda$	4-tensor of grain elasticity stiffness	(8.27), p.151
$\mathbb{D}_\Lambda^{\text{eff}}$	Effective grain stiffness tensor	(7.16), p.126
$\mathfrak{D}_\Lambda \equiv \mathbb{D}_M^{-1}\,[\mathbb{D}_\Lambda]$	Relative jump of grain stiffness	(7.18), p.127
$D_t\varepsilon^p_{eq}$	Equivalent plastic strain rate	(3.15), p.51
$\mathbf{D}_{P\omega}$	Damage-plastic stretching tensor	(2.5), p.35
E_3	Euclidean three-dimensional space	p.10
$\mathbf{E}, \mathbf{E}_E, \mathbf{E}_P$	Lagrangian total, elastic and plastic strain tensors in intrinsic picture	(1.3)–(1.8), p.12
$\mathbf{E}_{\Lambda E}$	Elastic micro-strain	(7.14), p.125
$\mathbf{E}^{gen}$	Hill's generalized strain tensor	(1.24), p. 14
$\mathbf{E}^{Hill}$	Hill's logarithmic strain tensor	(1.27), p.15
$\mathcal{E}_0^{ABC}$	Ricci's permutation tensor in material coordinates	(1.95), p.31

$\mathbf{F}$	Total deformation gradient	p.19
$\mathbf{F}', \mathbf{F}^*, \mathbf{F}$	Absolute, relative and average deformation gradient in an RVE	(8.19), p.150
$\mathbf{F}_P$	Replacement invariant plastic distortion tensor	(1.64), p.23
$\mathcal{F} : \mathcal{X} \mapsto \mathcal{R}$	Yield functional	(2.8), p.39
$F(\mathbf{S}, \mathbf{V}_P, \mathbf{L}_P)$	Dynamic yield function	(3.11), p.50
$\mathbf{G}, \mathbf{G}_E, \mathbf{G}_P$	Total, elastic and plastic deformation gradient tensors in engineering picture	(1.17), p.12
$\mathbb{H}$	Nonassociativity tensor	(3.25), p.54
$\mathbb{I} = (\mathbf{1} \square \mathbf{1} + \mathbf{1} \overset{T}{\square} \mathbf{1})/2$	Unit 4-tensor	(3.22), p.52
$\mathbf{J}_{0\Lambda}$	Material grain microinertia tensor	(8.23), p.150
$\mathbf{J}_u \equiv \mathbf{q}$	Internal energy flux	Table 4.1, p.62
$\mathbf{J}_s$	Entropy flux	Table 4.1, p.62
$\mathbf{L}$	Velocity gradient tensor	(1.35), p. 16
$\mathbb{L}$	4-tensor of magnetostriction constants	(6.36), p. 109
$\mathcal{L}$	Set of all second-order tensors	p.36
$\mathcal{L}^+ \subset \mathcal{L}$	Subset of $\mathcal{L}$ with elements having positive determinant	p.36
$\mathbb{L}_1$	Coleman–Noll 4-tensor of viscosity coefficients	(4.14), p.67
$\mathfrak{L}_2$	6-tensor of viscosity coefficients	(4.14), p.67
L_3	Distant parallelism 3-space	p.26
$\mathbf{L}_{P\omega}$	Damage-plastic "velocity gradient" tensor	(2.5), p.35
$\mathcal{M}$	Universal viscoplastic constant	(5.15), p. 87
$\mathbf{M}$, $\mathcal{M}$	Moments of skew and symmetric parts of stress inside RVE	(8.50), p. 155
$\mathbf{N}^a, \mathbf{N}^a_E, \mathbf{N}^a_P$	Total, elastic and plastic strain principal eigenvectors	(1.24), p.14
$\mathbb{P}_1$	4-tensor of plastic viscosity coefficients	(2.27), p.43

$\mathfrak{P}_{11}$	6-tensor of plastic viscosity coefficients	(2.27), p.43
$\mathbf{P}^t_\omega(s)$	Relative inelastic (damage-plastic) deformation history	(2.7), p.36
$\mathcal{Q}$	Nonmetricity 3-tensor	(1.71), p.24
$\mathcal{R}$, $\mathcal{R}^+$	Sets of real and positive numbers	p.36
$\mathbf{R}, \mathbf{R}_E, \mathbf{R}_P$	Total, elastic and plastic rotation tensors	(1.21)–(1.22), p.13
$\mathbf{R}_{P\omega}$	Damage-plastic rotation tensor	(2.6), p.35
$(\mathbb{R})_{\dot{\alpha}\dot{\beta}\dot{\gamma}}^{\cdot\,\cdot\,\cdot\,\delta}$	Riemann–Christoffel curvature in structural coordinates of (ν_t^ω) space	(1.78), p.25
$(\mathbf{R})^{AB}$	Einstein's curvature 2-tensor in material coordinates	(1.95), p.31
$\mathbf{S} \equiv \mathbf{T}_\nu$	Piola–Kirchhoff stress tensor related to (ν_t) configuration	(2.10), p. 38
$\widetilde{\mathbf{S}} = \mathbf{S}/\mathrm{det}\boldsymbol{\Phi}$	Convective stress tensor	(3.8), p. 49
$(\mathbb{S})_{\dot{a}\dot{b}}^{\cdot\,\cdot\,cd}$	Eshelby's 4-tensor in spatial coordinates of (χ_t) configuration	(1.85), p.28
$\mathcal{S}_{\dot{\alpha}\dot{\beta}}^{\cdot\,\cdot\,\gamma}$	Torsion 3-tensor in structural coordinates of (ν_t^ω) space	(1.84), p.26
$\mathcal{S}$	Piola–Kirchhoff stress functional	(2.10), p.38
$\mathbf{S} := \mathcal{S}(\gamma^\#, \boldsymbol{\Pi}^t)$	Dynamic stress tensor	(2.21), p.40
$\mathbf{S}^\# := \mathcal{S}(\gamma^\#, \mathbf{0}^t)$	Static stress tensor	(2.21), p.40
$\mathbf{S}^\# = k^\#\mathbf{S}$	Static yield stress	(3.13), p.51
$\mathcal{T}$	Cauchy stress functional	(2.8), p.36
$T(X^K, t)$	Absolute temperature	p.10
T_0	Reference (room) temperature	p.10
$\mathbf{T} \equiv \mathbf{T}_\chi$	Cauchy stress tensor	(1.88), p.28
$\mathbf{T}_\kappa$	Piola–Kirchhoff stress tensor related to (κ) configuration	(6.4), p.103
$\mathbf{U}, \mathbf{U}_E, \mathbf{U}_P$	Right stretch total, elastic and plastic tensors	(1.21)–(1.22), p.13
$\mathcal{U}_x, \mathcal{U}_y$	Elastic region in stress and	p.39

	strain spaces	
$\mathcal{V}_x, \mathcal{V}_y$	Elastic-viscoplastic region in stress and strain spaces	p.39
$\mathbf{U}_{P\omega}$	Damage-plastic right stretch tensor	(2.6), p.35
$\mathbf{V}, \mathbf{V}_E, \mathbf{V}_P$	Left stretch total, elastic and plastic tensors	(1.21)–(1.22), p.13
$\mathbf{V}_{P\omega}$	Damage-plastic left stretch tensor	(2.6), p.35
W	Total work per unit volume of (χ_t)	(3.18), p.51
W_E	Elastic work per unit volume of (χ_t)	(3.19), p.52
W_P	Plastic work per unit volume of (χ_t)	(3.20), p.52
$\mathbf{W}_{P\omega}$	Damage-thermo-plastic spin tensor	(2.5), p.35
$\mathcal{X}$	Space of stress histories	p.39
X^K	Material coordinates	p.10
$\mathcal{Y}$	Space of strain histories	p.38
$Y,\ Y_0$	Initial equivalent dynamic yield stress and its static counterpart	(5.16), p. 88
$\mathbf{b}$	Body force	Table 4.1, p.62
$\mathbf{b}$	Burgers vector	(1.39), p.17
$c_\Lambda \equiv \Delta V_\Lambda / \Delta V$	Grain concentration factors	(8.4), p.146
$\mathrm{dev}\mathbf{A} \equiv \mathbf{A}_d \equiv \mathbf{A} - \frac{1}{3}\mathbf{1}\,\mathrm{tr}\mathbf{A}$	Deviatoric part of a 2-tensor $\mathbf{A}$	(3.10), p.50
$\mathrm{d}s$	Surface element	(1.39), p.17
$d_e s,\ d_i s$	External and internal entropy increase	p.64
$\mathbf{e}, \mathbf{e}_E, \mathbf{e}_P$	Eulerian total, elastic and plastic strain tensors in intrinsic picture	(1.3)–(1.8), p.12
$\langle \mathbf{e}^u \rangle, \langle \mathbf{e} \rangle$	Effective Eshelby's unconstrained and constrained strain	(1.86), p.28
$f(t) = \mathcal{F}(\mathbf{X}^t)$	Yield functional	(2.8), p.39
$f^\#$	Static yield function	(3.13), p.51
g	Free energy density	(4.3), p.62
$\mathbf{g}_k, \mathbf{g}_0^M$	Spatial and material base vectors	(8.2), p.146

$\mathbf{h}_\alpha$	Structural lattice vectors	(1.59), p.21
$\mathbf{h}_\alpha^\omega$	Structural lattice vectors in damaged natural state (ν_t^ω)	(1.68), p.24
$p \in \mathcal{P}$	Inelastic path length	(2.14), p. 39
$\mathbf{q}$, $\mathfrak{D}_t q$	Heat flux and its time increment	Table 4.1, p.62
$\mathbf{s}_\gamma(\mathbf{\Pi}^t)$	Overstress tensor	(2.21), p.40
s	Entropy density	Table 4.1, p.62
$(\text{skw}\mathbf{A})_{mn} \equiv A_{[mn]}$	Skew symmetric part of 2-tensor $\mathbf{A}$	p.25
$(\text{sym}\mathbf{A})_{mn} \equiv A_{(mn)}$	Symmetric part of 2-tensor $\mathbf{A}$	p.25
t	Time	p.10
u	Internal energy density	Table 4.1, p.62
$\mathbf{x}', \mathbf{x}^*, \mathbf{x}$	Absolute, relative and center position inside a grain in (χ_t)	(8.18), p.149
x^k	Spatial coordinates	p.19

Greek letters

$\mathbf{\Delta}_E$	Thermoelastic distortion tensor	(2.1), p. 34
$\mathbf{\Delta}_{P\omega}$	Thermo-damage-plastic distortion tensor	(2.4), p.35
$\Delta\mathbf{e}^c = \mathbf{e} - \langle\mathbf{e}\rangle$	Eshelby's constrained strain	(1.85), p.28
$\Delta\mathbf{e}^u = \mathbf{e}^u - \langle\mathbf{e}^u\rangle$	Eshelby's unconstrained strain	(1.85), p.28
$\nabla \equiv \nabla_\chi$	Gradient operator in spatial coordinates of (χ_t)	(1.39), p.17
$\nabla_0 \equiv \nabla_\kappa$	Gradient operator in material coordinates of (κ)	(1.3), p.19
∇_ν	Gradient operator in structural coordinates of (ν_t)	(1.71), p.24
$\mathbf{\Phi}$	Elastic distortion tensor	(1.16), p.12
$\mathbf{\Pi}$	Plastic distortion tensor	(1.16), p.12
$\mathbf{\Pi}_D$	Dashner's relative plastic distortion	(1.63), p.22
$\mathbf{\Pi}^d$	Quasi-plastic damage distortion tensor	(2.2), p.34

$\boldsymbol{\Pi}^t$	Relative thermo-inelastic history	(2.7), p. 36
$\boldsymbol{\Pi}^\theta \equiv \boldsymbol{\Pi}^\omega(\theta)$	Thermal quasi-plastic distortion tensor	(1.82), p.26
$\boldsymbol{\Pi}^\omega$	Quasi-plastic distortion tensor	(1.66), p.23
$\boldsymbol{\Sigma}_\Lambda$	Micro-stress	(7.14), p.125
$\Omega_{\alpha\beta}^{\cdot\cdot\gamma}$	Anholonomic object	(1.75), p.25
$\alpha_1, \alpha_2, \alpha_3$	Anisotropic coefficients of thermal expansion	(1.80), p.26
γ_s	Entropy production density	(4.7), p.63
$\boldsymbol{\varepsilon}_P$ & $\boldsymbol{\varepsilon}^P$	Hill's logarithmic strain tensor	(5.1), p.82 & (11.2), p.204
ε_{eq}^p	Equivalent plastic strain	(3.16), p.51
$\zeta(t)$	Vakulenko's thermodynamic time	(4.43), p.74
η	Kröner's incompatibility tensor	(1.95), p.31
$\theta \equiv T - T_0$	Temperature increment	p.10
$\vartheta, \quad D_t\vartheta$	Empirical temperature and its rate	p.70
$\theta^t(s)$	Relative temperature history	(2.7), p. 36
(κ)	Global reference configuration (made of real or ideal crystal)	p.10
$\lambda_a, \lambda_{Ea}, \lambda_{Pa}$	Total, elastic and plastic principal extension ratios	(1.26), p.15
$D_t\lambda$	Inelastic mechanical power	(6.60), p.113
$D_t\lambda_{\mathrm{mag}}$	Irreversible magnetic power	(6.69), p.115
μ	Forgetfulness measure	p.40
(ν_t)	Natural state reference elements at constant temperature θ_0	p.10
(ν_t^ω)	Natural state damaged elements	p.23
$(\nu_t^{d\theta})$	Isothermally relaxed natural state damaged elements	p.34
$(d\xi)^\lambda$	Anholonomic structural coordinates	(1.70), p.24
$\pi_{(1)}, \ldots, \pi_{(6)}$	Plastic slip generators	(1.23), p.14
ϱ	Mass density	(4.1), p.61

$\boldsymbol{\sigma}, \boldsymbol{\tau}$	Symmetric and skew symmetric part of Cauchy stress	(8.30), p.152
σ_{eq}	Mises equivalent stress	(3.17), p.51
$\sigma_u,\ \sigma_s$	Internal energy and entropy source density	Table 4.1, p.62
$\varphi(\mathbf{x}')$	Mass distribution function	(8.17), p.149
$(\chi), (\chi'), (\chi''), (\chi_t)$	Global actual deformed configuration (made of real crystal)	Fig.1.1, p.7 Fig.1.4, p.11
(χ_0)	Initial real crystal configuration	Fig.1.4, p.11
(χ_t^{fict})	Murakami's fictitious undamaged configuration	p.36
(χ_t^{res})	Global "unloaded" Euclidean configuration in E_3	Fig.1.4, p.11
$\omega_{\Lambda E}$	Microelastic spin	(8.16), p.149

Hebrew letters

$\aleph, \aleph^T, \aleph^P$	Total, thermal and plastic dissipation	(7.27), p. 130
$\aleph^{PM}$	Magnetoplastic dissipation	(6.65), p. 114

Abbreviations

CTIP, RTIP	Classical and rational thermodynamics of irreversible processes	p.62
EndoTIP	Endochronic irreversible thermodynamics	p.74
ETIP	Extended irreversible thermodynamics	p.62
PCR	Perzyna–Chaboche–Rabotnov extended constitutive model	p.88
MAM	Micunovic–Albertini–Montagnani constitutive model	p.88
RVE	Representative volume element	p.76

Index

Printed in the United States of America